London Mathematical Society Lecture Note Series

Managing Editor:
Professor N.J. Hitchin,
Mathematical Institute, 24–29 St. Giles, Oxford OX1 3DP, UK

All the titles listed below can be obtained from good booksellers or from
Cambridge University Press. For a complete series listing visit
http://publishing.cambridge.org/stm/mathematics/lmsn/

London Mathematical Society Lecture Note Series: 316

Linear Logic in Computer Science

Edited by

Thomas Ehrhard
Institut de Mathématiques de Luminy, Marseille

Jean-Yves Girard
Institut de Mathématiques de Luminy, Marseille

Paul Ruet
Institut de Mathématiques de Luminy, Marseille

Philip Scott
Department of Mathematics and Statistics, University of Ottawa

CAMBRIDGE
UNIVERSITY PRESS

CAMBRIDGE
UNIVERSITY PRESS

University Printing House, Cambridge CB2 8BS, United Kingdom

Cambridge University Press is part of the University of Cambridge.

It furthers the University's mission by disseminating knowledge in the pursuit of
education, learning and research at the highest international levels of excellence.

www.cambridge.org
Information on this title: www.cambridge.org/9780521608572

First published 2004

A catalogue record for this publication is available from the British Library

ISBN 978-0-521-60857-2 Paperback

Contents

Contents

Preface

This volume is published in honour of the Azores summer school on Linear Logic and Computer Science held August 30 – Sept. 7, 2000 in St. Miguel, Azores. It can be considered as the third in a series, following a volume dedicated to the Cornell Linear Logic workshop of 1993 published as vol. 222 in the LMSLNS series of Cambridge University Press, and volumes dedicated to the Tokyo meeting of 1996 published as vol. 227(1-2) and 294(3) of Theoretical Computer Science. The summer school was attended by students and researchers from the different sites of the EU Training and Mobility of Researchers project "Linear Logic in Computer Science" (ERBFMRX-CT-97-0170, 1998 - 2002). The Organizing Committee consisted of: V. Michele Abrusci (University of Rome), Nuno Barreiro (University of Lisbon), and Jose Luiz Fiadeiro (University of Lisbon). The school included a series of tutorials, together with thematic sessions covering applications and new directions.

The main purpose of this book is twofold: to give a detailed overview of some well-established developments of Linear Logic under the guise of four tutorials, and to present some of the more recent advances and new directions in the subject through refereed contributions and invited papers. This book does not pretend to exhaustively cover the field of Linear Logic. In the spirit of the TMR "Linear" network, of which the Azores summer school was the climax, we decided to pay particular attention in this volume to the connections of Linear Logic with Computer Science.

We thank the authors of the various contributions for their wide-ranging and accessible presentations.

October 2003

Thomas Ehrhard
Jean-Yves Girard
Paul Ruet
Philip Scott

Contributors

J.-M. Andreoli
Xerox Research Centre Europe, Grenoble, and Institut de Mathématiques de Luminy. 163 avenue de Luminy, Case 907, 13288 Marseille Cedex 9, France.
andreoli@iml.univ-mrs.fr

R. Blute
Department of Mathematics and Statistics, University of Ottawa. Ottawa, Ontario, K1N 6N5, Canada.
rblute@mathstat.uottawa.ca

C. Faggian
Dipartimento di Matematica Pura ed Applicata, Università degli Studi di Padova. Via Belzoni 7, 35131 Padova, Italy.
claudia@math.unipd.it

M.-R. Fleury-Donnadieu
Institut de Mathématiques de Luminy. 163 avenue de Luminy, Case 907, 13288 Marseille Cedex 9, France.
mrd@iml.univ-mrs.fr

J.-Y. Girard
Institut de Mathématiques de Luminy. 163 avenue de Luminy, Case 907, 13288 Marseille Cedex 9, France.
girard@iml.univ-mrs.fr

S. Guerrini
Dipartimento di Informatica, Università degli Studi di Roma La Sapienza. Via Salaria, 113, 00198 Roma, Italy.
guerrini@dsi.uniroma1.it

J. Lambek
Department of Mathematics and Statistics, McGill University. 805 Sherbrooke W, Montreal, Québec, H3A 2K6, Canada.
`lambek@math.mcgill.ca`

O. Laurent
Preuves, Programmes et Systèmes, Université Denis Diderot. Case 7014, 2 Place Jussieu, 75251 Paris Cedex 05, France.
`Olivier.Laurent@pps.jussieu.fr`

P.-A. Melliès
Preuves, Programmes et Systèmes, Université Denis Diderot. Case 7014, 2 Place Jussieu, 75251 Paris Cedex 05, France.
`Paul-Andre.Mellies@pps.jussieu.fr`

D. Miller
INRIA - Futurs & Laboratoire d'Informatique LIX, École Polytechnique. Rue de Saclay, 91128 Palaiseau Cedex France.
`dale.miller@inria.fr`

M. Quatrini
Institut de Mathématiques de Luminy. 163 avenue de Luminy, Case 907, 13288 Marseille Cedex 9, France.
`quatrini@iml.univ-mrs.fr`

Ph. Scott
Department of Mathematics and Statistics, University of Ottawa. Ottawa, Ontario, K1N 6N5, Canada.
`phil@mathstat.uottawa.ca`

L. Tortora de Falco
Dipartimento di Filosofia, Università "Roma Tre". Via Ostiense, 234, 00146 Roma, Italy.
`tortora@uniroma3.it`

G. Winskel
Computer Laboratory, University of Cambridge. William Gates Building, JJ Thomson Avenue, Cambridge CB3 0FD, UK.
`Glynn.Winskel@cl.cam.ac.uk`

Part one
Tutorials

1

Category Theory for Linear Logicians

Richard Blute and Philip Scott

Department of Mathematics and Statistics
University of Ottawa

Abstract

This paper presents an introduction to category theory with an emphasis on those aspects relevant to the analysis of the model theory of linear logic. With this in mind, we focus on the basic definitions of category theory and categorical logic.

An analysis of cartesian and cartesian closed categories and their relation to intuitionistic logic is followed by a consideration of symmetric monoidal closed, linearly distributive and *-autonomous categories and their relation to multiplicative linear logic. We examine nonsymmetric monoidal categories, and consider them as models of noncommutative linear logic. We introduce traced monoidal categories, and discuss their relation to the geometry of interaction. The necessary aspects of the theory of monads is introduced in order to describe the categorical modelling of the exponentials. We conclude by briefly describing the notion of *full completeness*, a strong form of categorical completeness, which originated in the categorical model theory of linear logic.

No knowledge of category theory is assumed, but we do assume knowledge of linear logic sequent calculus and the standard models of linear logic, and modest familiarity with typed lambda calculus.

1.0 Introduction

Category theory arose as an organizing framework for expressing the *naturality* of certain constructions in algebraic topology. Its subsequent applicability, both as a language for simply expressing complex relationships between mathematical structures and as a mathematical theory in its own right, is remarkable. Categorical principles have been put to good use in virtually every branch of mathematics, in most cases leading to profound new understandings.

Roughly a *category* is an abstraction of the principle that the morphisms between objects of interest are just as important as the objects themselves. So a category will consist of two classes, the class of objects and the class of morphisms between objects. One must have a composition law, and each object must come equipped with a specified identity morphism. This data must satisfy some evident axioms. From this simple definition, an enormous theory follows. For example, one next defines morphisms between categories; these are *functors*. One can go on to define morphisms between functors; these are *natural transformations*, and on and on. There is a remarkably rich interaction between these structures. As expositions of this theory, we highly recommend [53, 21].

Categorical logic begins with the idea that, given a logic, one can form a category whose objects are formulas and whose morphisms are (equivalence classes of) proofs. The question of the proper notion of equivalence is extremely important and delicate. We will examine it in some detail below. There are several benefits to the formation of this category. First, under this interpretation, the logic's connectives are naturally exhibited as functors, and the logic's inference rules are exhibited as natural transformations. Then models of the logic can be simply defined as structure-preserving functors from this syntactic category to a category with the appropriate structure. Second, the category so formed will typically be freely generated in a certain sense, and can thus be used to derive general information about all categories of the same structure. The most well-developped examples of this idea are the relations between intuitionistic logic and cartesian closed categories, and between linear logic and *-autonomous categories. Both of these relationships will be described below.

The goal of this paper is to establish sufficient categorical background to understand these relationships and their consequences. We will introduce cartesian closed categories (cccs) and describe the translation between cccs and intuitionistic logic. This is the most well-established example of categorical logic, and is the subject of the book [51]. This is followed by a consideration of monoidal, symmetric monoidal closed, linearly distributive and *-autonomous categories and the translation between these structures and multiplicative linear logic. One of the most intriguing aspects of linear logic is that it is sufficiently flexible as a logical system to allow one to define noncommutative versions. With this in mind, we examine nonsymmetric monoidal categories, and consider them as models of noncommutative linear logic. We will especially focus on examples arising from the representation theory of *Hopf algebras*.

We also introduce traced monoidal categories, which arose independently of linear logic, but were subsequently seen to provide the appropriate framework for the analysis of Girard's geometry of interaction. Computationally, the most important fragment of linear logic is the exponential fragment, and its categorical structure leads one to the notion of *Seely model*. The necessary aspects of the theory of monads is introduced in order to describe the categorical modelling of the exponentials. We conclude by briefly describing the notion of *full completeness*, a strong form of categorical completeness, which originated in the categorical model theory of linear logic [3]. Full completeness is an excellent example of the influence of categorical principles on logical semantics, not just for linear logic, but for general logics.

No knowledge of category theory is assumed, but we do assume knowledge of linear logic sequent calculus and the standard models of linear logic. Also it would help to have a modest familiarity with typed lambda calculus (as in Girard's [39]). This paper may be considered a companion to the article [59], but stressing the linear logic aspects. We note that we only focus on aspects of category theory of immediate relevance to linear logic. So important topics like limits and colimits are omitted.

1.1 Categories, Functors, Natural Transformations

1.1.1 Basics of Categories

A *category C* consists of two classes, *Objects* and *Arrows*, together with two functions *Arrows* $\overset{dom}{\underset{cod}{\rightrightarrows}}$ *Objects* satisfying the following properties (we write $A \xrightarrow{f} B$ for: $f \in Arrows$, $dom(f) = A$ and $cod(f) = B$):

- There are *identity* arrows $A \xrightarrow{id_A} A$, for each object A,
- There is a partially-defined binary *composition* operation on arrows, denoted by juxtaposition,

$$\frac{A \xrightarrow{f} B \quad B \xrightarrow{g} C}{A \xrightarrow{gf} C}$$

(defined only when $dom(g) = cod(f)$) satisfying the following equations:

(i) $f id_A = f = id_B f$, where $A \xrightarrow{f} B$,

(ii) $h(gf) = (hg)f$, where $A \xrightarrow{f} B \xrightarrow{g} C \xrightarrow{h} D$.

A category is called *large* or *small* depending upon whether its class of objects is respectively a proper class or a set, in the sense of Gödel-Bernays set theory. We denote by $C(A, B)$ the collection of arrows $A \to B$ in the category C. A category is *locally small* if $C(A, B)$ is a set, for all objects A, B. It is convenient to represent arrows graphically. Equations in categories are also typically represented graphically and are called *commutative diagrams*.

Many familiar classes of structures in mathematics and logic can be organized into categories. Here are some basic examples. Verification that the set of arrows is closed under composition as well as satisfying the axioms of a category is left as an exercise.

Set: This (large) category has the class of all sets as Objects, with all set-theoretic functions as Arrows. Identity arrows and composition of arrows are defined in the usual way.

Rel: This has the same objects as **Set**, but an arrow $A \xrightarrow{R} B$ is a binary relation $R \subseteq A \times B$. Here composition is given by *relational product*, i.e. given $A \xrightarrow{R} B \xrightarrow{S} C$,

$$A \xrightarrow{SR} C = \{(a, c) \in A \times C \mid \exists b \in B \text{ such that } (a, b) \in R \& (b, c) \in S\}$$

while the identity arrows $A \xrightarrow{id_A} A$ are given by the diagonal relations: $id_A = \{(a, a) \mid a \in A\}$.

Universal Algebras: Here Objects can be any equational class of algebras (e.g. semigroups, monoids, groups, rings, lattices, heyting or boolean algebras, $\cdots$). Arrows are *homomorphisms*, i.e. set-theoretic functions preserving the given structure. Composition and identities are induced from **Set**. We use boldface notation for the names of the associated categories, e.g. **Group, Lat, Heyt,** for the categories of groups, lattices, and Heyting algebras, resp.

Vec$_k$: Here Objects are vector spaces over a field **k**, and Arrows are k-linear maps. We usually omit the subscript **k**, and write **Vec** for short. An important subcategory of **Vec** is the category **Vec**$_{fd}$ of finite dimensional vector spaces and linear maps. Of course, one can also consider various classes of topological vector spaces and normed spaces, with appropriate notions of map.

Top: Here objects are topological spaces and morphisms are continuous maps. One can also consider various homotopy categories, i.e.

where the morphisms are homotopy equivalence classes of continuous maps. It is from this sort of example that category theory originally arose.

Poset: Here the objects are partially-ordered sets and morphisms are monotone maps. A particularly important example arising in theoretical computer science is the category ω-**CPO** of posets in which ascending countable chains $\cdots a_i \leq a_{i+1} \leq a_{i+2} \leq \cdots$ have suprema, and in which morphisms are poset maps preserving suprema of countable chains. Composition and identities are inherited from **Set**. For an introduction to this and other aspects of domain theory, see [8].

Of course, those areas of mathematics that heavily use category theory, e.g. algebraic topology, algebraic geometry, and homological algebra, are replete with many more sophisticated examples.

The previous examples were large categories, i.e. in which the collections of objects form proper classes in the sense of set theory. We now present some "small" categories, based on much smaller collections of objects and arrows:

One: The category with one object and one (identity) arrow.

Discrete categories: These are categories where the only arrows are identities. A set X becomes a discrete category, by letting the objects be the elements of X, and adding an identity arrow $x \to x$ for each $x \in X$. All (small) discrete categories arise in this way.

A monoid: A monoid M gives a category with one object, call it $\mathcal{C}_M$, as follows: if the single object is $*$, we define $\mathcal{C}_M(*, *) = M$. Composition of maps is multiplication in the monoid. Conversely, note that every category $\mathcal{C}$ with one object corresponds to a monoid, namely $\mathcal{C}(*, *)$.

A preorder: A preordered set $\mathbb{P} = (P, \leq)$ (where $\leq$ is a reflexive & transitive relation) may be considered as a category, whose objects are just the elements of $\mathbb{P}$ and in which we define $\mathbb{P}(a, b) = \{*\}$ if $a \leq b$ and $\mathbb{P}(a, b) = \emptyset$ if $a \nleq b$. Thus, given two objects $a, b \in \mathbb{P}$, there is at most one arrow from a to b; moreover, there is an arrow $a \to b$ in $\mathbb{P}$ exactly when $a \leq b$. In this case, the category laws are exactly the preorder conditions.

Graphs and finite categories: A *graph* (more precisely, a directed multigraph), consists of a pair of sets, called *Objects* and *Arrows*, together with two functions $Arrows \overset{dom}{\underset{cod}{\rightrightarrows}} Objects$. Every (small) category has an underlying graph, obtained by simply ignoring the other data beyond *dom, cod*. In particular, any finite category can be represented by simply drawing its underlying graph and assuming the existence of all well-defined compositions of arrows. Notice that all vertices in the underlying graph of a category have loops (given by identity arrows). Indeed, another way of looking at a category is as a kind of graph with additional structure (i.e. identity edges, a composition law and equations).

Graphs form a category **Graph** whose objects are graphs and whose arrows are pairs of functions $Arrows_0 \xrightarrow{f} Arrows_1$ and $Objects_0 \xrightarrow{g} Objects_1$ such that $gdom_0 = dom_1 f$ and $gcod_0 = cod_1 f$.

1.1.2 Deductive systems as categories

In the 1960's, Lambek introduced the novel idea of using Gentzen's methods in category theory and linguistics. His new approach involved the use of proof-theoretical methods in constructing free categories and for solving coherence (i.e. decision) problems. At the same time he emphasized a fundamental new idea: arrows in (freely generated) categories are equivalence classes of proofs. Lambek's work raises a question of particular relevance to linear logicians: what should the equations between proofs be? There is no ultimate answer except that Lambek's work would seem to say that the equations should be elegant and natural from the viewpoint of category theory. This section will follow [51] closely. For more on the history of this area, see [51] and the references therein.

Definition 1.1 A *deductive system* is a labelled directed graph (whose objects are called *formulas* and whose arrows are called *labelled sequents*). There are certain specified arrows (called *axioms*) among which are arrows $A \xrightarrow{id_A} A$, for all formulas A, and certain specified rules (called "inference rules") for generating new arrows from old ones, among which

$$\frac{A \xrightarrow{f} B \quad B \xrightarrow{g} C}{A \xrightarrow{gf} C} \; cut$$

is the composition rule called "cut":   , for all formulas A, B, C.

A deductive system freely generates "labelled proof trees" by the following procedure:

- Axioms are proof trees.
- The set of proof trees must be closed under the inference rules.

The root of a proof tree is called a "provable sequent", or "proof" for short, while the leaves of the tree are axioms.

Example 1.2 Let $\mathcal{G}$ be a graph. The *deductive system freely generated from* $\mathcal{G}$ is defined as follows:

(i) The formulas are the objects of $\mathcal{G}$ (also called *atomic formulas*).

(ii) The axioms consist of a distinguished *identity* axiom $A \xrightarrow{id_A} A$, for each formula A, together with all the arrows of $\mathcal{G}$ (the latter are sometimes called *nonlogical axioms*).

(iii) Cut is the only rule of inference.

A deductive system freely generated from $\mathcal{G}$ forms a category $F(\mathcal{G})$, *the category freely generated from* $\mathcal{G}$, whose objects are all the formulas and whose arrows are equivalence classes of proofs. Namely, we impose equations between proof trees by taking the congruence relation generated by the following equations:

$$\frac{A \xrightarrow{f} B \quad B \xrightarrow{id_B} B}{A \xrightarrow{id_B f} B} \quad = \quad A \xrightarrow{f} B$$

$$\frac{A \xrightarrow{id_A} A \quad A \xrightarrow{f} B}{A \xrightarrow{f id_A} B} \quad = \quad A \xrightarrow{f} B$$

$$\frac{A \xrightarrow{f} B \quad \dfrac{B \xrightarrow{g} C \quad C \xrightarrow{h} D}{B \xrightarrow{hg} D}}{A \xrightarrow{(hg)f} D} \quad = \quad \frac{\dfrac{A \xrightarrow{f} B \quad B \xrightarrow{g} C}{A \xrightarrow{gf} C} \quad C \xrightarrow{h} D}{A \xrightarrow{h(gf)} D}$$

An important special case is the following:

Example 1.3 (Deductive systems generated by discrete graphs) A graph $\mathcal{G}_0$ is *discrete* if it has no arrows: it may be identified with a set (of objects). The deductive system generated from the set $\mathcal{G}_0$ has only atomic formulas (objects of $\mathcal{G}_0$) and for axioms it has only identity axioms $A \xrightarrow{id_A} A$, for atomic formulas A. $F(\mathcal{G}_0)$ is called the *free category generated from the set of objects* $\mathcal{G}_0$.

We will later consider freely generated categories with additional structure (i.e. with additional operations on formulas, and additional axioms and rules of inference). It is possible conversely to *define* a category as a certain kind of deductive system, although in that case it will not necessarily be freely generated: the class of objects may simply be specified and the class of arrows merely closed under appropriate operations and equations (see [51]). Moreover, if the category is large, like **Set**, the objects and arrows form proper classes, which is not exactly what logicians are familiar with.

1.1.3 Operations on categories

There are many ways of forming new categories out of old ones. Two basic operations are the following:

Dualization: If C is a category, so is its *dual* C^{op}, with the same objects, but whose arrows are reversed (i.e. interchange *dom* and *cod*). Clearly $(C^{op})^{op} = C$ and reversing all arrows changes commutative diagrams in C to commutative diagrams in C^{op}. In other words, we have the following bijective correspondence:

$$\frac{f \colon A \to B \text{ in } C}{f \colon B \to A \text{ in } C^{op}}$$

Products: If C, D are categories, so is their cartesian product $C \times D$, with the obvious structure: objects are pairs of objects, arrows are pairs of arrows, composition and identities are defined componentwise.

Finally, we end with a useful notion:

Definition 1.4 A *subcategory* C of a category B is any category whose class of objects and arrows are contained in those of B, respectively, and which is closed under the "operations" in B of *domain, codomain, composition, and identity*. C is a *full* subcategory of B if for all objects $A, B \in C$, $C(A, B) = B(A, B)$. In other words, a full subcategory is determined by just restricting the class of objects, since the arrows are predetermined by B.

For example, we often consider small subcategories whose objects are of "bounded size" within the large examples above: e.g. the full subcategories of (i) finite sets and (ii) finite dimensional vector spaces, and more generally, for a fixed infinite cardinal κ, sets (resp. vector spaces) of cardinality (resp. dimension) bounded by κ.

1.1.4 Functors

If $\mathcal{C}, \mathcal{D}$ are categories, a *functor* $F : \mathcal{C} \to \mathcal{D}$ is a pair $F = (F_{ob}, F_{arr})$, where $F_{ob} : Objects(\mathcal{C}) \to Objects(\mathcal{D})$, and similarly for arrows, satisfying the following (we omit the subscripts ob, arr): if $A \xrightarrow{f} B$ then $FA \xrightarrow{F(f)} FB$ with: $F(gf) = F(g)F(f)$ and $F(id_A) = id_{FA}$. A functor $F : \mathcal{C}^{op} \to \mathcal{D}$ is sometimes called *contravariant*. From the definition of the opposite category, a contravariant functor F preserves the identity arrows, but reverses composition: $F(gf) = F(f)F(g)$.

Examples 1.5

1. *Forgetful (also called Underlying) Functors.* These include forgetful functors $U : \mathbf{Posets} \to \mathbf{Set}$, $U : \mathbf{Top} \to \mathbf{Set}$, $U : \mathbf{Alg} \to \mathbf{Set}$ (where $\mathbf{Alg}$ is any category of universal algebras and homomorphisms between them). U maps objects and arrows to their underlying set (omitting the other structure).

   Sometimes, one only forgets part of the structure, e.g. there are several forgetful functors on $\mathbf{TopGrp}$; we have $U_1 : \mathbf{TopGrp} \to \mathbf{Grp}$ which maps a topological group and a continuous group homomorphism to its underlying group (and the underlying group homomorphism), and similarly there is $U_2 : \mathbf{TopGrp} \to \mathbf{Top}$.

2. *Representable (or Hom) Functors.* If $A \in \mathcal{C}$, we have the dual co- and contravariant homs:

   1. *Covariant hom* : $\mathcal{C}(A, -) : \mathcal{C} \to \mathbf{Set}$ given by:

      $$B \mapsto \mathcal{C}(A, B)$$
      $$B \xrightarrow{f} C \mapsto \mathcal{C}(A, f) : \mathcal{C}(A, B) \to \mathcal{C}(A, C)$$
      $$\text{where } \mathcal{C}(A, f)(g) = fg.$$

   2. *Contravariant hom:* $\mathcal{C}(-, A) : \mathcal{C}^{op} \to \mathbf{Set}$ given by:

      $$B \mapsto \mathcal{C}(B, A)$$
      $$B \xrightarrow{f} C \mapsto \mathcal{C}(f, A) : \mathcal{C}(C, A) \to \mathcal{C}(B, A)$$
      $$\text{where } \mathcal{C}(A, f)(g) = gf.$$

3. *Powerset Functors.* There are co- and contravariant powerset functors on $\mathbf{Set}$:

(i) *Covariant* $\mathcal{P}$: **Set** $\to$ **Set** given by:

$$A \mapsto \mathcal{P}(A) = \{X \mid X \subseteq A\}$$
$$A \xrightarrow{f} B \mapsto \mathcal{P}(f) : \mathcal{P}(A) \to \mathcal{P}(B)$$
$$\text{where } \mathcal{P}(f)(X) = f[X]$$

(ii) *Contravariant* $\mathcal{P}$: **Set**^{op} $\to$ **Set** given by:

$$A \mapsto \mathcal{P}(A) = \{X \mid X \subseteq A\}$$
$$A \xrightarrow{f} B \mapsto \mathcal{P}(f) : \mathcal{P}(B) \to \mathcal{P}(A)$$
$$\text{where } \mathcal{P}(f)(Y) = f^{-1}[Y]$$

4. *Free Algebra Functors.* F : **Set** $\to$ **Alg**, where $F(X) =$ the free algebra generated by set X (e.g. **Alg** can be **Mon**, **Grp**, **Vec**, etc.)

5. *Identity and Inclusion Functors*: For example, Id : **Set** $\to$ **Set**, and the evident inclusion Inc : **Vec**$_{fd}$ $\hookrightarrow$ **Vec** of finite dimensional vector spaces among all vector spaces.

6. *Dual Spaces*: Let $V \in$ **Vec** and $V^{\perp} = Lin(V, \mathbf{k})$, the dual space of V. Exercise: show there are two functors: $(-)^{\perp}$: **Vec**^{op} $\to$ **Vec** and $(-)^{\perp\perp}$: **Vec** $\to$ **Vec**.
   Typically, this functor would be denoted V^*, but we will suggest that in some settings, this notion of linear dual actually models linear negation quite successfully, hence our choice of notation.

7. Let P and P' be posets, viewed as categories. We leave it to the reader to verify that a functor $F: P \to P'$ is the same thing as an order-preserving function from P to P'.

1.1.5 Natural Transformations

Given functors $F, G : \mathcal{C} \to \mathcal{D}$, a *natural transformation* is a family of arrows, indexed by the objects of $\mathcal{C}$, $\{\theta_C : FC \to GC \mid C \in \mathcal{C}\}$ such that for every $f : C \to D$, the following diagram commutes:

$$
\begin{array}{ccc}
FC & \xrightarrow{\theta_C} & GC \\
\downarrow{\scriptstyle Ff} & & \downarrow{\scriptstyle Gf} \\
FD & \xrightarrow{\theta_D} & GD
\end{array}
$$

Given n-ary functors $F, G : \mathcal{C}^n \to \mathcal{D}$, a family of arrows, indexed by n-tuples of objects of $\mathcal{C}$, $\alpha_{A_1,\cdots,A_n} : F(A_1, \cdots, A_n) \to G(A_1, \cdots, A_n)$ is said to be *natural in* A_i if fixing all the other arguments $A_j, j \neq i$, the resultant family $\alpha_{\dots,A_i\dots} : F(\cdots,A_i,\cdots) \to G(\cdots,A_i,\cdots)$ determines a natural transformation between functors $\mathcal{C} \to \mathcal{D}$ with respect to the ith argument as variable.

Examples 1.6

1. *Double Dual:*   Define $\theta : Id \to (-)^{\perp\perp} : \mathbf{Vec} \to \mathbf{Vec}$, where $\theta_V : V \to V^{\perp\perp}$ is given by:

$$\theta_V(x)(f) = f(x) \quad \text{for} \ f \in V^{\perp}, x \in V.$$

Exercise: Verify that θ is well-defined, and that the appropriate natural transformation diagram commutes. It may be shown that θ_V is an isomorphism if and only if V is finite dimensional. However, note that if V is indeed finite dimensional, there is no *natural* isomorphism $\eta : Id \to (-)^{\perp}$, even though for each V, $V \cong V^{\perp}$ in this case. The reason is that any such isomorphism depends on a choice of basis.

2. *Functor Categories:* Let $\mathcal{C}, \mathcal{D}$ be categories, with $\mathcal{C}$ small. Let $Funct(\mathcal{C}, \mathcal{D})$ be the category whose objects are functors from $\mathcal{C}$ to $\mathcal{D}$, and whose arrows are natural transformations between them, where we compose natural transformations as follows: given $F, G, H \in Funct(\mathcal{C}, \mathcal{D})$, define

$$FA \xrightarrow{(\psi\theta)_A} HA = FA \xrightarrow{\theta_A} GA \xrightarrow{\psi_A} HA$$

for each object $A \in \mathcal{C}$. In particular, if $\mathcal{C}$ is small, and $\mathcal{D} = \mathbf{Set}$, the category $Funct(\mathcal{C}^{op}, \mathcal{D}) = \mathbf{Set}^{\mathcal{C}^{op}}$ is called the category of *presheaves on* $\mathcal{C}$.

   If $\mathcal{C}$ is the small category with two objects and two non-identity arrows, $\bullet \rightrightarrows \bullet$, one can identify $\mathbf{Set}^{\mathcal{C}^{op}}$ with the category **Graph** of small graphs.

3. There is a category **Cat** of small categories and functors between them. There is a forgetful functor $U : \mathbf{Cat} \to \mathbf{Graph}$ which associates to every small category $\mathcal{C}$ its underlying graph.

1.1.6 Adjoints and Equivalences

An arrow in a category is an *isomorphism* or *iso* if it has a two-sided inverse. This corresponds to the usual mathematical notion of "isomorphism" in most familiar categories. In the case of functor categories, we obtain the following related notions:

- *Natural Isomorphisms*: A natural transformation $F \xrightarrow{\theta} G$ is a natural isomorphism if, for each A, $FA \xrightarrow{\theta_A} GA$ is an iso. We often write $F \cong G$ (leaving θ implicit) to denote such a natural isomorphism.

- *Isos of Categories*: A pair of functors $C \underset{G}{\overset{F}{\rightleftarrows}} D$ is an isomorphism of categories if $GF = Id_C$ and $FG = Id_D$. This is much too strong and rarely occurs in mathematics. A much more reasonable notion is the following:

- *Natural Equivalence*: A pair of functors $C \underset{G}{\overset{F}{\rightleftarrows}} D$ is a natural equivalence of categories if there are natural isomorphisms $GF \cong Id_C$ and $FG \cong Id_D$. We shall see many examples of this notion below.

Most mathematical duality theories, as in the case of the famous representation theorems of Stone, Gelfand, and Pontrjagin, amount to "contravariant" natural equivalences $C \cong D^{op}$. Barr's book [10] on *-autonomous categories, which analyzes such duality theories, is an important source of concrete models for (fragments of) linear logic. We shall discuss this later.

One of the most important concepts in category theory is that of *adjoint functors*. Given functors $D \underset{U}{\overset{F}{\rightleftarrows}} C$, we say F *is left adjoint to* U (denoted $F \dashv U$) if there is a natural isomorphism

$$\mathcal{D}(FC, D) \cong \mathcal{C}(C, UD) .$$

That is, there is a family of arrows $\{\alpha_{C,D} : \mathcal{D}(FC, D) \to \mathcal{C}(C, UD)\}$ which determines a natural isomorphism of functors (natural in C and D), $\alpha_{-,-} : \mathcal{D}(F-, -) \xrightarrow{\cong} \mathcal{C}(-, U-)$. This isomorphism determines a natural bijection of arrows

$$\frac{FC \to D \quad \text{in } \mathcal{D}}{C \to UD \quad \text{in } \mathcal{C}.}$$

Indeed, the statement that F is left adjoint to U is equivalent to the following *universal mapping property* of the functor F: for each object $C \in \mathcal{C}$, there is an object $FC \in \mathcal{D}$ and an arrow $\eta_C : C \to UFC$, such that for any arrow $f : C \to UD \in \mathcal{C}$, there is a unique $f^* : FC \to D \in \mathcal{D}$ satisfying: $U(f^*)\eta_C = f$, i.e.

Exercise 1.7 (Adjoints)

(i) Prove the equivalence of F being left adjoint to U and the universal mapping property above.

(ii) Given adjoints $F \dashv U$ as above, show there are natural transformations $\eta : Id_C \to UF$ and $\varepsilon : FU \to Id_D$ satisfying $(U\varepsilon)(\eta U) = id_U$ and $(\varepsilon F)(F\eta) = id_F$, where id denotes an identity natural transformation.

(iii) Show that there is a one-one correspondence between solutions $(F, \eta, (-)^*)$ of the universal mapping problem in part (i) with quadruples $(F, U, \eta, \varepsilon)$ satisfying the equations in part (ii).

Notions defined by universal mapping properties are unique up to isomorphism. For example, adjoint functors determine each other uniquely up to natural isomorphisms.

Adjoint functors abound in mathematics. Lawvere has used this in an attempted axiomatic foundation for large parts of mathematics.

Examples 1.8 (Adjoints)

1. *Galois Correspondences:* Consider two pre-orders as categories, with a pair of adjoint functors (order-preserving maps) between them : $(P, \leq) \overset{F}{\underset{G}{\rightleftarrows}} (Q, \leq)$. Then $F \dashv G$ means: $F(a) \leq b$ iff $a \leq G(b)$, for all $a \in P, b \in Q$. Let $j = GF : Q \to Q$. This gives a monotone *closure operator* satisfying: (i) $a \leq j(a)$ and (ii) $j^2(a) \leq j(a)$, for all $a \in P$

2. *Free Algebras:* In categories of universal algebras, the left adjoint to U exists, where $\mathbf{Alg} \overset{F}{\underset{U}{\rightleftarrows}} \mathbf{Set}$. Here $F(X)$ is the free (universal) algebra generated by the set X (for a still more concrete example, replace $\mathbf{Alg}$ by the category $\mathbf{Mon}$ of monoids. Then $F(X) = X^*$, the free monoid on the set X.) In general, $\eta_X : X \to UFX$ is the "inclusion of generators" which maps the

set X into the underlying set of the free algebra $F(X)$. The universal property of adjoint functors reduces to the familiar universal property of free algebras.

3. *Topological examples* There is an evident forgetful functor $U: \mathbf{Top} \to \mathbf{Set}$. This functor has both left and right adjoints. We leave it as an exercise to find them.

4. *Free Structures and Free Categories*: Generalizing the previous examples, left adjoints to forgetful functors typically determine "free" structures. A special case fundamental to categorical logic is the construction of free (structured) categories on graphs (see Exercise 1.9 below). We have an adjoint situation $\mathbf{Cat} \underset{U}{\overset{F}{\rightleftarrows}} \mathbf{Graph}$ in which the forgetful functor U has a left adjoint F, where $F(\mathcal{G})$ is the free (small) category generated by a (small) graph $\mathcal{G}$. More generally, we will later introduce free cartesian, cartesian closed, and $*$-autonomous categories. The point is that categorical logic allows us to construct such free categories directly from the formulas and proofs of certain logics.

Exercise 1.9 Prove $F \dashv U$, where $\mathbf{Cat} \underset{U}{\overset{F}{\rightleftarrows}} \mathbf{Graph}$ and $F(\mathcal{G})$ is the free (small) category generated by a (small) graph $\mathcal{G}$. In particular describe the η and the ε of the adjointness.

The next exercise is important: it illustrates Lawvere's slogan: *many categorical notions arise as adjoints to previously defined functors.*

Exercise 1.10 (Categorical Structure via Adjoints)

1. For any category $\mathcal{C}$, there is a unique functor to the one-object category: $\mathcal{C} \overset{!}{\longrightarrow} \mathbf{One}$. Postulating that $\mathcal{C}$ has a left (resp. right) adjoint corresponds to saying $\mathcal{C}$ has an *initial object* $\bot$ (resp. a *terminal object* $\top$). The universal properties say: for any object $C \in \mathcal{C}$ there is a unique arrow $C \overset{!_C}{\longrightarrow} \top$ (resp. a unique arrow $\bot \overset{0_C}{\longrightarrow} C$). Letting $\{*\}$ be any one-element set, this says: $\mathcal{C}(C, \top) \cong \{*\}$ and $\mathcal{C}(\bot, C) \cong \{*\}$.
   E.g. In $\mathbf{Set}$, $\bot = \emptyset$ and $\top = \{*\}$ (any one element set). In $\mathbf{Vec}$, $\bot = \top = \{0\}$, the trivial space.

2. *Products and Coproducts* : For any category $\mathcal{C}$, there is a diagonal functor $\mathcal{C} \overset{\Delta}{\longrightarrow} \mathcal{C} \times \mathcal{C}$. If we postulate a right adjoint $\Delta \dashv R$, then for all $C, A, B \in \mathcal{C}$, $\mathcal{C} \times \mathcal{C}(\Delta(C), (A, B)) \cong \mathcal{C}(C, R(A, B))$. Writing

$R(A, B) = A \times B$, show that we have a natural isomorphism $\mathcal{C}(C, A) \times \mathcal{C}(C, B) \cong \mathcal{C}(C, A \times B)$. We say $A \times B$ is the *cartesian product* of A and B. Dually, postulating a left adjoint $L \dashv \Delta$ determines a *coproduct* (= product in $\mathcal{C}^{op}$). Writing $L(A, B) = A + B$, show that this satisfies $\mathcal{C}(A + B, C) \cong \mathcal{C}(A, C) \times \mathcal{C}(B, C)$. E.g. In **Set**, $A \times B$ exists and is the usual cartesian product, while $A + B$ exists and is the disjoint union. In **Vec**$_{fd}$ and Abelian Groups, products and coproducts exist and coincide: $V \times W \cong V \oplus W$. In **Top**, we obtain the usual product topology.

3. *Exponentials*: If $\mathcal{C}$ has products, prove that there is an induced functor $\mathcal{C} \times \mathcal{C} \xrightarrow{\times} \mathcal{C}$. Fix an object $A \in \mathcal{C}$. Consider the induced functor $\mathcal{C} \xrightarrow{-\times A} \mathcal{C}$, given on objects by $C \mapsto C \times A$. Suppose that $- \times A$ has a right adjoint $(-)^A$. Show that this means there is a natural isomorphism $\mathcal{C}(C \times A, B) \cong \mathcal{C}(C, B^A)$. This property is called *cartesian closedness*, and will be considered in the next section. Verify that in **Set**, the exponentials B^A exist, where B^A is the set of all functions from A to B. Investigate cartesian closedness in some of the other categories we have mentioned. What does it mean for a poset, considered as a category, to have products and exponentials? For a difficult problem, which categories of topological spaces have exponentials? For a discussion of the existence of topological cartesian closed categories, see [53].

1.1.7 Cartesian and Cartesian Closed Categories

We shall now begin a process of equationally axiomatizing categories with products and function spaces, as introduced in Exercise (1.10) above. These categories have significant connections to the proof theory of certain intuitionist propositional calculi.

Definition 1.11 A *cartesian category* is a category with finite products, i.e. binary products together with a terminal object.

Thus we have natural bijections:

$$\mathcal{C}(C, \top) \cong \{*\} \tag{1.1}$$
$$\mathcal{C}(C, A) \times \mathcal{C}(C, B) \cong \mathcal{C}(C, A \times B) \tag{1.2}$$

We shall be interested in categories with a *specified* cartesian structure. The following is a standard technique in categorical logic. Clearly we must postulate that there is a terminal object $\top$ and a binary operation on objects denoted $A \times B$. What about arrows? We shall chase the identity arrow starting from the RHS of isomorphism (1.2), since this is the only distinguished structure we have at hand. So, letting $C = A \times B$, the identity arrow $id_{A \times B}$ on the RHS maps to a pair of arrows on the LHS, which we call *projections* $A \times B \xrightarrow{\pi_1} A, A \times B \xrightarrow{\pi_2} B$. Conversely, going from the LHS to the RHS of (1.2), we wish to internalize pairing: given a pair of arrows $C \xrightarrow{f} A$ and $C \xrightarrow{g} B$, define a "pairing" $C \xrightarrow{\langle f,g \rangle} A \times B$. Using these operations, we will then impose equations specifying the bijections (1.1), (1.2).

So, on arrows we postulate the following distinguished structure, for all objects A, B, C:

- *Terminal*:   An arrow $C \xrightarrow{!_C} \top$
- *Projections*: $A \times B \xrightarrow{\pi_1^{A,B}} A$, $A \times B \xrightarrow{\pi_2^{A,B}} B$

  $$C \xrightarrow{f} A \quad C \xrightarrow{g} B$$
- *Pairing*: $C \xrightarrow{\langle f,g \rangle} A \times B$

The isomorphisms (1.1) , (1.2) above may be given equationally by imposing the following identities (for all objects A, B, C):

$$!^C = f, \quad \text{for any } f : C \to \top \tag{1.3}$$

$$\pi_1 \langle f, g \rangle = f , \quad \pi_2 \langle f, g \rangle = g , \quad \langle \pi_1 h, \pi_2 h \rangle = h \tag{1.4}$$

$$\text{where } f : C \to A, \; g : C \to B, \; h : C \to A \times B$$

Thus, a *cartesian category with specified structure* is given by the above data: an object $\top$, a binary operation $\times$ on objects, distinguished families of arrows $!_C, \pi_i^{A,B}, \langle f, g \rangle$ for all objects A, B, C, satisfying the above equations.

Exercise 1.12

(i) Work out specified cartesian structure for **Set**, **Vec**$_{fd}$ and **Top**.

(ii) Show that a poset $\mathbb{P}$, considered as a category, is a cartesian category iff it is a $\wedge$-semilattice with top element.

Example 1.13 (Deductive system for $\{\wedge, \top\}$ generated by $\mathcal{G}_0$)

Let $\mathcal{G}_0$ be a discrete graph (cf. Exercise 1.3). We will now give an explicit description of the deductive system generated by $\mathcal{G}_0$. Formulas are freely generated from atoms (i.e. objects of $\mathcal{G}_0$), using $\{\wedge, \top\}$. Sequents are freely generated from the following "axioms" and "rules":

Axioms: $A \xrightarrow{id} A$, $A \wedge B \xrightarrow{\pi_1} A$, $A \wedge B \xrightarrow{\pi_2} B$, $C \xrightarrow{!^C} \top$.

Rules:
$$\frac{A \xrightarrow{f} B \quad B \xrightarrow{g} C}{A \xrightarrow{gf} C} \; cut \qquad \frac{C \xrightarrow{f} A \quad C \xrightarrow{g} B}{C \xrightarrow{\langle f,g \rangle} A \wedge B} \; pairing$$

Finally, we obtain a cartesian category $F(\mathcal{G}_0)$ (*the free cartesian category generated by the discrete graph $\mathcal{G}_0$*) by letting the objects be the formulas and letting arrows be equivalence classes of proofs, where we impose the smallest congruence relation forcing the equations (1.3), (1.4) of a cartesian category to hold.

We generalize this example to arbitrary graphs in Exercise 1.15 below.

In general, categorical constructions given by universal mapping properties are only determined up to isomorphism. However in categorical logic and proof theory, it is natural to consider categories with specified structure (as above) and "strict" functors, i.e. those preserving the structure on the nose [51].

Definition 1.14 Cart is the category of cartesian categories and *strict* cartesian functors, i.e. those preserving the structure on-the-nose: $F(A \times B) = F(A) \times F(B), F(\top) = \top$.

There is a forgetful functor **Cart** $\xrightarrow{U}$ **Graph**. This functor has a left adjoint **Graph** $\xrightarrow{F}$ **Cart**, where for any graph $\mathcal{G}$, $F(\mathcal{G})$ is *the cartesian category freely generated from $\mathcal{G}$*.

Exercise 1.15 (Free Cartesian Categories) Let $\mathcal{G}$ be a graph. Following Examples 1.2 and 1.13, construct $F(\mathcal{G})$ as a deductive system and prove that $F \dashv U$. [**Hint:** Objects of $F(\mathcal{G})$ are formulas in the language $\{\top, \wedge\}$ freely generated from the objects of $\mathcal{G}$ (i.e. consider the objects of $\mathcal{G}$ as atomic formulae). Proofs are freely generated from the Axioms in Example 1.13 along with all the arrows $A \xrightarrow{f} B \in \mathcal{G}$ (considered as nonlogical axioms) using the rules. Finally, impose the smallest congruence relation on proofs making $F(\mathcal{G})$ a cartesian category.]

Definition 1.16 A *cocartesian* category is the dual of a cartesian category, i.e. a category with binary coproducts and an initial object. A *cocartesian category with distinguished structure* is obtained by dualizing the cartesian case, i.e. we postulate the following distinguished structure, for all objects A, B, C:

- *Initial:*   An arrow $\perp \xrightarrow{O_C} C$
- *Injections:* $A \xrightarrow{in_1^{A,B}} A+B$, $B \xrightarrow{in_2^{A,B}} A+B$

$$A \xrightarrow{f} C \quad B \xrightarrow{g} C$$

- *Copairing:* $A+B \xrightarrow{[f,g]} C$

The relevant equations specifying the isomorphisms in Exercise 1.10(2) are obtained by dualizing equations (1.3), (1.4). Free cocartesian categories may be obtained by setting up an appropriate deductive system for $\{\perp, \vee\}$ using the above structure (cf. Exercise 1.15). It is common to denote initial objects by 0, rather than $\perp$.

Cartesian Closed Categories. We now wish to equationally axiomatize those cartesian categories with specified exponentials (cf. Exercise 1.10).

Definition 1.17 A category C is *cartesian closed* (or a *ccc*) if it is cartesian and, for each object A, the endofunctor $C \xrightarrow{-\times A} C$ has a right adjoint, denoted $(-)^A$.

The adjointness says that for each A, we have an isomorphism $C(C \times A, B) \cong C(C, B^A)$, natural in C and B.

To axiomatize *specified* exponential structure (on top of specified cartesian structure) we specify: (i) there is also a binary operation B^A on objects; (ii) on arrows we postulate the following arrow schema and unary rule-schema for generating new arrows from old, in addition to the cartesian structure (for all objects A, B, C).

- *Evaluation:* $ev_{AB} : B^A \times A \longrightarrow B$

$$C \times A \xrightarrow{f} B$$

- *Currying:* $C \xrightarrow{f^*} B^A$

Finally, we impose the following equations in addition to the cartesian equations:

- (*Beta*) $ev\langle f^*\pi_1, \pi_2 \rangle = f : C \times A \to B$
- (*Eta*) $(ev\langle g\pi_1, \pi_2\rangle)^* = g : C \to B^A$.

Thus, a *cartesian closed category with specified structure* is given by the following data: a specified object $\top$, two binary operations $\times$ and $(-)^{(-)}$ on objects, the basic arrows

$$A \xrightarrow{id_A} A, \quad A \xrightarrow{!_A} \top, \quad A \times B \xrightarrow{\pi_1^{A,B}} A, \quad A \times B \xrightarrow{\pi_2^{A,B}} B, \quad B^A \times A \xrightarrow{ev_{A,B}} B,$$

the unary rule of Currying, and two binary rules of composition and pairing. Finally we postulate the equations of cartesian categories with (Beta) and (Eta).

Exercise 1.18 Check that the equations guarantee the bijection $\mathcal{C}(C \times A, B) \cong \mathcal{C}(C, B^A)$, and this bijection is natural in C and B.

The category **CCC** is defined as follows: its objects are cartesian closed categories with specified structure and its morphisms are functors preserving the structure on the nose. There is a forgetful functor **CCC** $\xrightarrow{U}$ **Graph**.

Examples 1.19

1. **Set** (see Exercise 1.10) and more generally functor categories (presheaves) **Set**$^{\mathcal{C}^{op}}$. In the case of **Set**, B^A is the set of all functions from A to B, ev is *evaluation*: $ev(g, a) = g(a)$, and currying is: $f^*(c)(a) = f(c, a)$, for all $g \in B^A, f \in B^{C \times A}, a \in A, c \in C$.

2. ω-**CPO**: *Objects* are posets in which ascending ω-chains have suprema. *Arrows* are functions preserving suprema of chains (hence, monotone). Products are cartesian products, with pointwise order structure, $\top = \{*\}$, $B^A = \omega$-$\mathrm{CPO}(A, B)$ with order and sups defined pointwise (e.g. $(\bigvee_n f_n)(a) = \bigvee f_n(a)$). The rest of the structure is induced from **Set**.

3. **Heyting Semilattices**: A cartesian closed poset $(P, \leq, \top, \wedge, \Rightarrow)$ is a poset satisfying, for all $a, b, c \in P$,

$$a \leq \top \qquad\qquad \frac{c \leq a \quad c \leq b}{c \leq a \wedge b}$$

$$a \wedge b \leq a \quad a \wedge b \leq b, \qquad c \wedge a \leq b \text{ iff } c \leq a \Rightarrow b$$

So, $b^a = a \Rightarrow b$ is the largest element whose meet with a is less than or equal to b. A *Heyting Algebra* $(P, \leq, \top, \wedge, \Rightarrow, \vee, \bot)$ is a cartesian closed poset with finite coproducts and an initial object. These are the posetal models of intuitionistic propositional calculus.

   The canonical example is due to Stone and Tarski: Let $X \in$ **Top** be a topological space. Then $\mathcal{O}(X)$, the poset of open subsets of X, is a Heyting algebra: for $U, V \in \mathcal{O}(X)$, $U \wedge V = U \cap V$, $U \vee V = U \cup V$, $U \Rightarrow V = int((X \setminus U) \cup V)$, $\top = X$, $\bot = \emptyset$.

4. **Cat**: The category of (small) categories in Example 1.6 is cartesian closed. We have already introduced the notion of product

of two categories, and we leave it as an exercise to verify that the appropriate functor category acts as an exponential in this setting.

5. **Deductive Systems for $\{\wedge, \Rightarrow, \top\}$ and free ccc's:** In general, ccc's will correspond to labelled deductions in intuitionistic $\{\wedge, \Rightarrow, \top\}$-logic. We add to the cartesian $\{\wedge, \top\}$-fragment one new axiom schema $ev_{A,B}$ (evaluation) and one new rule of inference (*Currying*):

$$(A \Rightarrow B) \wedge A \xrightarrow{ev_{A,B}} B \qquad \text{and} \qquad \frac{C \wedge A \xrightarrow{f} B}{C \xrightarrow{f^*} (A \Rightarrow B)} \text{ Curry} \quad .$$

We form $F(\mathcal{G})$, *the free ccc generated from graph $\mathcal{G}$*, as follows. Formulas are generated from the atomic formulas (i.e. objects of $\mathcal{G}$) using $\{\top, \wedge, \Rightarrow\}$. Proofs are generated from the nonlogical axioms (i.e. arrows of $\mathcal{G}$) together with the axioms (identity), (terminal), (projections), (evaluation) using the rules: (pairing), and (currying). We impose the equations of ccc's between proofs. The operation $F(-)$ is functorial. Indeed, the forgetful functor U has a left adjoint F, $\mathbf{CCC} \underset{U}{\overset{F}{\rightleftarrows}} \mathbf{Graph}$, with $F(\mathcal{G})$ the free ccc as described above.

Labels on proofs may be encoded by typed lambda terms, in the familiar manner. This is detailed in [51]. For example, in the currying rule above, $f^* = \lambda_{x:A} f(\langle z, x \rangle)$ where $z : C$.

Finally, the universal property of $F(\mathcal{G})$ says the following: for any ccc $\mathcal{C}$ and graph morphism $J : \mathcal{G} \to U(\mathcal{C})$, there is a unique extension to a strict ccc-functor $[\![-]\!]_J : F(\mathcal{G}) \to \mathcal{C}$.

Exercise 1.20 (For λ-calculus hackers) Verify in what sense the equations (Beta) and (Eta) above correspond to their λ-calculus counterparts. Actually, (Beta)–as written–corresponds to a restricted version of β-conversion, where we substitute a *variable* rather than an arbitrary term.

6. **CCC's = Typed Lambda Calculi:** This example is basic to categorical logic and proof theory. Cartesian closed categories are equivalent to typed lambda calculi (with product types) in a strong sense. Let $\mathbf{CCC}$ be the category of ccc's with specified structure and strict ccc functors. Similarly, we may define the category of typed λ-calculi, whose objects are (not necessarily freely generated) typed lambda-calculi, and whose morphisms are *translations*, i.e. interpretations strictly preserving the lambda structure (see

[51]). There is a natural equivalence of categories:

$$\textbf{CCC} \underset{L}{\overset{C}{\rightleftarrows}} \textbf{Typed } \lambda\textbf{-Calculus}$$

Here, associated to every ccc C there is a typed lambda calculus $L(C)$, the *internal language* of C. Roughly speaking, the types of $L(C)$ are the objects of C and the terms are freely generated, using the arrows of C as new term-forming operations (where currying corresponds to λ-introduction). The equations are generated by the equalities in C. Conversely, $C(\mathcal{L})$, the ccc *syntactically generated by a lambda theory* $\mathcal{L}$, is essentially the closed term model, viewed as a ccc (for details, see [51]). We remark that for this to go through, we require that our languages (in this case typed λ-calculi) need not be freely generated (in the same sense that deductive systems can be generalized). Moreover, $F(\mathcal{G})$, the free ccc generated by graph $\mathcal{G}$, is equivalent to $C(L(\mathcal{G}))$, where $L(\mathcal{G})$ is the typed lambda calculus generated by the graph $\mathcal{G}$ (analogous to $L(C)$).

This categorical equivalence of ccc's, typed lambda calculi, and equivalence classes of proofs in intuitionistic deductive systems is the ultimate categorical form of the Curry-Howard isomorphism, and is due essentially to Lambek.

7. **Presheaves.** $Set^{C^{op}}$, the category of *presheaves on* C, is the functor category whose objects are contravariant functors $C^{op} \to$ **Set** and whose maps are natural transformations between them. **Set** is the special case when $C = $ **One**. The ccc structure of presheaves is given as follows:

$\top(A) = \{*\}$
$(F \times G)(A) = F(A) \times G(A)$
$G^F(A) = nat(C(-, A) \times F, G)$

$G^F \times F \xrightarrow{ev} G$ is defined by:
$ev_C(\theta, c) = \theta_C(id_C, c), c \in F(C)$.
$\theta^* : H \to G^F$ is defined by:
$\theta_A^*(a)_C(h, c) = \theta_C(H(h)(a), c)$,
where $h : C \to A, a \in H(A)$.

For some purposes, it is slightly more convenient to consider "covariant" presheaf categories $Set^{\mathcal{D}}$, which of course are included in the previous case, by observing that $Set^{\mathcal{D}} = Set^{(\mathcal{D}^{op})^{op}}$.

8. **Special Case.** G-Sets as presheaves.

Let G be a group, X a set. Let $Sym(X) = $ the group of all bijections of X. A *G-set* is a group homomorphism $G \to Sym(X)$.

Equivalently, a G-set is a *left action map* $\cdot : G \times X \to X$, denoted $(g, x) \mapsto g \cdot x$ satisfying:

(i) $!_G \cdot x = x$, for all $x \in X$;

(ii) $g_1 \cdot (g_2 \cdot x) = (g_1 g_2) \cdot x$, for all $g_i \in G$, $x \in X$.

The category G-**Set** of G-sets and G-set maps is defined as follows. Objects are G-sets. A G-set arrow $X \to Y$ is an *equivariant function*, i.e. a **Set**-function $f : X \to Y$ such that for all $g \in G$, $x \in X$, $f(g \cdot x) = g \cdot f(x)$.

Exercise: G-**Set** $\cong$ **Set**^{G}, where in the right-hand-side, the group G is considered as a category with one object (in which all arrows are isos).

Hence G-**Set** is a ccc. The *ccc*-structure can be described as follows. Let X, Y be two G-sets.

Product: $X \times Y$, with action $g \cdot (x, y) = (g \cdot x, g \cdot y)$.

Exponentials: Y^X (all set maps), with action $(g \cdot f)(a) = g \cdot (f(g^{-1} \cdot a))$. In particular, we have $ev(g \cdot (f, x)) = g \cdot ev(f, x)$, for all $f \in Y^X, x \in X, g \in G$.

9. **Per($\mathbb{N}$)**. A *per* (partial equivalence relation) is a symmetric, transitive relation on a set. We shall consider the category of pers on a functionally complete partial combinatory algebra. For example, consider the Kleene algebra $(\mathbb{N}, \cdot)$, in which $m \cdot n = \{m\}(n)$ is the application of the m partial recursive function to input n. We form the category **Per($\mathbb{N}$)** as follows: the *objects* of **Per($\mathbb{N}$)** are the pers on $\mathbb{N}$, denoted $R, S, T, \cdots$. The *arrows* of **Per($\mathbb{N}$)** are equivalence classes of certain partial recursive functions, denoted by their gödel number. Given a partial recursive function $\{e\} : \mathbb{N} \rightharpoonup \mathbb{N}$, e represents an arrow $R \to S$ iff $\quad \forall m, n[mRn \Rightarrow e \cdot m \downarrow, e \cdot n \downarrow$ and $e \cdot m S e \cdot n]$. Two indices representing arrows $e, e' : R \to S$ are equivalent, denoted $e \sim e' : R \to S$, iff $\quad \forall m, n[mRn \to e \cdot m \downarrow, e' \cdot n \downarrow$ and $e \cdot m S e' \cdot n]$.

This structure forms a ccc. For products, the recursive bijection $\mathbb{N} \times \mathbb{N} \cong \mathbb{N}$, induces a pairing function $\langle -, - \rangle$. Define $\langle a, b \rangle R \times S \langle a', b' \rangle$ iff $a R a'$ and $b S b'$. For exponentials, define $(S^R, \sim_{S^R}) = (\mathbf{Per}(\mathbb{N})(R, S), \sim)$, where $\sim$ is the above equivalence relation on indices. Getting the ccc structure, notably the operation of Currying, requires some elementary recursion theory (Kleene's s-m-n theorem) [9]. This example admits many generalizations.

10. **Coherence Spaces and Stable Maps.** A *coherence space* $\mathcal{A}$ is a family of sets satisfying:

- (i) $a \in \mathcal{A}$ and $b \subseteq a$ implies $b \in \mathcal{A}$.
- (ii) if $B \subseteq \mathcal{A}$ and if $\forall c, c' \in B(c \cup c' \in \mathcal{A})$ then $\cup B \in \mathcal{A}$.

In particular, $\emptyset \in \mathcal{A}$. Morphisms are *stable maps*, i.e. monotone maps preserving pullbacks and filtered colimits. That is, $f : \mathcal{A} \to \mathcal{B}$ is a stable map if (i) $b \subseteq a \in \mathcal{A}$ implies $f(b) \subseteq f(a)$, (ii) $f(\cup_{i \in I} a_i) = \cup_{i \in I} f(a_i)$, for I directed, and (iii) $a \cup b \in \mathcal{A}$ implies $f(a \cap b) = f(a) \cap f(b)$. This gives a category **Coh**. Every coherence space $\mathcal{A}$ yields a reflexive-symmetric (undirected) graph $(|\mathcal{A}|, \bigcirc)$ where $|\mathcal{A}| = \{a \mid \{a\} \in \mathcal{A}\}$ and $a \bigcirc b$ iff $\{a, b\} \in \mathcal{A}$. Moreover, there is a bijective correspondence between such graphs and coherence spaces.

Given two coherence spaces $\mathcal{A}, \mathcal{B}$ their product $\mathcal{A} \times \mathcal{B}$ is defined via the associated graphs as follows: $(|\mathcal{A} \times \mathcal{B}|, \bigcirc_{\mathcal{A} \times \mathcal{B}})$, with $|\mathcal{A} \times \mathcal{B}| = |\mathcal{A}| + |\mathcal{B}| = (\{1\} \times |\mathcal{A}|) \cup (\{2\} \times |\mathcal{B}|)$ where $(1, a) \bigcirc_{\mathcal{A} \times \mathcal{B}} (1, a')$ iff $a \bigcirc_{\mathcal{A}} a'$, $(2, b) \bigcirc_{\mathcal{A} \times \mathcal{B}} (2, b')$ iff $b \bigcirc_{\mathcal{B}} b'$, and $(1, a) \bigcirc_{\mathcal{A} \times \mathcal{B}} (2, b)$ for all $a \in |\mathcal{A}|, b \in |\mathcal{B}|$. The function space $\mathcal{B}^{\mathcal{A}} = \mathbf{Coh}(\mathcal{A}, \mathcal{B})$ of stable maps can be given the structure of a coherence space, ordered by Berry's order: $f \preceq g$ iff for all $a, a' \in \mathcal{A}$, $a' \subseteq a$ implies $f(a') = f(a) \cap g(a')$. For details, see [39]. This class of domains led to the discovery of linear logic (cf. Example 1.37).

A *bicartesian closed category* (biccc) is a ccc with binary coproducts and an initial object (often denoted by 0). It corresponds to the proof theory of full intuitionistic propositional logic, i.e. of the connectives $\{\wedge, \vee, \Rightarrow, \top, \bot\}$.

Exercise 1.21 A bicartesian closed category satisfies $A^{B+C} \cong A^B \times A^C$, $A^0 \cong 1$ and the distributive law: $(A + B) \times C \cong (A \times C) + (B \times C)$.

Observe that until now we have been discussing the proof theory of *intuitionistic* logics. What can we say about the proof theory of classical logic? Writing $\neg A = 0^A = A \Rightarrow 0$, notice that in any biccc there is a canonical arrow or proof $A \to \neg\neg A$. A naive guess for a model for classical logic is to demand that this arrow should be an isomorphism, so $A \cong \neg\neg A$. Let us call such biccc's *Boolean categories*.

The following surprising theorem about biccc's also characterizes Boolean categories. For a proof and discussion of this theorem, see [51].

Theorem 1.22 (Joyal) *In any biccc, there is at most one arrow $A \to 0$. In particular, in the associated intuitionistic propositional calculus, there is at most one proof of $A \to \perp$ and hence at most one proof of $\neg A$, up to equivalence of proofs.*

Thus Boolean categories are necessarily preorders and, up to equivalence of categories, the only such are boolean algebras!

So to understand the proof theory of classical logic requires a more sophisticated approach. It turns out that this involves categorical versions of Parigot's $\lambda\mu$-calculus and ideas arising from the notion of continuations in programming language theory. The appropriate categorical framework, called *control categories*, was developed by P. Selinger in [61].

1.2 Monoidal and *-Autonomous Structures

Definition 1.23 A *monoidal (or tensored) category* $(\mathcal{C}, I, \otimes, \alpha, \ell, r)$ is a category $\mathcal{C}$, with functor $\otimes : \mathcal{C} \times \mathcal{C} \to \mathcal{C}$, unit object $I \in ob(\mathcal{C})$, and specified isos: $\alpha_{ABC} : (A \otimes B) \otimes C \xrightarrow{\cong} A \otimes (B \otimes C)$, $\ell_A : I \otimes A \xrightarrow{\cong} A$, $r_A : A \otimes I \xrightarrow{\cong} A$ satisfying the following: $\quad \ell_I = r_I : I \otimes I \to I$, as well as:

$$
\begin{array}{ccc}
A \otimes (I \otimes C) & \xrightarrow{\alpha} & (A \otimes I) \otimes C \\
{\scriptstyle 1 \otimes \ell_C} \downarrow & & \downarrow {\scriptstyle r_A \otimes 1} \\
A \otimes C & = & A \otimes C
\end{array}
$$

$$
\begin{array}{ccc}
A(B(CD)) \xrightarrow{\alpha} (AB)(CD) \xrightarrow{\alpha} ((AB)C)D \\
{\scriptstyle 1 \otimes \alpha} \downarrow \qquad\qquad\qquad\qquad \uparrow {\scriptstyle \alpha \otimes 1} \\
A((BC)D) \xrightarrow{\hspace{2cm} \alpha \hspace{2cm}} (A(BC))D
\end{array}
$$

where we omit $\otimes$'s in the second diagram for typographical reasons. This diagram is known as the Mac Lane pentagon.

A monoidal category is a very basic structure. There are any number of additional structures one may add to this basic definition. The structures of relevance to this paper are *symmetric* structure, *closed* structure, or *traced* structure. We now begin the description of these structures.

Suppose first that there is a natural isomorphism $s_{AB} : A \otimes B \to B \otimes A$ satisfying the following three diagrams:

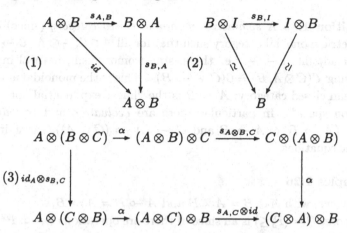

$\mathcal{C}$ is *symmetric* if diagrams (1), (2), and (3) commute.

Examples 1.24

1. Any cartesian category, with $\otimes = \times$.

2. Any co-cartesian category ($=$ finite coproducts), with $\otimes = +$

3. **Rel**$_\times$. This is the category **Rel** whose objects are sets and whose arrows are binary relations. The functor $\otimes : \textbf{Rel} \times \textbf{Rel} \to \textbf{Rel}$ is defined as follows. On objects, $\otimes = \times$, while on maps, $A \otimes B \xrightarrow{R \otimes S} C \otimes D$ is given by: $(a, b) R \otimes S (c, d)$ iff aRc & bSd. The tensor unit $I = \{*\}$, any one element set.

4. **Rel**$_+$. This is again the category **Rel**, except $\otimes = +$ (disjoint union), $I = \emptyset$, and where $A \otimes B \xrightarrow{R \otimes S} C \otimes D$ is given by:

$$(a, 0) R \otimes S (c, 0) \text{ if and only if } aRc$$

$$(b, 1) R \otimes S (d, 1) \text{ if and only if } bSd$$

where disjoint union in **Set** is given by: $X + Y = X \times \{0\} \cup Y \times \{1\}$

5. Two important monoidal subcategories of **Rel**$_+$ are:

   (i) **Pfn**: Sets and partial functions.

   (ii) **PInj**: Sets and partial injective functions.

6. **Vec**$_{fd}$ and **Vec**: (finite dimensional) vector spaces over **k**, where **k** is a field. Here $V \otimes W$ is taken to be the usual tensor product, and $I = \textbf{k}$.

Next, it is natural to ask that the tensor product have an appropriate adjoint, and this leads us to our next definition.

Definition 1.25 A *symmetric monoidal closed category* (smcc) C is a symmetric monoidal category such that for all $A \in C$, $- \otimes A : C \to C$ has a right adjoint $A \multimap -$, i.e. there is an isomorphism, natural in B, C, satisfying $C(C \otimes A, B) \cong C(C, A \multimap B)$. This is the monoidal analog of cartesian closed category; $A \multimap B$ is the "linear exponential" or "linear function space". In particular there are *evaluation* and *coevaluation* maps $(A \multimap B) \otimes A \to B$ and $C \to (A \multimap (C \otimes A))$, satisfying the adjoint equations.

Examples 1.26

1. Any ccc, with $A \otimes B = A \times B$ and $A \multimap B = A \Rightarrow B$.
2. A poset $\mathcal{P} = (P, \le)$ is an smcc iff there are operations $\otimes$, $\multimap$: $P^2 \to P$, $1 \in P$ satisfying:

   (i) $(P, \otimes, 1)$ is a commutative monoid.
   (ii) $\otimes$, $\multimap$ are functorial in the posetal sense: i.e. $x \le x', y \le y'$ implies $x \otimes y \le x' \otimes y'$ and $x' \multimap y \le x \multimap y'$
   (iii) (Closedness) $x \otimes y \le z$ iff $x \le y \multimap z$.

3. *Girard's Phase Semantics*: This is a posetal smcc, in the sense of Example 2 above. Let $M = (M, \cdot, e)$ be a commutative monoid. Consider the poset $\mathcal{P}(M)$, the powerset of M. We view $\mathcal{P}(M)$ as a poset ordered by inclusion. For $X, Y \in \mathcal{P}(M)$, define

$$
\begin{aligned}
X \otimes Y &= \{x \cdot y \mid x \in X, y \in Y\} =_{def} X \cdot Y \\
X \multimap Y &= \{z \in M \mid z \cdot X \subseteq Y\} \\
I &= \{e\}
\end{aligned}
$$

4. **Vec**, where $V \otimes W$ is the usual algebraic tensor product and $V \multimap W = Lin(V, W)$. More generally, consider $\mathcal{R}$-Modules over a commutative ring $\mathcal{R}$, with the standard algebraic notions of $V \otimes_{\mathcal{R}} W$ and $V \multimap W = Hom(V, W)$.

5. $\mathcal{MOD}(G)$. This example extends groups acting on sets to groups acting linearly on vector spaces. Let G be a group and V a vector space. A *representation of G on V* is a group homomorphism $\rho : G \to Aut(V)$; equivalently, it is a left G-action $G \times V \longrightarrow V$ (satisfying the same equations as a G-set) such that $v \mapsto g \cdot v$ is a linear automorphism, for each $g \in G$. The pair (ρ, V) is called a *G-module* or *G-space*. $\mathcal{MOD}(G)$ has as objects the G-modules

and as morphisms the linear maps commuting with the G-actions. Define the smcc structure of $\mathcal{MOD}(G)$ as follows:

$$V \otimes W \;=\; \text{the usual tensor product, with action determined by}$$
$$g \cdot (v \otimes w) = g \cdot v \otimes g \cdot w$$
$$V \multimap W \;=\; Lin(V, W), \text{ with action } (g \cdot f)(v) = g \cdot f(g^{-1} \cdot v),$$
$$\text{the contragredient action.}$$

We recommend [32] as an introduction to group representation theory.

Exercise 1.27 Formulate intuitionistic linear logic (ILL) as a deductive system, and show (with appropriate equations between proofs) it forms an smcc.

We now come to the fundamental definition, which will correspond to the proof theory of MLL. To model classical linear logic, we need an involutive negation. *In what follows, $\perp$ should not be confused with its use in the previous chapter (as an initial object).* The idea is that one chooses an object, which will be called $\perp$, and then defines (linear) negation via the formula $A^\perp = A \multimap \perp$. However to make this negation involutive, we should have $A = A^{\perp\perp}$, or since we are approaching this categorically, $A \cong A^{\perp\perp}$. This leads to the definition of *-autonomous category*.

Definition 1.28 A *-autonomous category* $(\mathcal{C}, \otimes, I, \multimap, \perp)$ is an smcc with a distinguished *dualizing* object $\perp$, such that (letting $A^\perp = A \multimap \perp$), the canonical map $\mu_A : A \to A^{\perp\perp}$ is an iso, for all A (i.e. "all objects are reflexive").

Facts about *-autonomous categories $\mathcal{C}$:

- We get a dualizing functor $\mathcal{C}^{op} \xrightarrow{\;()^\perp\;} \mathcal{C}$ s.t. $\mathcal{C}(A, B) \cong \mathcal{C}(B^\perp, A^\perp)$ which is a natural iso.

- $\mathcal{C}$ is closed under duality of categorical constructions: e.g. $\mathcal{C}$ has products iff it has coproducts, pullbacks iff pushouts, $\mathcal{C}$ is complete iff co-complete, etc.

- $(A \multimap B)^\perp \cong A \otimes B^\perp$ and $I \cong \perp^\perp$ Also $A \multimap B \cong B^\perp \multimap A^\perp$.

- We may define $A \,⅋\, B = (A^\perp \otimes B^\perp)^\perp$. In general, $\otimes \neq ⅋$, and (in general) there is not even a $\mathcal{C}$-morphism $A \otimes B \to A \,⅋\, B$.

- As we shall see below, categorical models of **MALL** (multiplicative, additive linear logic) will be ∗-autonomous categories with products (and hence coproducts).

Example 1.29 *Sets and relations.* The category $\mathbf{Rel}_\times$, with its usual monoidal structure, is probably the simplest ∗-autonomous category. The dualizing object is any one element set. We leave the details as an exercise. We will consider this example further below when we introduce compact closed categories.

Example 1.30 *Finite-dimensional vector spaces.* The category $\mathbf{Vec}_{fd}$, with its usual monoidal structure, is also a ∗-autonomous category. The dualizing object is the base field. We will also consider this example further when we introduce compact closed categories.

Example 1.31 *∗-autonomous posets and lattices.* Girard's phase semantics [33] gives examples of ∗-autonomous lattices, i.e. structures $(P, \le, \otimes, I, \multimap, \bot, \wedge, \vee)$ which are posetal ∗-autonomous categories. One method of construction is to consider closure operators $j : P \to P$ such that $j(x) \otimes j(y) \le j(x \otimes y)$. We consider the j-closed elements, i.e. the fixed points $Fix(j) = \{p \in P \mid j(p) = p\}$. We then seek to define a ∗-autonomous structure on $Fix(j)$.

For example, consider Girard's phase semantics (Example 1.26). Observe that the powerset of a monoid $\mathcal{P}(M)$ has both a lattice as well as a phase semantics structure. Pick an arbitrary $\bot \in \mathcal{P}(M)$, and consider $(-)^\perp : \mathcal{P}(M) \to \mathcal{P}(M)$ given by: $X^\perp = \{p \mid p \cdot X \subseteq \bot\}$. Let $j = (-)^{\perp\perp}$. On the set $Fix(j) = (-)^{\perp\perp}$-closed elements of $\mathcal{P}(M)$ (Girard calls them *facts*), we define: $\quad G \otimes H = (G \cdot H)^{\perp\perp}, \quad G \,\invamp\, H = (G^\perp \cdot H^\perp)^\perp,$ $G \wedge H = G \cap H, \quad G \vee H = (G \cup H)^{\perp\perp}.$

Example 1.32 *Finiteness spaces.* This example is due to T. Ehrhard [30]. It is an elaboration of the category **Rel**. Let X be a set and $u, v \subseteq X$ subsets of X. Say that u and v are *orthogonal*, written $u \perp v$, if $u \cap v$ is finite. If $\mathcal{F}$ is a set of subsets of X, write $\mathcal{F}^\perp = \{v \subseteq X \mid u \perp v \text{ for all } u \in \mathcal{F}\}$.

Ehrhard defines a *finiteness space* to be a pair $(X, \mathcal{F})$ where X is a set and $\mathcal{F}$ is a set of subsets of X such that $\mathcal{F}^{\perp\perp} = \mathcal{F}$. A morphism $R \colon (X, \mathcal{F}) \to (Y, \mathcal{G})$ is a subset $R \subseteq X \times Y$ such that for all $u \in \mathcal{F}$ we have $R(u) \in \mathcal{G}$, and for all $v \in \mathcal{G}^\perp$ we have $R^{op}(v) \in \mathcal{F}^\perp$. Here R^{op} is the reciprocal of R. It is straightforward to verify that this is indeed a category, with composition the usual relational composition. Then define:

- $(X, \mathcal{F})^{\perp} = (X, \mathcal{F}^{\perp})$
- $(X, \mathcal{F}) \otimes (Y, \mathcal{G}) = (X \times Y, \{u \times v | u \in \mathcal{F}, v \in \mathcal{G}\}^{\perp\perp})$
- $\perp = (\emptyset, \{\emptyset\})$

The rest of the details that we indeed have a $*$-autonomous category are straightforward.

Example 1.33 *Poset-valued sets.* This is a class of models constructed by de Paiva and Schalk [56], which can also be thought of as a generalization of **Rel**. One considers a $*$-autonomous poset, P, for example a Girard quantale or a phase space, as described above. Then a *P-valued set* is defined to be a pair (A, f) where A is a set and $f: A \rightarrow P$ is a function. A morphism between P-valued sets $R: (A, f) \rightarrow (B, g)$ is a relation $R: A \rightarrow B$ such that xRy implies $f(x) \leq g(y)$. Then one defines $(A, f) \otimes (B, g) = (A \times B, f \otimes g)$, where $f \otimes g$ is defined using the monoidal structure of P. The rest of the $*$-autonomous structure of P similarly lifts to the category.

Example 1.34 *Topological vector spaces.* We have already mentioned that the category of finite-dimensional vector spaces is $*$-autonomous, with the usual notion of dual space acting as negation. If one only wishes to consider discrete vector spaces, this is the best one can do. Indeed it is a standard result that a vector space is isomorphic to its second dual if and only if it is finite-dimensional, the problem being that the second dual of an infinite-dimensional space is substantially larger than the original space. If one wishes to consider infinite-dimensional spaces, one must add an additional topological structure.

So one passes to a category in which the objects are topological vector spaces and the morphisms are linear *continuous* maps. The hope in doing this is that in requiring continuity, one will decrease the size of the dual space to such an extent that one will be able to obtain additional objects isomorphic to their second dual and still retain the closed structure. It was the consideration of such spaces by Barr that led to the original axiomatization of $*$-autonomy. One appropriate notion of topology, introduced by Lefschetz, is the *linear topology*. This is the notion that led Barr to his axiomatization. See [10, 13] for the details of the following.

Definition 1.35 Let V be a vector space. A topology τ on V is a *linear topology* if it satisfies the following three properties:

- τ is Hausdorff.
- The topology τ makes V a topological vector space, i.e. addition and scalar multiplication are continuous.
- The origin has a neighborhood basis of open linear subspaces.

We get a category **TVec** when one takes as morphisms the linear, continuous maps. It can be shown that this is a symmetric monoidal closed category. The tensor product in **TVec** is given by an appropriate topology on the tensor of the underlying spaces, and the internal hom is given by the space of linear, continuous maps, again with an appropriate topology. This notion of topology is ideal in that one can show that the usual embedding of V into its second dual is always a bijection. So we have indeed reduced the second dual space to the appropriate size. If one restricts to the category of objects for which the embedding is also a homeomorphism, one obtains a category **RTVec**, and Barr demonstrates:

Theorem 1.36 RTVec *is a complete, cocomplete $*$-autonomous category.*

Example 1.37 *Coherence spaces and linear maps.* This example led to linear logic. Recall the ccc **Coh** of coherence spaces and stable maps was discussed at the end of Section 1.1.7, Example 10. A morphism $f : \mathcal{A} \to \mathcal{B}$ in **Coh** is *linear* if for any $X \subseteq \mathcal{A}$ such that for all $b, c \in X$, $b \cup c \in \mathcal{A}$, we have $f(\bigcup X) = \bigcup \{ f(b) \mid b \in X \}$. Let **Coh**$_{lin}$ be the subcategory of **Coh** consisting of coherence spaces and linear maps. This is $*$-autonomous, via the familiar constructions [39].

Example 1.38 *And many more....* The above list is by no means comprehensive. $*$-autonomous categories appear in many guises, in many branches of mathematics. There are at least three additional examples which should certainly be mentioned.

- Game semantics. An extremely important class of examples arises from game theory, with important computational properties. We recommend [6] as an introduction.
- The Chu construction. This is a simple construction, which applied to a symmetric monoidal closed category (with pullbacks) canonically yields a $*$-autonomous category. Despite being

Arrow-generating Rules	Equations
$A \xrightarrow{id} A$ $\qquad$ $\dfrac{A \xrightarrow{f} B \quad B \xrightarrow{g} C}{A \xrightarrow{gf} C}$	equations of a category
$\dfrac{A \xrightarrow{f} B \quad A' \xrightarrow{g} B'}{A \otimes A' \xrightarrow{f \otimes g} B \otimes B'}$	$\otimes$ is a functor : $ff' \otimes gg' = (f \otimes g)(f' \otimes g')$ $\qquad id \otimes id = id$
$A \otimes (B \otimes C) \xrightarrow{\alpha} (A \otimes B) \otimes C$ $A \otimes B \xrightarrow{s} B \otimes A$ $I \otimes A \xrightarrow{\ell} A$	α, s, ℓ are natural isos and equations for smcc's
$\dfrac{A \otimes B \xrightarrow{f} C}{A \xrightarrow{f^*} (B \multimap C)}$ $\multimap$R $(A \otimes B) \otimes A \xrightarrow{ev} B$	equations for monoidal closedness
$\dfrac{\Gamma \xrightarrow{f} A \quad \Gamma \xrightarrow{g} B}{\Gamma \xrightarrow{\langle f, g \rangle} A \times B}$ $A \times B \xrightarrow{\pi_1} A \quad A \times B \xrightarrow{\pi_2} B$ $A \xrightarrow{!_A} \top$	cartesian products

Fig. 1.1. SMCC's with Products.

straightforward, it would seem to have a number of remarkable properties. The construction is due to Barr and Chu [10], and has been studied extensively by Pratt. See [27] for one example of the applicability of Chu spaces.

• Recent work of Ehrhard and Regnier on Köthe spaces and the differential lambda-calculus suggests a whole new avenue to explore in the categorical semantics of linear logic. See [29, 31].

Summary of necessary structure. To aid comparison with proof theory, let us finally sum up the situation so far.

Figure 1.1 gives us symmetric monoidal closed categories (smcc's) with products. At this point we could also add coproducts, denoted + (or in linear logic ⊕), and their associated equations, dual to products. But

Arrow-generating Rules	Equations
$\dfrac{A \xrightarrow{f} B}{B^{\perp} \xrightarrow{f^{\perp}} A^{\perp}}$	$(-)^{\perp}$ is contravariant functor
$A^{\perp} \longrightarrow (A \multimap \perp)$ $(A \multimap \perp) \longrightarrow A^{\perp}$	These are natural isos
$(A \multimap B) \to (B^{\perp} \multimap A^{\perp})$	Natural strength iso
$A \to ((A \multimap \perp) \multimap \perp)$	natural iso

Fig. 1.2. Adding Negation.

as mentioned previously, once we have *-autonomous categories, we get duality for free, essentially by De Morgan duality.

The equations in Figure 1.2 specify that the action of the functor $(-)^{\perp}$ is given by a dualizing object $\perp$, and a natural iso $(-)^{\perp\perp} \cong id$.

The next notion is much more familiar mathematically, although logically it corresponds to a rather degenerate case of linear logic: the case where $\otimes = \wideparen{\gamma}$:

Definition 1.39 A *compact closed category* [49] is a symmetric monoidal category such that for each object A there exists a dual object A^*, and canonical morphisms:

$$\nu: I \to A \otimes A^*$$
$$\psi: A^* \otimes A \to I$$

such that evident equations hold. In the case of a strict monoidal category, these equations reduce to the usual adjunction triangles.

Lemma 1.40

- *Compact closed categories are *-autonomous, with the tensor unit as dualizing object.*
- *As in any *-autonomous category in which the tensor unit is the dualizing object, there is a canonical morphism $A \otimes B \to A \wideparen{\gamma} B$. (This is an instance of a more general observation. Such categories validate the **Mix** rule, which states:*

$$\frac{\vdash \Gamma \quad \vdash \Delta}{\vdash \Gamma, \Delta} \; Mix$$

*This rule is not valid in linear logic, but the theory with **Mix** added is of great interest.) For compact closed categories, this map is an isomorphism.*

Examples 1.41

- **Rel**$_\times$ is compact. On objects, define $A \otimes B = A \multimap B = A \times B$, $I = \bot = \{*\}$ (any 1-element set) and $A^\bot = A$. On morphisms, if $A \xrightarrow{R} B$, define $B^\bot \xrightarrow{R^\bot} A^\bot = B \xrightarrow{R^{op}} A$. It is easy to check that

$$\mathbf{Rel}_\times(C \otimes A, B) \cong \mathbf{Rel}_\times(C, A \multimap B) \cong \mathcal{P}(C \times A \times B)$$

$$\mathbf{Rel}_\times(A, B) \cong \mathbf{Rel}_\times(B^\bot, A^\bot)$$

$$(A \otimes B)^\bot = A \times B = A^\bot \otimes B^\bot$$

- **Vec**$_{fd}$ (finite dimensional vector spaces over field **k**) is also compact: Here $\bot = \mathbf{k}$, $V \multimap W = Lin(V, W)$, so $V^\bot = V \multimap \bot = Lin(V, \mathbf{k}) = V^*$, the dual space of V. There is a natural isomorphism $V \cong V^{**}$ given by the canonical map $V \to V^{**}$. Indeed, an arbitrary vector space is finite dimensional iff this canonical map $V \to V^{**}$ is an isomorphism.

- Let P and P' be posets. An *order ideal* from P to P' is a relation $R \subseteq P \times P'$ satisfying

$$x_1 \leq_P x_2 \; \& \; x_2 R y_2 \; \& \; y_2 \leq_{P'} y_1 \implies x_1 R y_1$$

One readily verifies that order ideals do indeed form a category with the inequality itself, viewed as a binary relation, acting as identity. The compact closed structure of **Rel**$_\times$ extends readily, except now $P^\bot = P^{op}$. See [57] for a detailed discussion of this category.

What are monoidal functors between monoidal categories? Here there may be several notions. Let us pick an important one:

Definition 1.42 A *monoidal functor* between monoidal categories is a 3-tuple (F, m_I, m) where $F : \mathcal{C} \to \mathcal{D}$ is a functor, together with two natural transformations $m_I : I \longrightarrow F(I)$ and $m_{UV} : F(U) \otimes F(V) \longrightarrow$

$F(U \otimes V)$ satisfying some coherence diagrams (which we omit). F is *strict* if m_I, m_{UV} are identities. A monoidal functor is *symmetric* if m commutes with the symmetries: $m_{B,A} s_{FA,FB} = F(s_{A,B}) m_{A,B}$, for all A, B.

Finally, we need an appropriate notion of natural transformation for monoidal functors.

Definition 1.43 A natural transformation $\alpha : F \to G$ is *monoidal* if it is compatible with both m_I and m_{UV}, for all U, V, in the sense that the following equations hold:

$$\alpha_I \circ m_I \;=\; m_I$$
$$m_{UV} \circ (\alpha_U \otimes \alpha_V) \;=\; \alpha_{U \otimes V} \circ m_{UV}$$

Let $*$-**Aut**$_{st}$ be the category of $*$-autonomous categories and strict $*$-autonomous functors. We wish to construct *free* such categories, i.e. to find a left adjoint F to the forgetful functor $U : *$-**Aut**$_{st} \to$ **Graph**, so $F(\mathcal{G})$ will be the free $*$-autonomous category generated by the graph $\mathcal{G}$. The procedure is now familiar: one sets up an appropriate deductive system (generated by $\mathcal{G}$) for MLL (cf. Example 5 in Section 1.1.7) and imposes the relevant equations between proofs. A related, but more delicate issue is to set up the fundamental equivalence of categorical logic, as in Example 6 in Section 1.1.7 , between $*$-autonomous categories and their *internal logics*, which are calculi of proof-terms. Thorny categorical questions like dealing with the units and coherence equations must also be taken into account.

These and related issues are discussed in the work of Cockett, Seely and others, see [23, 24, 14, 15, 16], which begins with an alternate approach to $*$-autonomous categories and the model theory of linear logic. The starting point is the notion of *linearly distributive category*[†]. Roughly, *LDC*s axiomatize multiplicative linear logic in terms of tensor and par, as opposed to tensor and negation. So an *LDC* is a category with two monoidal structures which interact via a linear distribution. One may then add negation as an additional structure.

Definition 1.44 A *symmetric linearly distributive category* (SLDC) is a category $\mathcal{C}$ equipped with

- Two bifunctors $\otimes, \invamp \colon \mathcal{C} \times \mathcal{C} \to \mathcal{C}$, together with objects and isomorphisms endowing $\mathcal{C}$ with two monoidal structures.

[†] Originally referred to as weakly distributive categories.

- Linear distributivity natural transformations

  (i) $\omega^L_{ABC} : A \otimes (B \,\wp\, C) \to (A \otimes B) \,\wp\, C$

  (ii) $\omega^R_{ABC} : A \otimes (B \,\wp\, C) \to (A \otimes C) \,\wp\, B$

- a number of coherence conditions.

A *symmetric linearly distributive category with negation* is an SLDC together with an object function $(-)^\perp$ on $ob(\mathcal{C})$ and natural maps $A \otimes A^\perp \xrightarrow{\gamma_A} \perp$, $I \xrightarrow{\tau_A} A \,\wp\, A^\perp$ satisfying:

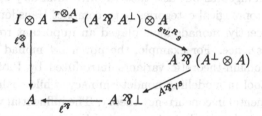

Theorem 1.45 (Cockett-Seely) *SLDC's with negation are the same as ∗-autonomous categories.*

Let us give an example from the above reference.

Example 1.46 A *shift monoid* $(M, 0, +, a)$ is a commutative monoid with an invertible element a. Let $x \cdot y = x + y - a$. Then note that $x \cdot (y + z) = (x \cdot y) + z$, which is an instance of a linear distributivity. Shift monoids are exactly discrete linearly distributive categories. *Shift groups*, i.e. shift monoids which are groups, are exactly discrete ∗-autonomous categories. In the latter case, linear negation is defined by $x^\perp = a - x$.

The references cited above, beginning with the notion of linearly distributive category, extend the categorical analysis of linear logic in several directions. Using (two-sided) proof nets, a natural deduction system for various fragments of linear logic, the authors give explicit constructions of free weakly distributive and ∗-autonomous structures [14], and extend this idea to the exponential fragment [23]. In [24, 16], these ideas are extended to include functors between linearly distributive categories and a logic for the analysis of such functors is presented.

1.3 Monads and Exponentials

1.3.1 Monads

Monads (also called triples or standard constructions) arose in the 1960's from the theory of adjoint functors and have played a central role in category theory ever since. It turns out that many categories of interest are "monadic", i.e. equivalent to categories of (Eilenberg-Moore) algebras of a monad, and thus arise from a pair of adjoint functors, as we discuss below. The general theory has many consequences, for example "monadicity" is a far-reaching generalization of the concept of "equationality" which includes not only traditional equational varieties of universal algebras but also theories with infinitary operations as well as certain topological categories (e.g. compact Hausdorff spaces).

More recently, monads have played an important role in theoretical computer science. For example, the power set monad discussed below (and its domain-theoretic variants introduced by Plotkin) are now a standard tool in modelling nondeterminacy, while coalgebraic methods are fundamental in concurrency theory. The influential work of E. Moggi [55] introduced monads and comonads into programming language semantics as a kind of structuring tool: they permit a modular treatment of such important programming features as exceptions, side-effects, nondeterminism, resumptions, dynamic allocation, etc.

Definition 1.47 A *monad* on a category C is a 3-tuple (T, η, μ), where $T : C \to C$ is a functor, $\eta : Id \longrightarrow T$ (unit) and $\mu : T^2 \longrightarrow T$ (multiplication) are two natural transformations satisfying the following equations:

$$
\begin{array}{ccc}
T & \xrightarrow{\ T\eta\ } & T^2 \\
{\scriptstyle \eta T}\downarrow & \searrow{\scriptstyle id} & \downarrow{\scriptstyle \mu} \\
T^2 & \xrightarrow[\ \mu\ ]{} & T
\end{array}
\qquad\qquad
\begin{array}{ccc}
T^3 & \xrightarrow{\ \mu T\ } & T^2 \\
{\scriptstyle T\mu}\downarrow & & \downarrow{\scriptstyle \mu} \\
T^2 & \xrightarrow[\ \mu\ ]{} & T
\end{array}
$$

A *comonad* on D is a monad on D^{op}. Thus a comonad is a functor $G : D \to D$, together with natural transformations $\varepsilon : G \to Id$ (counit) and $\delta : G \to G^2$ (comultiplication) satisfying the dual of the diagrams above.

Examples 1.48

1. *Power Set Monad:* The covariant power set functor $\mathcal{P} : \mathbf{Set} \to \mathbf{Set}$ determines a monad $(\mathcal{P}, \mu, \eta)$, where $\eta : Id \to \mathcal{P}$ is given by:

$\eta_X(x) = \{x\}$ and $\mu : \mathcal{P}^2 \to \mathcal{P}$ is given by: $\mu_A(\mathcal{F}) = \cup \mathcal{F}$, where $\mathcal{F} \subseteq \mathcal{P}(A)$

2. *Adjoint Functors*: The canonical examples of monads and comonads arise from a pair of adjoint functors $\mathcal{D} \underset{U}{\overset{F}{\rightleftarrows}} \mathcal{C}$, where $F \dashv U$. Let $T = UF : \mathcal{C} \to \mathcal{C}$. Following Exercise 1.7, there are natural transformations $\eta : Id \to UF$ and $\mu = U\varepsilon F : UFUF \to UF$. We leave it as an exercise to check the monad equations.

Continuing this example, we also obtain a comonad $G = FU : \mathcal{D} \to \mathcal{D}$, where $\varepsilon : FU \to Id$ and $\delta = F\eta U : FU \to FUFU$.

In fact, *every* monad $T : \mathcal{C} \to \mathcal{C}$ arises from a pair of adjoint functors $\mathcal{D} \underset{U}{\overset{F}{\rightleftarrows}} \mathcal{C}$. Although the category $\mathcal{D}$ is not unique, there are two natural choices:

Theorem 1.49 (Kleisli, Eilenberg-Moore) Every monad (T, η, μ) on $\mathcal{C}$ arises from a pair of adjoint functors $\mathcal{D} \underset{U}{\overset{F}{\rightleftarrows}} \mathcal{C}$ for two choices of $\mathcal{D}$, now called the Kleisli and the Eilenberg-Moore categories, respectively.

- *The Kleisli Category* of T :
  $\mathcal{D} = Kleisli(T)$ is defined as follows: the objects of $\mathcal{D}$ are the same as the objects of $\mathcal{C}$. The hom-sets are defined as $\mathcal{D}(A, B) = \mathcal{C}(A, TB)$. We define the categorical structure of $\mathcal{D}$ as follows:

  1. *Identity arrows* are defined by setting $id_A : A \to A$ in $\mathcal{D} = \eta_A : A \to TA$ in $\mathcal{C}$.

  2. *Composition* in $\mathcal{D}$ is defined via composition in $\mathcal{C}$ as:

$$\frac{A \xrightarrow{f} B \quad B \xrightarrow{g} C}{A \xrightarrow{gf} C} \text{ in } \mathcal{D} \quad = \quad \frac{\dfrac{A \xrightarrow{f} TB \quad \dfrac{B \xrightarrow{g} TC}{TB \xrightarrow{Tg} T^2C}}{A \xrightarrow{(Tg)f} T^2C} \quad T^2C \xrightarrow{\mu_C} TC}{A \xrightarrow[\mu_C(Tg)f]{} TC}$$

We leave it as an exercise to verify that the associated pair of adjoint functors $\mathcal{D} \underset{U}{\overset{F}{\rightleftarrows}} \mathcal{C}$ are given by: $U(A) = T(A)$, $U(h) = \mu_B T(h)$ for any object A and any arrow $h : A \to B$ in $\mathcal{D} = Kleisli(T)$ and $F(A) = A$, $F(f) = \eta_B f$ for objects A and arrows $f : A \to B$ in $\mathcal{C}$.

- *The Eilenberg-Moore Category*
  The Eilenberg-Moore category $\mathcal{D} = \mathcal{C}^T$ of a monad T is defined as

follows: its objects are arrows $TA \xrightarrow{\alpha} A$ (called T-*algebras*) satisfying: $\alpha\eta_A = id_A$ and $\alpha\mu_A = \alpha T(\alpha)$. Morphisms of T-algebras are arrows $A \xrightarrow{f} B \in C$ commuting with the T-algebra structure. The definition of objects and arrows in C^T is illustrated by the following commuting diagrams:

We leave it to the reader to check that C^T is a category, with functors $C^T \underset{U}{\overset{F}{\rightleftarrows}} C$ given by: $U(TA \xrightarrow{\alpha} A) = A$ and $U(h) = h$ for objects and arrows in C^T, and $F(A) = T^2(A) \xrightarrow{\mu_A} TA$ and $F(f) = T(f)$ for objects and arrows in C.

Exercise 1.50

(i) The Kleisli category of the power set monad P on **Set** has sets for objects and functions $A \xrightarrow{r} P(B)$ for arrows. The arrows may be identified with relations $R \subseteq A \times B$. Check that composition in the Kleisli category corresponds to relational composition. Conclude that $Kleisli(P) = $ **Rel**.

(ii) The Eilenberg-Moore category of P is exactly the category of sup-complete lattices and sup-preserving maps.

It may be shown that the Kleisli category of a monad is equivalent to the full subcategory of the Eilenberg-Moore category consisting of all free algebras, where free algebras in the Eilenberg-Moore category are those of the form $\mu_A \colon T^2A \rightarrow TA$. .

We can dualize the entire discussion above and speak of the co-Kleisli category of a comonad $G : D \rightarrow D$, of Eilenberg-Moore categories of coalgebras, etc. In this case, $Co - Kleisli(G)$ will be equivalent to the full subcategory of cofree coalgebras of the Eilenberg-Moore category of G. This will be relevant for linear logicians, as we now show.

1.3.2 Adding Exponentials to Linear Logic

The deductive system for MALL, and the equations between proofs we postulated previously, correspond to the theory of *-autonomous categories with products and coproducts. Although minor variations are possible (e.g. weak vs ordinary products), the story so far seems to yield a natural and satisfying categorical modelling of MALL proof theory.

Unfortunately, the exponentials are less clear: the structure seems less canonical. Work by many categorical logicians has refined the original Seely model (e.g. see [60, 11, 15]), resulting in interesting and reasonable equations between proofs. We begin with seven basic derivations and postulate equations which arise directly from the categorical point of view.

$$A \xrightarrow{f} B$$

Functoriality $\quad !A \xrightarrow{!f} !B$

Monoidalness $\quad I \xrightarrow{m_I} !I \qquad !A \otimes !B \xrightarrow{m_{AB}} !(A \otimes B)$

Products $\quad I \xrightarrow{n_I} !\top \qquad !A \otimes !B \xrightarrow{n_{AB}} !(A \& B)$

Dereliction $\quad !A \xrightarrow{\varepsilon_A} A$

Weakening $\quad !A \xrightarrow{\varepsilon'_A} I$

Contraction $\quad !A \xrightarrow{\delta'_A} !A \otimes !A$

Digging $\quad !A \xrightarrow{\delta_A} !!A$
(Storage)

Let C be a model of MALL proofs, i.e. a $*$-autonomous category with products (and hence coproducts). We postulate the following additional equational data:

- $(!, m_I, m_{AB}) : C \to C$ is a monoidal endofunctor
- $!A \xrightarrow{\varepsilon_A} A$ and $!A \xrightarrow{\delta_A} !!A$ are monoidal natural transformations.
- $(!, \delta, \varepsilon)$ is a monoidal comonad.
- n_I, n_{AB} are natural isomorphisms.
- The associated adjunction structure $\langle F, U, \eta, \varepsilon \rangle$ between the co-Kleisli category of $!$ and C is monoidal.
- Various coherence equations [15, 46].

Having products and the canonical isomorphism $!(A \& B) \cong !A \otimes !B$ gives added features that must be postulated in weaker fragments (cf [11, 15]). For example, the following are a consequence of the above properties:

Examples 1.51

(i) The endofunctor $!$ establishes an isomorphism of the following cocommutative comonoids: $I \xleftarrow{\epsilon'_A} !A \xrightarrow{\delta'} !A \otimes !A \cong !(\top \longleftarrow A \xrightarrow{\Delta} A \& A)$

(ii) The forgetful functor from the category of $\otimes$-comonoids in $\mathcal{C}$, say $\otimes$-Comonoids$(\mathcal{C}) \xrightarrow{U} \mathcal{C}$ has a right adjoint $U \dashv$! rendering $(!A, \epsilon'_A, \delta'_A)$ a cofree,cocommutative $\otimes$-comonoid object in $\mathcal{C}$.

Finally, we remark that the essence of Girard's original translation of intuitionistic logic into CLL is the following observation:

Proposition 1.52 (Seely) *The co-Kleisli category of the comonad* $(!, \delta, \varepsilon)$ *is a cartesian closed category, in which* $A \Rightarrow B = \; !A \multimap B$

Indeed, recall that in the co-Kleisli category of !, $Hom(A, B)$ is defined to be $\mathcal{C}(!A, B)$.

Examples 1.53

(i) The category **Rel** has exponentials. Let X be a set, and define $!X$ to be the set of all finite multisets on X. We leave the remaining details to the reader. But we note that this is an instance of a more general construction. Finite multisets on X could be written as follows:

$$!X = 0 \oplus X \oplus X \otimes_s X \oplus X \otimes_s X \otimes_s X \ldots$$

Here 0 is the initial object, $\oplus$ is disjoint union, which acts as both product and coproduct in this category, and $\otimes_s$ is the *symmetrized tensor*, and is expressed as a certain quotient (coequalizer). For example, $X \otimes_s X$ is the coequalizer of the identity and the symmetry map. In general, the n'th symmetric group acts on the n-fold tensor of X with itself, and the symmetrized tensor is the coequalizer of all of these maps.

It was an observation of Barr that this formula works frequently, but certainly not always. We recommend [46] as a reference which considers these issues.

(ii) The inclusion **Coh**$_{lin} \hookrightarrow$ **Coh** has a left adjoint ! $\dashv \hookrightarrow$. Thinking of a coherence space as a graph, $!\mathcal{A} = (\mathcal{A}_{fin}, \bigcirc)$, where $\mathcal{A}_{fin}$ is the set of finite cliques in the graph $\mathcal{A}$ and where $a \bigcirc b$ iff $a \cup b$ is a clique. The co-Kleisli category induced by this comonad is equivalent to **Coh**. (see for example [63]).

(iii) The category of finiteness spaces described above also has an exponential, as observed in [30]. Let $(X, \mathcal{F})$ be a finiteness space, then define its exponential by

$$!(X, \mathcal{F}) = (M(X), M(\mathcal{F}))$$

where $M(X)$ is the set of all finite multisets on X, and $M(\mathcal{F}))$ is an appropriate set of subsets. See [30] for further details.

1.4 Traced monoidal categories and the geometry of interaction

Traced monoidal categories, introduced by Joyal, Street, and Verity [47], provide a convenient framework for discussing iteration, parametrized feedback and fixedpoints in computation, algebra of networks, and categorical aspects of Girard's Geometry of Interaction (GoI) program [4, 40, 43].

Definition 1.54 A *traced symmetric monoidal category* is a symmetric monoidal category $(\mathcal{C}, \otimes, I, s)$ with a family of functions $\mathrm{Tr}^U_{X,Y} : C(X \otimes U, Y \otimes U) \longrightarrow C(X, Y)$ pictured in Figure 1.3, called a *trace*, subject to the following conditions:

(i) **Natural** in X, $\mathrm{Tr}^U_{X,Y}(f)g = \mathrm{Tr}^U_{X',Y}(f(g \otimes 1_U))$, where $f : X \otimes U \longrightarrow Y \otimes U$, $g : X' \longrightarrow X$,

(ii) **Natural** in Y, $g\mathrm{Tr}^U_{X,Y}(f) = \mathrm{Tr}^U_{X,Y'}((g \otimes 1_U)f)$, where $f : X \otimes U \longrightarrow Y \otimes U$, $g : Y \longrightarrow Y'$,

(iii) **Dinatural** in U, $\mathrm{Tr}^U_{X,Y}((1_Y \otimes g)f) = \mathrm{Tr}^{U'}_{X,Y}(f(1_X \otimes g))$, where $f : X \otimes U \longrightarrow Y \otimes U'$, $g : U' \longrightarrow U$,

(iv) **Vanishing (I, II)**, $\mathrm{Tr}^I_{X,Y}(f) = f$ and $\mathrm{Tr}^{U \otimes V}_{X,Y}(g) = \mathrm{Tr}^U_{X,Y}(\mathrm{Tr}^V_{X \otimes U, Y \otimes U}(g))$, for $f : X \otimes I \longrightarrow Y \otimes I$ and $g : X \otimes U \otimes V \longrightarrow Y \otimes U \otimes V$.

(v) **Superposing**,

$$g \otimes \mathrm{Tr}^U_{X,Y}(f) = \mathrm{Tr}^U_{W \otimes X, Z \otimes Y}(g \otimes f)$$

for $f : X \otimes U \longrightarrow Y \otimes U$ and $g : W \longrightarrow Z$.

(vi) **Yanking**, $\mathrm{Tr}^U_{U,U}(\sigma_{U,U}) = 1_U$.

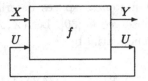

Fig. 1.3. The trace $\mathrm{Tr}_{X,Y}^{U}(f)$

We think of $\mathrm{Tr}_{X,Y}^{U}(f)$ as "feedback along U", as in Figure 1.3. Similarly, the axioms of traced monoidal categories have a geometrical representation given in [4] (Appendix 1).

Examples 1.55

1. The category $\mathbf{Rel}_\times$ is traced. Let $R : X \times U \longrightarrow Y \times U$ be a morphism in $\mathbf{Rel}_\times$. Then $\mathrm{Tr}_{X,Y}^{U}(R) : X \longrightarrow Y$ is defined by: $\mathrm{Tr}_{X,Y}^{U}(R)(x,y) = \exists u.R(x,u,y,u)$.

2. The category $\mathbf{Vec}_{fd}$ is traced. Let $f : V \otimes U \longrightarrow W \otimes U$ be a linear map, where U, V, W are finite dimensional vector spaces with bases $\{u_i\}, \{v_j\}, \{w_k\}$. We define $\mathrm{Tr}_{V,W}^{U}(f) : V \longrightarrow W$ by:

$$\mathrm{Tr}_{V,W}^{U}(f)(v_i) = \sum_{j,k} a_{ij}^{kj} w_k \qquad \text{where } f(v_i \otimes u_j) = \sum_{k,m} a_{ij}^{km} w_k \otimes u_m.$$

This reduces to the usual trace of $f : U \longrightarrow U$ when V and W are one dimensional.

3. Note that both $\mathbf{Rel}$ and $\mathbf{Vec}_{fd}$ are compact closed categories. More generally [47], every compact closed category has a canonical trace:

$$Tr_{A,B}^{U}(f) = A \cong A \otimes I \overset{id \otimes \eta}{\longrightarrow} A \otimes U \otimes U^{\perp} \overset{f \otimes id}{\longrightarrow} B \otimes U \otimes U^{\perp} \overset{id \otimes ev'}{\longrightarrow} B \otimes I \cong B$$

where $ev' = ev \circ s$.

4. The category $\omega\text{-}\mathbf{CPO}_{\perp}$ consists of objects of $\omega\text{-}\mathbf{CPO}$ with a smallest element $\perp$, and maps of $\omega\text{-}\mathbf{CPO}$ that do not necessarily preserve $\perp$. Here $\otimes = \times$, $I = \{\perp\}$. The (dinatural) family of least-fixed-point combinators $Y_U : U^U \to U$ induces a trace, given as follows (using informal lambda calculus notation): for any $f : X \times U \to Y \times U$, $Tr_{X,Y}^{U}(f)(x) = f_1(x, Y_U(\lambda u.f_2(x,u)))$, where $f_1 = \pi_1 \circ f : X \times U \to Y$, $f_2 = \pi_2 \circ f : X \times U \to U$ and $Y_U(\lambda u.f_2(x,u)) = $ the smallest element u' of U such that $f_2(x,u') = u'$.

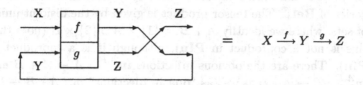

Fig. 1.4. Generalized Yanking

In the above examples, $\otimes$ is based on (cartesian) product. Unfortunately, these examples do not really illustrate the notion of feedback as data flow: the movement of tokens through a network. This latter view, emphasized in work of Abramsky and later Haghverdi and Hines (cf. [4, 40, 43]), is illustrated by examples based on coproduct-like traces, given below.

Exercise 1.56 (Generalized Yanking) Let $\mathcal{C}$ be a traced symmetric monoidal category, with arrows $f : X \to Y$ and $g : Y \to Z$. Then $g \circ f = Tr_{X,Z}^Y(s_{Y,Z} \circ (f \otimes g))$. Find an algebraic proof of this fact. Geometrically, the reader should stare at the diagram in Figure 1.4 , and do a "string-pulling" argument (cf. [47])

The next examples of traced monoidal categories arise in considering "coproduct-like" traces, and are related to dataflow interpretations of graphical networks. We illustrate this view with categories connected to **Rel**.

Examples 1.57

1. **Rel$_+$** , the category **Rel** with $\times = +$, disjoint union. Suppose $X + U \xrightarrow{R} Y + U$ is a relation. The coproduct injections induce four restricted relations : $R_{UU}, R_{UY}, R_{XY}, R_{XU}$ (for example, $R_{XY} \subseteq X \times Y$ is such that $R_{XY}(x, y) = R(in_1^{X,U}(x), in_1^{Y,U}(y))$. Let R^* be the reflexive, transitive closure of the relation R. A trace can be defined as follows:

$$Tr_{X,Y}^U(R) = R_{XY} \cup \bigcup_{n \geq 0} R_{UY} \circ R_{UU}^n \circ R_{XU}$$

$$= R_{XY} \cup R_{UY} \circ R_{UU}^* \circ R_{XU}. \tag{1.5}$$

2. Consider the categories **Pfn** and **PInj** of sets and partial functions

(resp. sets and partial injective functions), as monoidal subcategories of $\mathbf{Rel}_+$. The tensor product is given by the disjoint union of sets, where we identify $A + B = \{1\} \times A \cup \{2\} \times B$ (note that this is not a coproduct in $\mathbf{PInj}$, although it is a coproduct in $\mathbf{Pfn}$). There are the obvious injections $in_1^{A,B} : A \to A + B$ and $in_2^{A,B} : B \to A + B$ as well as "quasiprojections" $\rho_1 : A + B \longrightarrow A$ given by $\rho_1((1, a)) = a$ (where $\rho_1((2, b))$ is undefined) and similarly for $\rho_2 : A + B \longrightarrow B$.

Given a morphism $f : X + U \longrightarrow Y + U$, we may consider its four "components" $f_{XY} : X \to Y$, $f_{XU} : X \to U$, $f_{UX} : U \to X$, and $f_{UU} : U \to U$ obtained by pre- and post-composing with injections and quasiprojections: for example, $f_{XY} = X \xrightarrow{in_1} X + U \xrightarrow{f} Y + U \xrightarrow{\rho_1} Y$, (See Figure 1.5).

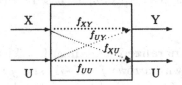

Fig. 1.5. Components of $f : X + U \to Y + U$

Both $\mathbf{Pfn}$ and $\mathbf{Pinj}$ are traced, the trace being given by the following iterative formula

$$\mathrm{Tr}_{X,Y}^U(f) = f_{XY} + \sum_{n \in \omega} f_{UY} f_{UU}^n f_{XU}, \tag{1.6}$$

which we interpret as follows: a family $\{h_i\}_{i \in I} : X \longrightarrow Y$ is said to be *summable* if the h_i's have pairwise disjoint domains and codomains. In that case, we define their sum

$$(\sum_{i \in I} h_i)(x) = \begin{cases} h_j(x), & \text{if } x \in \mathrm{Dom}(h_j) \text{ for some } j \in I; \\ \text{undefined}, & \text{else.} \end{cases}$$

From a dataflow view, particles enter through X, travel around a loop on U some number n of times, then exit through Y. Numerous other examples of such "coproduct-like" traces are studied in [4].

The iterative trace formulas (1.5) and (1.6) are versions of Girard's *Execution Formula* from his GoI program. A general categorical framework for discussing such traces, and their connections to Girard's original work, is studied in Haghverdi's work [40].

On a more general level, starting with a traced monoidal category $\mathcal{C}$, we now describe a compact closed category $\mathbf{Int}(\mathcal{C})$ described in [47] (also called $\mathcal{G}(\mathcal{C})$ in [1]) which captures in abstract form many of the features of Girard's Geometry of Interaction program, as well as the general ideas behind game semantics. We follow the treatment in Abramsky [1].

Definition 1.58 (The Int Construction) Given a traced monoidal category $\mathcal{C}$ we define a compact closed category, $\mathbf{Int}(\mathcal{C})$, as follows [47, 1]:

- Objects: Pairs of objects (A^+, A^-) where A^+ and A^- are objects of $\mathcal{C}$.

- Arrows: An arrow $f : (A^+, A^-) \longrightarrow (B^+, B^-)$ in $\mathbf{Int}(\mathcal{C})$ is an arrow $f : A^+ \otimes B^- \longrightarrow A^- \otimes B^+$ in $\mathcal{C}$.

- Identity: $1_{(A^+, A^-)} = s_{A^+, A^-}$.

- Composition: Arrows $f : (A^+, A^-) \longrightarrow (B^+, B^-)$ and $g : (B^+, B^-) \longrightarrow (C^+, C^-)$ have composite $g \circ f : (A^+, A^-) \longrightarrow (C^+, C^-)$ given by:

$$g \circ f = Tr^{B^- \otimes B^+}_{A^+ \otimes C^-, A^- \otimes C^+}(\beta(f \otimes g)\alpha)$$

  where $\alpha = (1_{A^+} \otimes 1_{B^-} \otimes s_{C^-, B^+})(1_{A^+} \otimes s_{C^-, B^-} \otimes 1_{B^+})$ and $\beta = (1_{A^-} \otimes 1_{C^+} \otimes s_{B^+, B^-})(1_{A^-} \otimes s_{B^+, C^+} \otimes 1_{B^-})(1_{A^-} \otimes 1_{B^+} \otimes s_{B^-, C^+})$.
  Pictorially, $g \circ f$ is given by symmetric feedback:

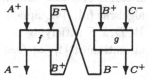

- Tensor: $(A^+, A^-) \otimes (B^+, B^-) = (A^+ \otimes B^+, A^- \otimes B^-)$ and for $(A^+, A^-) \longrightarrow (B^+, B^-)$ and $g : (C^+, C^-) \longrightarrow (D^+, D^-)$, $f \otimes g = (1_{A^-} \otimes s_{B^+, C^-} \otimes 1_{D^+})(f \otimes g)(1_{A^+} \otimes s_{C^+, B^-} \otimes 1_{D^-})$

- Unit: (I, I).

- Duality: The dual of (A^+, A^-) is given by $(A^+, A^-)^\perp = (A^-, A^+)$ where the unit $\eta : (I, I) \longrightarrow (A^+, A^-) \otimes (A^+, A^-)^\perp =_{def} s_{A^-, A^+}$ and counit $\epsilon : (A^+, A^-)^\perp \otimes (A^+, A^-) \longrightarrow (I, I) =_{def} s_{A^-, A^+}$.

- Internal Homs: As usual, $(A^+, A^-) \multimap (B^+, B^-) = (A^+, A^-)^\perp \otimes (B^+, B^-) = (A^- \otimes B^+, A^+ \otimes B^-)$.

Translating the work of [47] in our setting we obtain that $\mathbf{Int}(\mathcal{C})$ is a kind of "free compact closure" of $\mathcal{C}$ at the bicategorical level (for which the reader is referred to [47]):

Proposition 1.59 *Let $\mathcal{C}$ be a traced symmetric monoidal category*

   (i) $\mathbf{Int}(\mathcal{C})$ *defined above is a compact closed category. Moreover, $F_{\mathcal{C}} : \mathcal{C} \longrightarrow \mathbf{Int}(\mathcal{C})$ defined by $F_{\mathcal{C}}(A) = (A, I)$ and $F_{\mathcal{C}}(f) = f$ is a full and faithful embedding.*

   (ii) *The inclusion of 2-categories $\mathbf{CompCl} \hookrightarrow \mathbf{TraMon}$ of compact closed categories into traced monoidal ones has a left biadjoint with unit having component at $\mathcal{C}$ given by $F_{\mathcal{C}}$.*

Following Abramsky [1], we interpret the objects of $\mathbf{Int}(\mathcal{C})$ in a game-theoretic manner: A^+ is the type of "moves by Player (the System)" and A^- is the type of "moves by Opponent (the Environment)". The composition of morphisms in $\mathbf{Int}(\mathcal{C})$ is connected to Girard's execution formula . In [1] it is pointed out that $\mathcal{G}(\mathbf{Pinj})$ captures the essence of the original Girard model, while $\mathcal{G}(\omega\text{-}\mathbf{CPO}_\perp)$ is the model of GoI in [2].

Finally, we remark that in [4], a general analysis of such algebraic models of GoI is given. There it is shown how to use the above abstract GoI construction to obtain models of the $\{!, \multimap\}$ fragment of linear logic, presented in terms of *linear combinatory algebras*. These are certain combinatory algebras $(\mathcal{A}, \cdot)$ equipped with a map $! : \mathcal{A} \to \mathcal{A}$ and constants $B, C, I, K, W, D, \delta, F$ satisfying the combinatory identities for a Hilbert-style axiomatization of $\{!, \multimap\}$ (see also [63]). The method is sketched as follows.

Let $\mathcal{C}$ be a traced smc, with an endofunctor $T : \mathcal{C} \to \mathcal{C}$ and an object (called a *reflexive object*) $U \in \mathcal{C}$ with retractions $U \otimes U \lhd U$, $I \lhd U$, and $TU \lhd U$. Then if T satisfies some reasonable axioms and setting $V = (U, U)$ and $I = (I, I)$, it is shown in [4] how the homset $\mathbf{Int}(\mathcal{C})(I, V) = \mathcal{C}(U, U)$ naturally inherits the structure of a linear combinatory algebra. For example, in the case of $\mathcal{C} = \mathbf{Pinj}$, $\mathbb{N}$ is such a reflexive object, with endofunctor $T(-) = \mathbb{N} \times (-)$. This example underlies the original Girard GoI constructions. The model in [2] likewise arises from $\mathbf{Int}(\mathbf{CPO}_\perp)$. Moreover, Girard's original operator-theoretic models (in the category of Hilbert spaces), as well as Danos-Regnier's *small model* [26] are also captured in the above framework using some additional functorial structure (see [40], Section 6).

1.5 Nonsymmetric monoidal categories

One of the most appealing features of linear logic is its flexibility; one can readily define variants of linear logic which either have a full exchange rule or a very limited exchange rule. These variants correspond to the varying degrees of symmetry that one gives to the tensor. In short, just as there are nonsymmetric monoidal categories, there is nonsymmetric linear logic. The most interesting examples of nonsymmetric monoidal categories occur in the representation theory of *Hopf algebras*. Hopf algebras arise in many areas of physics, computer science and combinatorics. In this section, we review the basics of nonsymmetric monoidal categories, how Hopf algebras provide examples, and how these examples correspond to various types of linear logic.

If we drop the requirement that the tensor be symmetric, then one should consider categories with two internal HOM's. Thus we should have adjunctions of the form:

$$HOM(A \otimes B, C) \cong HOM(B, A \multimap C)$$
$$HOM(A \otimes B, C) \cong HOM(A, C \multimapinv B)$$

This is the definition of *biautonomous* category, an obvious generalization of the symmetric case. Of course, if the tensor happens to be symmetric, this will induce an isomorphism between the two HOM's.

Analogously, to define a nonsymmetric analogue of categories with dualizing objects one needs two duals, $A^{\perp}$ and $^{\perp}A$. (The dualizing object for each will be the same.) These will be subject to the isomorphisms:

$$^{\perp}(A^{\perp}) \cong (^{\perp}A)^{\perp} \cong A$$

More specifically, a biautonomous category has a canonical morphism:

$$A \longrightarrow {}^{\perp}(A^{\perp}) \cong (^{\perp}A)^{\perp}$$

and if this map is an isomorphism, then we have a *bi-*-autonomous* category. (In general, there will be no relationship between A and $A^{\perp\perp}$ in the nonsymmetric case.)

We now discuss a variant of these categories.

Definition 1.60 *If in a bi-*-autonomous category, the dualizing object, $\perp$, has the property that:*

$$^{\perp}A \cong A^{\perp}$$

or equivalently:

$$A \multimapdotinv \bot \cong \bot \multimapdot A$$

then $\bot$ *is said to be* cyclic. *A* ∗-*autonomous category with such a dualizing object is also said to be cyclic.*

In the posetal case, these are the Girard quantales, and were introduced by Yetter in [64] and studied by Rosenthal [58]. A notion of proof net for this theory is contained in [64].

Yetter's *cyclic linear logic* is obtained by replacing the usual exchange rule with:

$$\frac{\vdash A_1, A_2, \ldots, A_n}{\vdash A_{\sigma(1)}, A_{\sigma(2)}, \ldots, A_{\sigma(n)}}$$

where σ is a *cyclic* element of the symmetric group on n letters. It is straightforward to verify that a ∗-autonomous category with a cyclic dualizing object validates this rule.

Noncommutative linear logic, with the cyclic exchange rule would seem to be the optimal level of noncommutativity. The theory has an excellent semantics, sequent calculus and proof nets. Similarly well-behaved structures for fully noncommutative linear logic have proven to be much more problematic.

We note that the subject of noncommutative linear logic has not been explored as extensively as other aspects of linear logic. In addition to obtaining further noncommutative full completeness theorems, there are also a number of logical variants of cyclic linear logic that should be considered. Indeed, Ruet's recent variant, called simply *noncommutative logic*, ultimately suggests that the number of noncommutative variants may be almost endless. See [7] for an analysis of its syntax, as well as a notion of proof net for this logic.

Ruet's logic is a mix of commutative and noncommutative elements. In it, there are two sets of connectives, one an ordinary commutative tensor and par and the other a cyclic noncommutative tensor and par. Interaction between the two systems is mediated by a structural rule called *entropy*.

It is hoped that for any possible version of noncommutative linear logic, there is a corresponding notion of Hopf algebra (see below). For Ruet's logic, there is the notion of *entropic Hopf algebra* developed in [17].

1.5.1 *Representations of Hopf algebras*

We now introduce Hopf algebras as a means of constructing examples of nonsymmetric monoidal closed categories. Hopf algebras are best considered as a nonsymmetric generalization of the category $\mathcal{MOD}(G)$ of G-modules (see Examples 1.26, Number 5). We recommend [48, 54] as excellent introductions. We begin with some preliminaries. We assume throughout a fixed, but arbitrary field **k**.

Definition 1.61 An *(associative) algebra* is a k-vector space H equipped with maps $m: H \otimes H \to H$ and $\eta: k \to H$ which are called the multiplication and unit, and these must satisfy the evident equations for associativity and unit. Dually one may define a *(coassociative) coalgebra* as a space with maps $\Delta: H \to H \otimes H$ and $\varepsilon: H \to k$ satisfying the dual axioms. Then a *Hopf algebra* is a k-vector space H equipped with an algebra structure, a compatible coalgebra structure and a map $S: H \to H$ called the antipode satisfying appropriate equations. The following chart summarizes the necessary structure. All maps shown are linear.

Structure		Equations
Algebra	$m: H \otimes H \to H$ (multiplication) $\eta: k \to H$ (unit)	Associativity and Unit: $m \circ (m \otimes id) = m \circ (id \otimes m)$ and $\eta(1)$ is 2-sided unit for m.
Coalgebra	$\Delta: H \to H \otimes H$ (comultiplication) $\varepsilon: H \to k$ (counit)	Coassociativity with counit for comultiplication (dual to algebra structure).
Bialgebra	Algebra + Coalgebra	Δ and ε are algebra homs. (Equivalently m, η are coalgebra homs.)
Antipode	$S: H \to H$	Inverse to $id_H : H \to H$ under convolution

Here convolution refers to the operation on $Hom_k(H, H)$ defined by $(f * g)(c) = m((f \otimes g)(\Delta c))$. The identity for the convolution operation is given by $\eta\varepsilon : H \to H$. We say a Hopf algebra is *(co)commutative* if the (co)multiplication is (co)commutative (i.e. the appropriate diagram or its dual commutes [48, 54].) $\qquad\qquad \square$

Example 1.62 A particular Hopf algebra which provides the semantics of cyclic linear logic is known as the *shuffle algebra*. It is an example of an *incidence algebra* and is of fundamental importance in several areas of mathematics. The terminology below is motivated by thinking of shuffling a deck of cards.

Let X be a set and X^* the free monoid generated by X. We denote words (= strings) in X^* by $w, w', \cdots$ and occasionally $z, z' \ldots$. Elements $x, y, \cdots \in X$ are identified with words of length 1, the empty word (= unit of the monoid) is denoted by ϵ, and the monoid multiplication is given by concatenation of strings. We denote the length of word w by $|w|$. Let $\mathbf{k}[X^*]$ be the free k-vector space generated by X. We consider $\mathbf{k}[X^*]$ endowed with the following Hopf algebra structure (cf. [20]):

(i) $\mathcal{A} = \mathbf{k}[X^*]$ is an *algebra*, i.e. comes equipped with an associative k-linear multiplication (with unit) $m : \mathcal{A} \otimes \mathcal{A} \to \mathcal{A}$:

$$w \otimes w' \;\mapsto\; w{\cdot}w' = \sum_{u \in Sh(w,w')} u \qquad (1.7)$$

where $Sh(w, w')$ denotes the set of "shuffled" words of length $|w| + |w'|$ obtained from w and w'. Here, a *shuffle* of $w = a_1 \cdots a_m$ and $w' = a'_1 \cdots a'_n$ is a word of length $m + n$, say $w'' = c_1 \cdots c_{m+n}$ such that each of the a_i and a'_j occurs once in w''; moreover, within w'', a_i and a'_j occur in their original sequential order. For example, if $w = aba$ and $w' = bc$, we obtain the following set of shuffled words (where the letters from w' are underlined)

$$abab\underline{c},\; ab\underline{b}a\underline{c},\; ab\underline{b}a\underline{c},\; \underline{b}aba\underline{c},\; ab\underline{bc}a,\; ab\underline{bc}a,\; \underline{b}ab\underline{c}a,\; a\underline{bc}ba,\; \underline{ba}\underline{c}ba,\; \underline{bc}aba$$

Thus the summation $w{\cdot}w'$ is equal to

$$ababc + 2abbac + babac + 2abbca + babca + abcba + bacba + bcaba$$

Note that we always denote the shuffle multiplication with $\cdot$, as opposed to the monoid multiplication, for which we use concatenation.

The unit $\eta : \mathbf{k} \to \mathcal{A}$ arises by mapping $1 \mapsto \epsilon$.

(ii) $\mathcal{A} = \mathbf{k}[X^*]$ is a *coalgebra*, i.e. comes equipped with a coassociative comultiplication (with counit) $\Delta : \mathcal{A} \to \mathcal{A} \otimes \mathcal{A}$, defined as:

$$\Delta(w) = \sum_{w_1 w_2 = w} w_1 \otimes w_2 \qquad (1.8)$$

Note that in the equation $w_1 w_2 = w$ we are using the original monoid multiplication of X^*. The above pair $w_1 w_2$ is called a *cut* of w.

The counit $\varepsilon : \mathcal{A} \to \mathbf{k}$ is defined by:

$$\varepsilon(w) = \begin{cases} 1 & \text{if } w = \epsilon \\ 0 & \text{else} \end{cases} \tag{1.9}$$

Finally, there is an *antipode* defined as

$$S(w) = (-1)^{|w|}\bar{w} \tag{1.10}$$

where $\bar{w}$ denotes the word w written backwards.

Proposition 1.63 $\mathcal{A} = \mathbf{k}[X^*]$ *with the above structure forms a Hopf algebra with involutive antipode.*

1.5.2 H-*Modules*

In analogy with the notion of G-space, we may speak of the action of a Hopf algebra H on a vector space V. This is a linear map $\rho : \mathsf{H} \otimes V \to V$ satisfying the analog of the action equations above:

Definition 1.64 Given a Hopf algebra H, a *module* over H is a vector space V, equipped with a linear map called an H-*action* $\rho \colon \mathsf{H} \otimes V \to V$ such that the following diagrams commute:

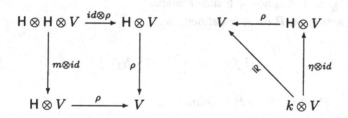

We will generally denote an H-action by concatenation, e.g. $\rho(h \otimes v) = hv$. Then the above diagrams translate, respectively, to: $(hh')v = h(h'v)$ and $\eta(1)v = v$, for all $h, h' \in \mathsf{H}, v \in V$. We shall frequently denote $\eta(1)$ by 1_{H}.

If (V, ρ) and (W, τ) are modules, then *a map of modules*, sometimes called an H-*map*, is a k-linear map $f \colon V \to W$ such that the following commutes:

$$H \otimes V \xrightarrow{id \otimes f} H \otimes W$$

$$\rho \downarrow \qquad\qquad \downarrow \tau$$

$$V \xrightarrow{\quad f \quad} W$$

i.e. in the above notation, $f(hv) = hf(v)$ for all $h \in H, v \in V$. We thus obtain a category $\mathcal{MOD}(H)$.

The above definition is a straightforward generalization from group representations; indeed, the latter arises as the special case $H = k[G]$. If U and V are modules, then $U \otimes V$ has a natural module structure given by:

$$H \otimes U \otimes V \xrightarrow{\Delta \otimes id} H \otimes H \otimes U \otimes V \xrightarrow{c_{23}} H \otimes U \otimes H \otimes V \xrightarrow{\rho \otimes \rho} U \otimes V$$

Theorem 1.65 (See [54, 48] for details.) $\mathcal{MOD}(H)$ *is a monoidal category. If the Hopf algebra is cocommutative, then the tensor product is symmetric. The unit for the tensor is given by the ground field with the module structure induced by the counit of* H.

Definition 1.66 Given an arbitrary Hopf algebra H with bijective antipode, and two H-modules, A and B, we will define two new H-modules, $A \multimap B$ and $B \multimapinv A$, as follows. In both cases, the underlying space will be $A \multimap_k B$, the space of k-linear maps.

The action on $B \multimapinv A$ is defined by:

$$(hf)(a) = \sum h_1 f(S(h_2)a) \tag{1.11}$$

and the action on $A \multimap B$ is defined by:

$$(hf)(a) = \sum h_2 f(S^{-1}(h_1)a) \tag{1.12}$$

where $\Delta(h) = \sum h_1 \otimes h_2$.

A proof of the following can be found in [54]. See that reference also for the history of these constructions, many of which are due to that author.

Theorem 1.67 *Let* H *be a Hopf algebra with bijective antipode. Then with the actions defined above,* $\mathcal{MOD}(\mathsf{H})$ *is a biautonomous category. The adjoint relation:*

$$HOM(A \otimes B, C) \cong HOM(B, A \multimap C)$$

holds whether or not the antipode is bijective. In the case of a cocommutative Hopf algebra, the two internal HOM's are equal.

So the representation theory of Hopf algebras provides us access to a wide variety of models of noncommutative (intuitionistic) linear logic. There are several ways to extend this to obtain classical models. One could restrict to finite-dimensional representations, or again use the topological category **RTVec** for representations. The expository paper [13] considers these ideas. In particular, we note the following as an example of the usefulness of Hopf algebras.

Theorem 1.68 *If* H *is a Hopf algebra with involutive antipode, then its category of finite-dimensional representations or representations in* **RTvec** *is a model of cyclic linear logic.*

As a corollary, we obtain from Proposition 1.63 that the *shuffle Hopf algebra* $\mathcal{A} = \mathbf{k}[X^*]$ models cyclic linear logic. Such Hopf algebras were used to obtain a *full completeness theorem* for cyclic multiplicative linear logic in [20]. This will be discussed in the next section.

1.6 Full Completeness and Representation Theorems

The most basic representation theorem of all is the Yoneda embedding:

Theorem 1.69 (Yoneda) *If* $\mathcal{A}$ *is locally small, the Yoneda functor* $\mathcal{Y} : \mathcal{A} \to \mathbf{Set}^{\mathcal{A}^{op}}$, *where* $\mathcal{Y}(A) = Hom_{\mathcal{A}}(-, A)$, *is a fully faithful embedding.*

Indeed, Yoneda preserves limits as well as cartesian closedness. This theorem, and its many variants, is critical to the development of category theory and categorical model theory.

However we seek mathematical models which fully and faithfully represent proofs. From the viewpoint of a logician, these are completeness theorems, but now at the level of proofs rather than provability. The results are known as *full completeness theorems*. The terminology arose in the work of Abramsky and Jagadeesan on full completeness for MLL + Mix in *-autonomous categories of games [3].

Definition 1.70 Let $\mathcal{F}$ be a free category. We say that a categorical model $\mathcal{M}$ is *fully complete for* $\mathcal{F}$ or that we have *full completeness of* $\mathcal{F}$ *with respect to* $\mathcal{M}$ if, with respect to some interpretation of the generators, the unique free functor $[\![-]\!] : \mathcal{F} \to \mathcal{M}$ is full. It is even better to demand that $[\![-]\!]$ is a fully faithful representation.

For example, suppose $\mathcal{F} = F(\mathcal{G}_0)$ is a free structured category (e.g. free ccc, *-autonomous, etc.) generated by the appropriate deductive system on a discrete graph $\mathcal{G}_0$. To say a categorical model $\mathcal{M}$ is fully complete for $\mathcal{F}$ means: any arrow $[\![A]\!] \to [\![B]\!] \in \mathcal{M}$ between definable objects is itself definable, i.e. it must be of the form $[\![f]\!]$ for some (equivalence class of a) proof $f : A \to B$ in $\mathcal{F}$. If the representation is fully faithful, then f is unique. Thus, by Curry-Howard-Lambek, any morphism in the model between definable objects is itself the image of a proof (or program); and this proof is unique if the representation is fully faithful.

Such results are mainly of interest when the models $\mathcal{M}$ are "genuine" mathematical models not a priori connected to the syntax. For example, an explicit use of the Yoneda embedding $\mathcal{Y} : \mathcal{F} \to \mathbf{Set}^{\mathcal{F}^{op}}$ is not what we want, since the target model $\mathbf{Set}^{\mathcal{F}^{op}}$ depends too much on $\mathcal{F}$.

Probably some of the earliest full completeness results were for free ccc's (i.e. for simply typed lambda-calculi). Plotkin in the 1970's and Statman in the 1980's studied lambda definability in terms of invariance under logical relations on set-theoretic Henkin models.

In the case of Linear Logic, the fundamental paper of Abramsky and Jagadeesan [3] proved full completeness for MLL + Mix, using categories of games with certain history-free winning strategies as morphisms. It is shown there that "uniform" history-free winning strategies are the denotations of unique proof nets. An alternate notion of game, developed by Hyland and Ong, permits eliminating the Mix rule in such game-theoretic full completeness theorems for the multiplicatives. These results paved the way for the most spectacular application of these game-theoretic methods: the solution of the full abstraction problem for PCF, by Abramsky, Jagadeesan, and Malacaria and by Hyland and Ong. See for example [45].

There have been a host of full completeness theorems for MLL + Mix, MLL, Yetter's CyLL, and recently for MALL. Very roughly speaking, we may distinguish two styles of fully-complete models in the literature:

- *Direct Models*: These are subcategories of some ambient *-autonomous category. The key idea is to impose an invariance or uniformity condition to restrict the class of arrows between definable types to those which are exactly the denotations of proofs. This is typical of the original game theoretic fully-complete models mentioned above (where proofs correspond to certain restricted kinds of winning strategies) as well as to Hamano's direct full completeness theorem for MLL + Mix, which uses a subcategory of **RTVec** restricted to certain Z-invariant maps.

- *Functorial Models*: This approach, and similar ones using relational transformers and Reynolds' parametricity, uses techniques of functorial polymorphism in [9]. The basic idea is to model formulas (i.e. types) as multivariant functors over some base monoidal category, and proofs as multivariant (dinatural) transformations. In the second case, one uses similar relational methods. In either approach, one often imposes additional uniformity requirements on (di)natural families to enable them to exactly correspond to proofs (e.g. in our previous work [19, 20], we supposed dinatural families (discussed below) are invariant under continuous group or Hopf-algebra actions.).

The functorial models and their variants provide a powerful and increasingly popular framework for full completeness proofs. Uniformity is now imposed over (di)natural families, with a much wider range of examples than the direct models approach. But this flexibility comes at a price: the functorial approach only applies to cut-free systems of LL, since dinatural transformations (as well as logical relations) do not compose in general. Hence functorial models, unlike direct models, are not a priori categories.

Since the work of Ralph Loader in the early 1990's, later generalized in work of Hyland and Tan, it has become increasingly important to lift known full completeness theorems to larger base categories. The techniques for doing this involve using a Chu space or Double Gluing construction on top of the base. The point of the Chu or Double-gluing construction is to eliminate various unwanted maps, e.g. moving from a compact base category or one satisfying Mix to a more general *-autonomous setting, and then to rebuild the whole functorial framework at this level. Haghverdi [40] introduces a new class of full completeness theorems for MLL by a 2-step process: first applying a GoI-model construction to certain traced monoidal categories, then applying modified Loader-Hyland-Tan techniques.

In the case of MALL, there are currently two full completeness theorems in the dinatural framework. Part of the difficulty here is that the associated notion of proof-nets for MALL is highly non-trivial. The

first model, by Abramsky and Melliès [5] uses dinaturals over a base
category of so-called *concurrent games*, which are themselves related to
a kind of double-gluing construction. The second, by the authors and
Hamano [18] uses the dinatural framework on a double gluing category
over Ehrhard's category of hypercoherence spaces.

As an example of how full completeness theorems work, we now give
a brief picture of the dinatural results in the authors' papers [19, 20].

Definition 1.71 Let C be a category, and $F, G : (C^{op})^n \times C^n \to C$
functors. A *dinatural transformation* $\theta : F \to G$ is a family of C-
morphisms $\theta = \{\theta_A : FAA \to GAA \mid A \in C^n\}$ satisfying (for any
n-tuple $f : A \to B \in C^n$):

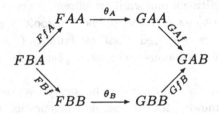

For a history of this notion, see [9].

Let C be a *-autonomous category. Given an MLL formula
$\varphi(\alpha_1, \ldots, \alpha_n)$ built from $\otimes, \multimap, (\ )^\perp$, with type variables $\alpha_1, \ldots, \alpha_n$,
we inductively define its *functorial interpretation* $[\![\varphi(\alpha_1, \ldots, \alpha_n)]\!]$:
$(C^{op})^n \times C^n \to C$ as follows (boldface letters are vectors of objects):

- $[\![\varphi]\!](\mathbf{AB}) = \begin{cases} B_i & \text{if } \varphi(\alpha_1, \ldots, \alpha_n) \equiv \alpha_i \\ A_i^\perp & \text{if } \varphi(\alpha_1, \ldots, \alpha_n) \equiv \alpha_i^\perp \end{cases}$
- $[\![\varphi_1 \otimes \varphi_2]\!](\mathbf{AB}) = [\![\varphi_1]\!](\mathbf{AB}) \otimes [\![\varphi_2]\!](\mathbf{AB})$.

It is readily verified that $[\![\varphi^\perp]\!] = [\![\varphi]\!]^\perp$ and $[\![\varphi_1 \multimap \varphi_2]\!](\mathbf{AB}) = [\![\varphi_1]\!](\mathbf{BA}) \multimap [\![\varphi_2]\!](\mathbf{AB})$, where $A \multimap B$ is defined as $(A \otimes B^\perp)^\perp$.

From now on, let $C = $ RTVec. The set $Dinat(F, G)$ of dinatural transformations from F to G is a vector space, under pointwise operations.
We call it the *proof space* associated to the sequent $F \vdash G$ (where we
identify formulas with definable functors.) If $\vdash \Gamma$ is a one-sided sequent,
then $Dinat(\Gamma)$ denotes the set of dinaturals from $\mathbf{k}$ to $[\![\mathbin{⅋} \Gamma]\!]$.

The following is proved in [19, 20]. A binary sequent is one where each
atom appears exactly twice, with opposite variances. A *diadditive dinatural transformation* is one which is a linear combination of substitution
instances of binary dinaturals.

Theorem 1.72 (Full Completeness for MLL + Mix) *Let F and G be formulas in MLL + Mix, interpreted as definable multivariant functors on* **RTVec**. *Then the proof space $Dinat(F, G)$ of diadditive dinatural transformations has as basis the denotations of cut-free proofs in the theory MLL + Mix.*

Example 1.73 The proof space of the sequent

$$\alpha_1, \alpha_1 \multimap \alpha_2, \alpha_2 \multimap \alpha_3, \ldots, \alpha_{n-1} \multimap \alpha_n \vdash \alpha_n$$

has dimension 1, generated by the evaluation dinatural. Thus any proof of this sequent must be a scalar multiple of the evaluation dinatural.

The proofs of the above results actually yield a fully faithful representation theorem for a free *-autonomous category with Mix, whose homsets are canonically enriched over vector spaces ([19]).

In the same paper we proved a similar Full Completeness Theorem and fully faithful representation theorem for Yetter's Cyclic Linear Logic. In this case one employs the category $\mathcal{RTMOD}(\mathsf{H})$ for a Hopf algebra H. The particular Hopf algebra used is the shuffle Hopf algebra. Once again we consider formulas as multivariant functors on **RTVec**, but restrict the dinaturals to so-called H -*uniform dinaturals* $\theta_{|V_1|, \cdots, |V_n|}$, i.e. those which are equivariant with respect to the H -action induced from the atoms, for H -modules $V_i \in \mathcal{RTMOD}(\mathsf{H})$. This is completely analogous to the techniques used in logical relations.

Theorem 1.74 (Full Completeness for CyLL + Mix) *Let F and G be formulas in MLL + Mix, interpreted as definable multivariant functors on* **RTVec**. *Let H be the shuffle Hopf algebra. Then the proof space of H -uniform diadditive dinatural transformations has as basis the denotations of cut-free proofs in the theory CyLL + Mix.*

From the large literature on MLL full completeness theorems, we end by discussing an interesting line of research stemming from seminal work of Ralph Loader [52], who early on proved full completeness theorems using a linear version of logical predicates. His work led M. Hyland and A. Tan to introduce the method of *double gluing* [46, 62] as a new categorical technique for generating fully complete functorial models.

Definition 1.75 Let $\mathcal{C} = (\mathcal{C}, \otimes, I, (-)^\perp)$ be a compact closed category. We define a new category, $\mathbf{G}\mathcal{C}$, the *double gluing category* of $\mathcal{C}$, whose objects are triples $\mathcal{A} = (A, \mathcal{A}_p, \mathcal{A}_{cp})$ where A is an object of $\mathcal{C}$, where

$A_p \subseteq \mathcal{C}(I, A)$ is called a set of *points* of A and where $A_{cp} \subseteq \mathcal{C}(A, I) \cong$ $\mathcal{C}(I, A^\perp)$ is called a set of *copoints* of A.

A morphism $f : A \longrightarrow B$ in **G$\mathcal{C}$** is a morphism $f : A \longrightarrow B$ in $\mathcal{C}$ such that $f(A_p) \subseteq B_p$ and $f^\perp(B_{cp}) \subseteq A_{cp}$. Composition and identities are induced from the underlying composition and identities in $\mathcal{C}$.

Proposition 1.76 *For any compact closed category $\mathcal{C}$, **G$\mathcal{C}$** is a *-autonomous category, in which*

$$A^\perp = (A^\perp, A_{cp}, A_p)$$
$$A \otimes B = (A \otimes B, (A \otimes B)_p, (A \otimes B)_{cp})$$
$$I_{\mathbf{GC}} = (I, \{id_I\}, \mathcal{C}(I, I))$$

*where $(A \otimes B)_p = \{\alpha \otimes \beta | \alpha \in A_p, \beta \in B_p\}$ and $(A \otimes B)_{cp} = \mathbf{GC}(A, B^\perp)$. The forgetful functor $U : \mathbf{GC} \to \mathcal{C}$ preserves the *-autonomous structure.*

We remark that in a logical setting one can think of an object $A \in \mathbf{GC}$ as a formula A in $\mathcal{C}$ together with a collection of proofs of A (the set A_p) and a collection of refutations of A (the set A_{cp}). Also, we remark that the double gluing construction works more generally for *-autonomous categories $\mathcal{C}$ (see [18, 46]). An important special case is:

Example 1.77 GRel denotes *the double gluing category over the category* **Rel**$_\times$. Its objects are triples $A = (A, A_p, A_{cp})$, where A is a set, $A_p \subseteq \mathbf{Rel}(I, A) = \mathcal{P}(A)$ and $A_{cp} \subseteq \mathbf{Rel}(A, I) = \mathcal{P}(A)$. A morphism $f : A \to B$ of **GRel** is a relation $R : A \to B$ of **Rel** such that:
(image condition:) $\forall \alpha \in A_p \; [\alpha]R := \{b \in \beta \mid \exists a \in \alpha (a, b) \in R\} \in B_p$
(co-image condition:) $\forall \beta \in B_{cp} \; R[\beta] := \{a \in \alpha \mid \exists b \in \beta (a, b) \in R\} \in A_{cp}$

There are many interesting full subcategories of **GRel**, e.g. Loader's *Linear Logical Predicates* and *Totality Spaces* [52], as well as **Coh**.

The Hyland-Tan approach to Loader's method is based on the following ideas. We start with a *compact closed* category $\mathcal{C}$ (e.g. $\mathcal{C} = $ **Rel**). We look at multivariant MLL-definable functors on the double gluing category **G$\mathcal{C}$** and dinatural transformations between them. Full completeness states that every such dinatural corresponds to a Danos-Regnier MLL proof net ρ_θ. The method is as follows:

(i) Using the forgetful functor $U : \mathbf{GC} \to \mathcal{C}$, the dinatural family θ on **G$\mathcal{C}$** is completely determined by arrows in $\mathcal{C}$. It thus suffices to prove a version of full completeness for compact categories $\mathcal{C}$. For **Rel** such a result holds, and it implies that every nontrivial such dinatural θ arises as a union of fixed-point-free involutions. Instantiating θ at appropriate subcategories of **GRel** determines axiom links of a proof structure ρ_θ.

(ii) One shows ρ_θ is a proof net, by showing it is acyclic and connected, using further instantiations of θ in **GRel**.

Haghverdi [40] applied these techniques to compact closed categories arising from GoI, e.g. of the form **Int**($\mathcal{C}$), for certain traced monoidal categories $\mathcal{C}$.

Finally, we should remark that **GRel** has products and coproducts, so is a model of MALL. But neither **GRel** nor *Dinat*(**GRel**) is fully complete for MALL. Instead, it turns out that one must move to *Dinat*(**GHCoh**), where **HCoh** is Ehrhard's category of *hypercoherences* [28] in order to get a full completeness theorem for MALL (see [18]).

Acknowledgements

Both authors would like to thank the entire Équipe de Logique de la Programmation (Luminy) and the TMR Network, along with the Directors Jean-Yves Girard and Laurent Regnier, for their kind hospitality and support. We also acknowledge support from operating grants from NSERC, Canada. Finally we thank Robert Seely and Mark Weber for helpful comments.

Bibliography

[1] S. Abramsky , Retracing Some Paths in Process Algebra. In *CONCUR 96*, Springer Lecture Notes in Computer Science 1119, pp. 1-17 (1996).

[2] S. Abramsky, R. Jagadeesan. New foundations for the geometry of interaction. *Information and Computation*, 111, pp. 53–119, (1994).

[3] S. Abramsky, R. Jagadeesan, Games and Full Completeness for Multiplicative Linear Logic, J. Symbolic Logic 59, pp. 543-574 (1994).

[4] S. Abramsky, E. Haghverdi, and P. Scott. Geometry of interaction and linear combinatory algebras. Mathematical Structures in Computer Science 12, pp. 625-665, (2002).

[5] S. Abramsky, P.-A. Melliès. Concurrent games and full completeness. *Proceedings, Logic in Computer Science 1999*. IEEE Press, (1999).

[6] S. Abramsky. Semantics of interaction: an introduction to game semantics. *in Semantics and logics of computation*, edited by A. Pitts and P. Dybjer, Cambridge University Press, (1997).

[7] V.M. Abrusci, P. Ruet. Non-commutative logic I : the multiplicative fragment. Annals of Pure and Applied Logic 101 pp.29-64, (2000).

[8] R. Amadio, P.L. Curien. *Domains and Lambda Calculi*. Cambridge University Press, (1998).

[9] E. Bainbridge, P. Freyd, A. Scedrov, P. Scott, Functorial polymorphism, Theoretical Computer Science 70, pp. 1403-1456, (1990).

[10] M. Barr. *-autonomous categories*. Springer Lecture Notes in Mathematics 752, (1980).

[11] G. Bierman. What is a categorical model of intuitionistic linear logic? *In* Proceedings of the Second International Conference on Typed Lambda Calculus and Applications. Lecture Notes in Computer Science 902, (1995).

[12] R. Blute. Linear logic, coherence and dinaturality. Theoretical Computer Science, 115:3-41, 1993.

[13] R. Blute, Hopf algebras and linear logic, Mathematical Structures in Computer Science 6, pp. 189-212, (1996).

[14] R. Blute, J. R. B. Cockett, R. A. G. Seely and T. Trimble. Natural deduction and coherence for weakly distributive categories. Journal of Pure and Applied Algebra 13, pp. 229–296, (1996)

[15] R. Blute, J. R. B. Cockett, R. A. G. Seely. ! and ?: Storage as tensorial strength. Mathematical structures in Computer Science 6, pp. 313-351, (1996).

[16] R. Blute, J. R. B. Cockett, R. A. G. Seely. The logic of linear functors. Mathematical structures in Computer Science 12, pp. 513-539, (2002).

[17] R. Blute, F. Lamarche, P. Ruet. Entropic Hopf algebras and models of non-commutative logic. Theory and Applications of Categories 10, pp. 424-460, (2002).

[18] R. Blute, M. Hamano, P. Scott. Softness of hypercoherences and MALL full completeness. In preparation, (2003).

[19] R. Blute, P. Scott. Linear Lauchli semantics, Annals of Pure and Applied Logic 77, pp. 101-142 (1996).

[20] R. Blute, P. Scott. The Shuffle Hopf algebra and noncommutative full completeness. Journal of Symbolic Logic 63, pp. 1413-1435, (1998).

[21] F. Borceux. *Handbook of Categorical Algebra* Cambridge University Press, (1993).

[22] Antonio Bucciarelli and Thomas Ehrhard. On phase semantics and denotational semantics in multiplicative-additive linear logic. Annals of Pure and Applied Logic 102, pp. 247-282, 2000.

[23] J. R. B. Cockett, R. A. G. Seely. Weakly distributive categories. Journal of Pure and Applied Algebra 114, pp. 133-173, (1997).

[24] J. R. B. Cockett, R. A. G. Seely. Linearly distributive functors. Journal of Pure and Applied Algebra 143, pp. 155-203, (1999).

[25] V. Danos, L. Regnier, The structure of multiplicatives, Arch. Math. Logic 28, pp.181-203, (1989).

[26] V. Danos, L. Regnier. Proof-nets and the Hilbert space, In *Advances in Linear Logic*, London Mathematical Society Lecture Notes Volume 222, (1995).

[27] H. Devarajan, D. Hughes, G. Plotkin, and V. Pratt. Full completeness of the multiplicative linear logic of Chu spaces. *in* Proceedings 14th Annual IEEE Symposium on Logic in Computer Science, LICS'99, Trento, Italy, July 1999.

[28] T. Ehrhard. Hypercoherences: a strongly stable model of linear logic. Mathematical Structures in Computer Science 3, pp. 365-385, (1993).

[29] T. Ehrhard. On Köthe sequence spaces and linear logic. Mathematical Structures in Computer Science 12, pp. 579-623, (2002).

[30] T. Ehrhard. Finiteness spaces, preprint, (2001).

[31] T. Ehrhard and L. Regnier. The differential lambda-calculus. *To appear in* Theoretical Computer Science. 2003.

[32] W. Fulton, J. Harris. Representation Theory: A First Course. Springer Verlag, (1991).

[33] J.-Y. Girard. Linear logic. *Theoretical Computer Science*, 50:1–102, 1987.

[34] J.Y. Girard. Geometry of interaction I: interpretation of system *F*. *Proceedings of the ASL Meeting, Padova, 1988*.

[35] J.Y. Girard. Linear Logic, its syntax and semantics. In *Advances in Linear Logic*, London Mathematical Society Lecture Notes Volume 222, (1995).

[36] J.Y. Girard. Geometry of Interaction III: accommodating the additives. *In* Advances in Linear Logic, London Mathematical Society Lecture Notes Volume 222, (1995).

[37] J.Y. Girard. Proof-nets: the parallel syntax for proof-theory, Logic and Algebra, eds Ursini and Agliano, Marcel Dekker, New York 1996.

[38] J.Y. Girard. Locus Solum. Mathematical Structures in Computer Science 11, pp. 301-506, (2001).

[39] J.Y. Girard, Y. Lafont, P. Taylor. *Proofs and Types*. Cambridge University Press, (1989)

[40] E. Haghverdi. Unique decomposition categories, geometry of interaction and combinatory logic. Math. Structures Comput. Sci. 10, pp. 205–230, (2000).

[41] M. Hamano. **Z**-modules and Full Completeness of Multiplicative Linear Logic, Annals of Pure Appl. Logic 107 , pp. 165-191 (2001).

[42] M. Hamano. Pontrjagin Duality and Full Completeness for Multiplicative Linear Logic (Without Mix), Math. Struct. in Comp. Science 10, pp. 231-259 (2000).

[43] P. Hines. The Algebra of Self-Similarity and its Applications. Thesis. University of Wales, (1997).

[44] J. M. E. Hyland and C.-H. L. Ong. Fair Games and Full Completeness for Multiplicative Linear Logic without the Mix-Rule. Preprint, 1993.

[45] J. M. E. Hyland and C.-H. L. Ong. On Full Abstraction for PCF. Information and Computation, Volume 163, pp. 285-408, December 2000

[46] J. M. E. Hyland and A. Schalk. Glueing and orthogonality for models of linear logic. Theoretical Computer Science 294, pp. 183-231, (2003).

[47] A. Joyal, R. Street and D. Verity. Traced monoidal categories. *Mathematical Proceedings of the Cambridge Philosophical Society*, 119:425–446, 1996.

[48] C. Kassel. *Quantum Groups*. Springer-Verlag, (1995)

[49] G. M. Kelly and M. Laplaza. Coherence for compact closed categories. Journal of Pure and Applied Algebra 19, pp. 193–213, (1980).

[50] Y. Lafont. Interaction nets. *In* Principles of Programming Languages (POPL 1990), p. 95-108, (1990).

[51] J. Lambek, P. Scott. *Introduction to Higher-Order Categorical Logic*. Cambridge University Press, (1988).

[52] R. Loader, Linear Logic, Totality and Full Completeness, *Symposium of Logic in Computer Science* (LICS), pp. 292-298 (1994)

[53] S. Mac Lane. *Categories for the Working Mathematician*, volume 5 of *Graduate texts in Mathematics*. Springer-Verlag, New York, 1971.

[54] S. Majid. *Foundations of Quantum Group Theory*. Cambridge University

Press, (2000).

[55] E. Moggi. Notions of computation and monads. Information And Computation 93, 1991.

[56] V. de Paiva, A. Schalk. Poset-valued sets, or, How to build models for linear logic, *to appear in* Theoretical Computer Science.

[57] R. Rosebrugh, R. Wood. Constructive complete distributivity IV, Applied categorical structures 2, pp. 119-144, (1994).

[58] K. Rosenthal. *Quantales and Their Applications*. Pitman Research Notes in Mathematics, (1990).

[59] P. Scott. Some aspects of categories in computer science, in *Handbook of Algebra* , Volume 2, *edited by* M. Hazewinkel, North-Holland, pp. 1-77, (2000).

[60] R.A.G. Seely. Linear logic, *-autonomous categories and cofree coalgebras. *Contemporary Mathematics*, Volume 92. American Mathematical Society, (1989).

[61] P. Selinger. Control categories and duality: on the categorical semantics of the lambda-mu calculus. Math. Structures Comput. Sci. 11 , pp. 207–260 (2001).

[62] A. Tan. Full completeness for models of linear logic. Thesis, Cambridge University, (1997).

[63] A. Troelstra. Lectures on Linear Logic, Cambridge University Press, (1992).

[64] D. Yetter, Quantales and (noncommutative) linear logic, Journal of Symbolic Logic 55, p. 41-64, (1990)

2
Proof Nets and the λ-Calculus[†]

Stefano Guerrini

Dipartimento di Informatica - Università Roma La Sapienza

Abstract

In this survey we shall present the main results on proof nets for the
Multiplicative and Exponential fragment of Linear Logic (MELL) and
discuss their connections with λ-calculus. The survey ends with a short
introduction to sharing reduction. The part on proof nets and on the
encoding of λ-terms is self-contained and the proofs of the main theorems
are given in full details. Therefore, the survey can be also used as a
tutorial on that topics.

2.1 Introduction

In his seminal paper on Linear Logic [13], Girard introduced proof nets
in order to overcome some of the limitations of the sequent calculus for
Linear Logic. At the price of the loss of the symmetries of sequents, proof
nets allowed to equate proofs that in sequent calculus differ by useless
details, and gave topological tools for the characterization and analysis
of Linear Logic proofs. Moreover, because of the encoding of λ-calculus
in the Multiplicative and Exponential fragment of Linear Logic (MELL),
it was immediately clear that proof nets might have become a key tool
in the fine analysis of the reduction mechanism, the dynamics, of λ-
calculus. Such a property was even clearer after that Girard introduced
GOI (Geometry of Interaction [14]) and after the work by Danos and
Regnier on the so-called local and asynchronous β-reduction [33, 11].
However, the key step towards the full exploitation of proof nets in
the analysis of λ-calculus dynamics was the discover by Gonthier et al.

[†] This work has been partially supported by the Italian MIUR Cofin Project "Reti
dimostrative: interazione e complessità".

65

[12] that Lamping's algorithm [25] for the implementation of λ-calculus optimal reduction [28] might be reformulated on proof nets using GOI. Such a correspondence was successively analyzed in more depth in [2], where it was shown that the paths definable in GOI coincide with the paths used in optimal reductions [4].

This paper is mainly a survey on the more relevant results on MELL proof nets and on their correspondence with λ-calculus. The main ideas of sharing graphs and optimal reductions are only sketched in the last part of the paper. For a detailed study of the latter topics, we refer the reader to [3, 22]. For a lack of space, many other interesting topics on proof nets will not be treated in the paper; for instance, we shall not consider the additives. Indeed, proof nets for the additives does not have the same nice properties of those for MELL. Several attempts have been tried in order to find a good formulation of additive proof nets (*e.g.*, see [17] or [36]), but none of that approaches is completely satisfactory yet. For more details on additives, see [35]. We shall not even consider quantifiers. On the contrary of what happens for the additives, the extension of proof nets to quantifiers does not pose any problem, see [15]

Another interesting issue that we shall not cover in the survey are the proof nets for classical calculi. For that purpose, we remark the polarized proof nets introduced by Laurent in [27]. Such proof nets allow to encode λμ-calculus and are a nice tool for the analysis of the computational content of classical logic.

Among the other topics that we shall not cover in the paper, we point out the non-commutative proof nets for Cyclic Linear Logic [1], or for Abrusci and Ruet's Non-Commutative Linear Logic [6].

Structure of the paper In section 2.2, we shall analyze multiplicative proof nets, starting with Multiplicative Linear Logic (MLL) in 2.2.1. In particular, we shall introduce proof structures (2.2.1.1), we shall give the Danos-Regnier correctness criterion (2.2.1.2) and the correctness criteria based on contractibility (2.2.1.3) and parsing (2.2.1.4), and we shall see that every Danos-Regnier correct proof net can be sequentialized (2.2.1.5); then, we shall define the notion of empire of a formula (2.2.1.6), we shall study how to compute empires (2.2.1.7), and we shall see the original sequentialization proof by Girard based on empires (2.2.1.8). In 2.2.2, we shall study the proof nets of Intuitionistic Multiplicative Linear Logic (IMLL); in particular, we shall see how to transform any MLL proof into an IMLL proof by orienting the corresponding net (2.2.2.1), and we shall see the so-called essential nets (2.2.2.2). In 2.2.3, we shall briefly discuss the time complexity of the algorithms implementing the criteria for proof net correctness. The section ends with the cut-elimination rules for MLL proof nets, in 2.2.4, and with an analysis of the problems that the introduction of unit cause in the definition of proof nets, in 2.2.5.

In section 5.3.5, we shall add the exponentials and define the proof nets for MELL. In particular, we shall introduce the key notion of exponential box (in 2.3.1), we shall extend the parsing algorithm already seen for MLL to MELL (in 2.3.2), and we shall see that weakening leads to the same problems caused by $\perp$ in the definition of proof net correctness (in 2.3.3). Finally, we shall define the rules for the elimination of exponential cuts (in 2.3.4).

In section 2.4, we shall study the correspondences between proof nets and λ-calculus. In particular, in 2.4.1, we shall start with linear λ-calculus; in 2.4.2, we shall see the $!A \multimap B$ encoding of λ-terms into proof nets; in 2.4.3, we shall define the correctness criterion for the proof nets that are images of λ-terms; in 2.4.4, we shall see that proof nets allow to define an interesting operational equivalence on λ-terms; in 2.4.5, we shall extend the encoding of λ-terms to the untyped case.

In section 2.5, we shall give a brief introduction to sharing reduction. In particular, in 2.5.1, we shall see how to represent boxes by means of indexes; in 2.5.2, we shall give an intuitive justification of the rules of sharing reduction; in 2.5.3, we shall point out that optimal reduction is a particular strategy of sharing reduction.

The parts on MELL nets and on the encoding of λ-terms (section 2.2, section 5.3.5 and part of section 2.4) are self-contained and the proofs of the main theorems are given in full details. Thus, the survey can be also used as a tutorial on these topics.

2.2 Multiplicative Proof Nets

2.2.1 MLL

The formulas of the multiplicative fragment without constants of Linear Logic, MLL for short, are defined by

$$p \mid A \,\mathbf{\mathcal{R}}\, B \mid A \otimes B$$

where p ranges over a set of atoms and A, B range over MLL formulas. The binary connective $\mathbf{\mathcal{R}}$ is named *par*, the binary connective $\otimes$ is named *tensor*. For every atom p, there is a dual atom $p^\perp$ s.t. $p^{\perp\perp} = p$. Duality extends to every formula by

$$A^{\perp\perp} = A \qquad\qquad (A \,\invamp\, B)^\perp = A^\perp \otimes B^\perp$$

Let us remark that $(.)^\perp$ is not a connective of the logic. Then, in the following, $A^\perp$ will denote the MLL formula that we can obtain from A by application of the above rules. In particular, $(A \otimes B)^\perp = A^\perp \,\invamp\, B^\perp$.

The rules of MLL are given in Figure 2.1, where Γ and Δ denote multisets of MLL formulas. If we define the linear implication $\multimap$ as $A \multimap B = A^\perp \,\invamp\, B$, we get modus ponens as a derived rule, namely, from a proof of $\vdash A \multimap B$ and a proof of $\vdash A$ we can get a proof of B. Then, if we denote by $A \sim B$ the fact that both $\vdash A \multimap B$ and $\vdash B \multimap A$ are provable (that corresponds to $\vdash A$ is provable iff $\vdash B$ is provable), it is readily seen that $A \,\invamp\, B \sim B \,\invamp\, A$ and $A \,\invamp\, (B \,\invamp\, C) \sim (A \,\invamp\, B) \,\invamp\, C$, and dually for the tensor, namely, par and tensor are commutative and associative. For more details on MLL (and Linear Logic), we refer the reader to [13, 10, 16].

$$\frac{}{\vdash A, A^\perp}\ \text{ax} \qquad\qquad \frac{\vdash \Gamma, A \qquad \vdash \Delta, A^\perp}{\vdash \Gamma, \Delta}\ \text{cut}$$

$$\frac{\vdash \Gamma, A, B}{\vdash \Gamma, A \,\invamp\, B}\ \invamp \qquad\qquad \frac{\vdash \Gamma, A \qquad \vdash \Delta, B}{\vdash \Gamma, \Delta, A \otimes B}\ \otimes$$

Fig. 2.1. MLL.

2.2.1.1 *Proof structures*

The main drawback of MLL sequent calculus is that it distinguishes proofs that differ by useless details. For instance, let us take

$$\frac{\dfrac{\vdots \Pi_1}{\vdash \Gamma, A, B, C} \qquad \dfrac{\vdots \Pi_2}{\vdash \Delta, D}}{\dfrac{\vdash \Gamma, \Delta, A, B, C \otimes D}{\vdash \Gamma, \Delta, A \,\invamp\, B, C \otimes D}} \qquad\qquad \frac{\dfrac{\dfrac{\vdots \Pi_1}{\vdash \Gamma, A, B, C}}{\vdash \Gamma, \Delta, A \,\invamp\, B, C} \qquad \dfrac{\vdots \Pi_2}{\vdash \Delta, D}}{\vdash \Gamma, \Delta, A \,\invamp\, B, C \otimes D}$$

That pair of proofs differ for the order in which the last two rules are applied, but, as the principal formulas of the rules (A, B and $A \,\invamp\, B$ in the $\invamp$-rule; C, D and $C \otimes D$ in the $\otimes$-rule) are disjoint, it would be better to have a syntax that allows to apply the last two rules in parallel, or at least, a syntax in which the two proofs have the same representation.

In order to find a parallel syntax, we move from sequents to graphs. In particular, we assume that each occurrence of a formula (in the following, we shall simply say formula in the place of occurrence of a formula) is the vertex of a graph and that each rule introduces a link between its principal formulas. In this way, the two proofs seen above become something like

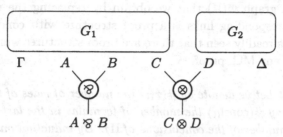

where G_1, G_2 are the graphs replacing Π_1 and Π_2, respectively. Let us notice that the conclusions of Π_1 and Π_2 become the roots of the graph, and that there is no ordering between the links replacing the last two rules of Π_1 and Π_2.

In the following, we shall write $\Gamma \rhd \Delta$, where Γ and Δ are disjoint sequences of formulas, to denote that there is a link between the *premises* Γ and the *conclusions* Δ—a link may have an empty set of conclusions or an empty set of premises, but Γ and Δ cannot be simultaneously empty. In MLL we have the following type of links (the type of the link can be written above the link symbol $\rhd$):

$$\overset{ax}{\rhd} A, A^\perp \qquad A, A^\perp \overset{cut}{\rhd} \qquad A, B \overset{\mathit{⅋}}{\rhd} A \,⅋\, B \qquad A, B \overset{\otimes}{\rhd} A \otimes B$$

Figure 2.2 gives then graphical representation of MLL links.

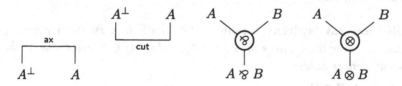

Fig. 2.2. MLL links.

Definition 2.1 (proof structure) *An* MLL *proof structure is a set of* MLL *links G where every formula is conclusion of one (and only one) link and is premise of at most one link. The formulas that are not premises of any link are the conclusions of G.*

Let $H \subseteq G$, where G is a proof structure; we say that H is a (proof) substructure of G. A formula is a premise of H if it is not conclusion of any link in H. We write, $H[\Delta] \vdash \Gamma$ to denote that Δ are the premises of H and Γ are its conclusions.

By the way, if Π is an MLL proof ending with the sequent $\vdash \Gamma$ (write $\Pi \vdash \Gamma$), the graph $\mathfrak{G}(\Pi)$ that we obtain by replacing the rules in Π with the corresponding links is a proof structure with conclusions Γ. However, it is readily seen that there are proof structures which are not the image of any MLL proof.

Remark 2.2 *Let us denote by $\#(x_\Pi)$ the number of rules of type x in a proof Π and by $\#(concl_\Pi)$ the number of formulas in the last sequent of Π (i.e., the number of the conclusions of Π). By induction on the length of Π, we can see that:*

$$\#(ax_\Pi) - 1 = \#(\otimes_\Pi) + \#(cut_\Pi)$$
$$\#(ax_\Pi) + 1 = \#(\mathscr{V}_\Pi) + \#(cut_\Pi) + \#(concl_\Pi)$$

2.2.1.2 Danos-Regnier correctness criterion

Let us denote by PS the set of the MLL proof structures. The image $\mathfrak{G}(MLL)$ of the MLL proofs is a proper subset of PS. In order to exploit the fact that we are using graphs to represent proofs, we want a geometric characterization of $\mathfrak{G}(MLL)$. Because of that, we shall associate a family of graphs to every $G \in$ PS, the so-called switches of G, and we shall say that G is a proof net when every switch of G is a tree. By induction, we shall see that every $G \in \mathfrak{G}(MLL)$ is a proof net. In a following section, see Theorem 2.13, we shall also see that every proof net is in $\mathfrak{G}(MLL)$.

Definition 2.3 (switch) *A switch of $G \in$ PS is a symmetric binary relation S over the vertices of G s.t. $(A,B),(B,A) \in S$ when one of the following cases holds:*

(i) $\overset{ax}{\triangleright} A, B \in G;$

(ii) $A, B \overset{cut}{\triangleright} \in G;$

(iii) $A, C \overset{\otimes}{\triangleright} B \in G$ or $C, A \overset{\otimes}{\triangleright} B \in G;$

(iv) $A, C \overset{\mathscr{V}}{\triangleright} B \in G$ or $C, A \overset{\mathscr{V}}{\triangleright} B \in G$, but $(C,B) \notin S.$

Let $\mathbb{SW}(G)$ be the set of the switches of G. If $S \in \mathbb{SW}(G)$, we shall write $A \frown_S B$ to denote that $(A,B) \in S$ and $A \frown_S B$ to denote that $(A,B) \notin S$. We see that S is antireflexive and that it defines a graph over the vertices of G (see Figure 2.3, where the thick lines are the switch edges introduced by the corresponding link), in which:

- every axiom and every cut introduces an edge between its premises or conclusions, respectively;
- every tensor link introduces two edges that connect the premises of the link to its conclusion;
- every par link introduces one (and only one) edge that connects one of its premises to its conclusion.

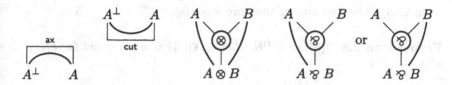

Fig. 2.3. Switch edges.

The switching of a par link (*i.e.*, the choice of which premise of the par link is connected to the conclusion) is the source of non-determinism in the construction of the switches of a proof structure. Because of this, it is readily seen that $|\mathbb{SW}(G)| = 2^{\#(\mathscr{8}_G)}$.

Definition 2.4 (proof net) $G \in$ PS *is Danos-Regnier correct or, equivalently, G is a proof net, say $G \in$ PN, when every switch of G is connected and acyclic.*

Remark 2.5 *Using the fact that in every tree there is one edge less than the number of the vertices of the tree, we can see that the equations in Remark 2.2 hold for every $G \in$ PN, if $\#(x_G)$ is the number of links of type x in G. In fact, the number of vertices in every switch of a proof structure G is equal to the number of formulas in G, that is equal to $2\#(\text{ax}_G) + \#(\mathscr{8}_G) + \#(\otimes_G)$ (count the conclusions of the links in G), but also equal to $2\#(\mathscr{8}_G) + 2\#(\otimes_G) + 2\#(\text{cut}_G) + \#(\text{concl}_G)$ (count the premises of the links in G); while the number of edges in every switch of G is $\#(\text{ax}_G) + \#(\mathscr{8}_G) + 2\#(\otimes_G) + \#(\text{cut}_G)$.*

Remark 2.6 *Let G' be the proof structure obtained by replacing a tensor for every cut in $G \in$ PS. It is readily seen that G is DR-correct iff G' is DR-correct. Because of this, studying proof net correctness, we shall frequently restrict w.l.o.g. to the case of cut free proof structures.*

Remark 2.7 *Given a proof structure G. Let $\preceq$ be the transitive and reflexive closure of the relation $\preceq_1$ defined by $A \preceq_1 B$ when $A \in \Gamma$ and $B \in \Delta$ and $\Delta \rhd \Gamma \in G$. Every ascending chain $A_1 \preceq \cdots \preceq A_k$ corresponds to a path in some switch of G. Now, let us assume that we can find an ascending chain $A \preceq_1 A_1 \preceq \cdots \preceq A_k \preceq A$; by the fact that every switch of a proof net is acyclic, G is not a proof net. Therefore, when G is a proof net, $\preceq$ is antisymmetric, that is, $\preceq$ is a partial order.*

We aim at proving that the *correctness criterion* in Definition 2.4 gives exactly the proof structures that correspond to MLL proofs (Theorem 2.13). The first step of that proof is easy.

Proposition 2.8 $\mathfrak{G}(\mathsf{MLL}) \subseteq \mathsf{PN}$, *that is, $\mathfrak{G}(\Pi)$ is a proof net for every MLL proof Π.*

Proof By induction on the length of Π. $\square$

The hard part in the proof of $\mathsf{PN} = \mathfrak{G}(\mathsf{MLL})$ is to show that $\mathsf{PN} \subseteq \mathfrak{G}(\mathsf{MLL})$. In fact, we have to prove that every proof net G has a *sequentialization*, that is, we have to show that there exists an order among the links in G that corresponds to the order of the rules in some MLL proof Π s.t. $\mathfrak{G}(\Pi) = G$. The original proof by Girard will be presented in 2.2.1.8. In the next sections, we shall give another characterization of proof nets in terms or graph rewriting that will lead to a simple proof of sequentialization and that will give an efficient implementation of DR-correctness.

2.2.1.3 Contraction criterion

The D(anos-)R(egnier) correctness criterion gives a clean and useful topological characterization of proof nets independent from the sequent calculus. However, it does not help in finding an efficient algorithm for the decision problem: "given $G \in \mathsf{PS}$, $G \in \mathfrak{G}(\mathsf{MLL})$?". In fact, a direct application of the DR-criterion leads to an exponential algorithm—we have to repeat a linear test for an exponential number of switches.

In this section and in the following one, we shall see two other correctness criteria—contraction and parsing—based on the idea that from the point of view of switches two correct substructures with the same premises and conclusions are equivalent. Both these criteria lead to quadratic time algorithms—or with some care in the implementation, to $n \log n$ algorithms. Moreover, they are the base for a linear algorithm for proof net correctness. By mean of the parsing algorithm, we shall also prove that $\mathfrak{G}(\mathsf{MLL}) = \mathsf{PN}$.

The definition of DR-correctness extends in the natural way to proof substructures. Moreover, let us assume that $H \subseteq G$ is DR-correct. In order to verify if $G \in \mathsf{PN}$ we do not need to check every switch of H, it suffices to choose one switch of H and to verify all switching possibility for the par links in $G \setminus H$. In fact, let us introduce a new kind of link, named *-link,

$$A_1, \ldots, A_k \overset{*}{\triangleright} B_1, \ldots, B_h$$

without any restriction on the number and the type of its premises and conclusions (however, by the definition of link, in every $\Gamma \overset{*}{\triangleright} \Delta$, $\Gamma \cup \Delta \neq \emptyset$ and $\Gamma \cap \Delta = \emptyset$). The switching of a *-link is the least graph s.t.

$$A_1 \frown \cdots \frown A_k \frown B_h \frown \cdots \frown B_1$$

(see Figure 2.4 also). Now, let us replace a DR-correct substructure $H[\Gamma] \vdash A$ with the link $\Gamma \overset{*}{\triangleright} \Delta$. The verification of G can be reduced to that of the new structure obtained by replacing H.

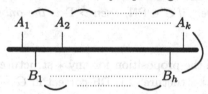

Fig. 2.4. *-link and the corresponding switching.

Definition 2.9 (*-structure) *A *-structure G is a set of MLL links and *-links s.t. every formula is conclusion of one (and only one) link and is premise of at most one link. The definitions of switch and of DR-correct *-structures are the natural extensions of Definition 2.3 and Definition 2.4.*

Lemma 2.10 *Let $H[\Gamma] \vdash \Delta$ be a DR-correct substructure of $G \in \mathsf{PS}$. $G \in \mathsf{PN}$ iff the *-structure $G[\Gamma \overset{*}{\triangleright} \Delta/H]$ obtained by replacing $\Gamma \overset{*}{\triangleright} \Delta$ for H in G is DR-correct.*

Proof Every $S \in \mathbb{SW}(G)$ is obtained by some $S' \in \mathbb{SW}(G[\Gamma \overset{*}{\triangleright} \Delta/H])$ by replacing a tree for the switching tree of the *-link $\Gamma \overset{*}{\triangleright} \Delta$. Moreover, for every S' there is such an S. Therefore, as S is a tree iff S' is a tree, we conclude. $\square$

Lemma 2.10 suggests the contraction algorithm defined by the rules in Figure 2.5 (this algorithm has been introduced with a different notation in [9]). By the way, if $R \overset{c}{\to} S$ is any instance of the rules in Figure 2.5 and $G = H; R$, then $G \overset{c}{\to} G'$, with $G' = H; S$; moreover, it is readily seen that, when G is a $*$-structure, G' is a $*$-structure too.

$$\overset{\text{ax}}{\triangleright} A, A^\perp \overset{c}{\to} \overset{\cdot}{\triangleright} A, A^\perp \tag{$\overset{\text{ax}}{\triangleright}$}$$

$$\Gamma \overset{\cdot}{\triangleright} \Delta, A, B;\ A, B \overset{\gamma}{\triangleright} A \,\gamma\, B \overset{c}{\to} \Gamma \overset{\cdot}{\triangleright} \Delta, A \,\gamma\, B \tag{$\overset{\gamma}{\triangleright}$}$$

$$A, B \overset{\otimes}{\triangleright} A \otimes B \overset{c}{\to} A, B \overset{\cdot}{\triangleright} A \otimes B \tag{$\overset{\otimes}{\triangleright}$}$$

$$A, A^\perp \overset{\text{cut}}{\triangleright} \overset{c}{\to} A, A^\perp \overset{\cdot}{\triangleright} \tag{$\overset{\text{cut}}{\triangleright}$}$$

$$\Gamma_1 \overset{\cdot}{\triangleright} \Delta_1, A;\ A, \Gamma_2 \overset{\cdot}{\triangleright} \Delta_2 \overset{c}{\to} \Gamma_1, \Gamma_2 \overset{\cdot}{\triangleright} \Delta_1, \Delta_2 \tag{$\overset{\cdot}{\triangleright}$}$$

Fig. 2.5. Contraction rules.

Proposition 2.11 *Let* $G \in$ PS *with* $G \vdash \Gamma$. *We have that* $G \in$ PN *iff* $G \overset{c}{\to}{}^* \overset{\cdot}{\triangleright} \Gamma$. *Moreover, if* $G \in$ PN, *then* $\overset{\cdot}{\triangleright} \Gamma$ *is the only* $\overset{c}{\to}$-*normal form of* G.

Proof Let us prove the proposition for any $*$-structure $G \vdash \Gamma$. Namely, let us prove that a $*$-structure G is DR-correct iff $G \overset{c}{\to}{}^* \overset{\cdot}{\triangleright} \Gamma$.
First of all, let us notice the following corollary of Lemma 2.10: if $G \overset{c}{\to} G'$, then G is DR-correct iff G' is DR-correct. Therefore, we can immediately conclude that $G \overset{c}{\to}{}^* \overset{\cdot}{\triangleright} \Gamma$ only if G is DR-correct. We left to prove that, if G is DR-correct, then $G \overset{c}{\to}{}^* \overset{\cdot}{\triangleright} \Gamma$ and that this is its only normal form. By the fact that contraction is strongly normalizing (every rule decreases the following measure: $\#(\text{ax}) + |G|$), it suffices to prove that, when G is DR-correct and contains more than one link, $G \overset{c}{\to} G'$ for some $G' \vdash \Gamma$. The statement trivially holds when G is correct and contains at least an axiom, or at least a tensor or at least a cut link. So, let us assume that G is correct and contains par and $*$-links only. By Remark 2.7, $\preccurlyeq$ is a partial order and, by hypothesis, any maximal formula w.r.t. that ordering is conclusion of a $*$-link $\overset{\cdot}{\triangleright} \Delta$. If $\Delta = \Gamma$, either G is not DR-correct or $\overset{\cdot}{\triangleright} \Delta$ is the only link in G. So, $\Delta \setminus \Gamma \neq \emptyset$. If some $A \in \Delta$ is premise of a $*$-link, we have a $\overset{\cdot}{\triangleright}$-redex and $G \overset{c}{\to} G'$ for some G'. If every $A \in \Delta \setminus \Gamma$ is premise of a par link, there must be $A, B \in \Delta$ s.t. $A, B \overset{\gamma}{\triangleright} C \in G$, otherwise we might construct a disconnected switch of G; therefore, $G \overset{c}{\to} G'$ for some G', in this case too. $\square$

2.2.1.4 Parsing

The sequent calculus of MLL suggests another contraction algorithm. The rules of this algorithm simulate the *parsing* of the proof structures in $\mathfrak{G}(\mathsf{MLL})$ (see [24] and [18]) and are given in Figure 2.6. Such rules ensure that $G \in \mathfrak{G}(\mathsf{MLL})$ implies $G \overset{p}{\to}{}^{*} \overset{\cdot}{\rhd} \Gamma$, if $G \vdash \Gamma$.

$$\overset{ax}{\rhd} A, A^{\perp} \overset{p}{\to} \overset{\cdot}{\rhd} A, A^{\perp} \tag{$\overset{ax}{\rhd}$}$$

$$\overset{\cdot}{\rhd} \Gamma, A, B\,;\, A, B \overset{\mathfrak{N}}{\rhd} A \,\mathfrak{N}\, B \overset{p}{\to} \overset{\cdot}{\rhd} \Gamma, A \,\mathfrak{N}\, B \tag{$\overset{\mathfrak{N}}{\rhd}$}$$

$$\overset{\cdot}{\rhd} \Gamma, A\,;\, \overset{\cdot}{\rhd} \Delta, B\,;\, A, B \overset{\otimes}{\rhd} A \otimes B \overset{p}{\to} \overset{\cdot}{\rhd} \Gamma, \Delta, A \otimes B \tag{$\overset{\otimes}{\rhd}$}$$

$$\overset{\cdot}{\rhd} \Gamma, A\,;\, \overset{\cdot}{\rhd} \Delta, A^{\perp}\,;\, A, A^{\perp} \overset{cut}{\rhd} \overset{p}{\to} \overset{\cdot}{\rhd} \Gamma, \Delta \tag{$\overset{cut}{\rhd}$}$$

Fig. 2.6. Parsing rules.

Proposition 2.12 *Let $G \in \mathsf{PS}$ with $G \vdash \Gamma$. Then, $G \in \mathsf{PN}$ iff $G \overset{p}{\to}{}^{*} \overset{\cdot}{\rhd}$ Γ. Moreover, if $G \in \mathsf{PN}$, then $\overset{\cdot}{\rhd} \Gamma$ is the only $\overset{p}{\to}$-normal form of G.*

Proof First of all, let us notice that the ∗-structures that we get by reduction of a proof structure contain ∗-links without premises only; let us say that a ∗-structure with that property is a parsing ∗-structure. Then, let us remark that the main point of the proof is to show that every DR-correct parsing ∗-structure that contains more than one link contains at least a $\overset{p}{\to}$-redex; the rest is similar to the proof of Proposition 2.11. Let $G \vdash \Gamma$ be a DR-correct parsing ∗-structure. W.l.o.g., we can restrict to the case in which G does not contain axioms. As a consequence, G contains at least a ∗-link.

By Remark 2.6 we can assume w.l.o.g. that G is cut free. Moreover, let us assume that G does not contain any $\overset{\mathfrak{N}}{\rhd}$-redex. We see that, if G contains more than one link, then every ∗-link in G has a conclusion that is premise of a tensor (otherwise, G would have a disconnected switch). In a cut and axiom free parsing ∗-structure, the first combinatorial equation in Remark 2.2 becomes $\#(\ast_G) = \#(\otimes_G) + 1$ (see Remark 2.5 also). Therefore, by the pigeon hole principle, there is at least an $A, B \overset{\otimes}{\rhd} C \in G$ s.t. A and B are both conclusions of ∗-links. As G is DR-correct, A and B cannot be conclusions of the same ∗-link. Therefore, G contains at least a $\overset{\otimes}{\rhd}$-redex. □

For more details on the parsing technique, we refer to [18].

2.2.1.5 Sequentialization

By Proposition 2.8, Proposition 2.12 and the fact that every proof structure that contracts to a $*$-link corresponds to an MLL proof, we get the
sequentialization theorem.

Theorem 2.13 (sequentialization) *Let $G \in$ PS with $G \vdash \Gamma$. Then,
G is a proof net iff there is an MLL proof Π s.t. $\mathfrak{G}(\Pi) = G$, that is,
$\mathfrak{G}(\text{MLL}) = \text{PN}$.*

2.2.1.6 Empires

Let A be a formula in the proof net G. We denote by $\text{PN}_G(A)$ the set
of the proof nets in G that have A as a conclusions, namely,

$$\text{PN}_G(A) = \{H \subseteq G \mid H \vdash \Gamma, A \text{ and } H \in \text{PN}\}$$

By the sequentialization theorem, $\text{PN}_G(A)$ is not empty. In fact, let us
take any proof Π s.t. $\mathfrak{G}(\Pi) = G$; there is at least a subproof Ξ of Π s.t.
A is in the final sequent of Ξ; then, $\mathfrak{G}(\Xi) \in \text{PN}_G(A)$.

Lemma 2.14 $\text{PN}_G(A)$ *is closed by intersection and union.*

Proof Let $H_u = M \cup N$ and $H_i = M \cap N$ with $M, N \in \text{PN}_G(A)$; it is
readily seen that $H_i, H_u \in$ PS. Let $S_i \in \mathbb{SW}(H_i)$ and $S_u \in \mathbb{SW}(H_u)$.
W.l.o.g., let us assume that S_i is the restriction of S_u to H_i; moreover,
let $S_M \in \mathbb{SW}(M)$ and $S_N \in \mathbb{SW}(N)$ be the restrictions of S_u to M and
N, respectively. As S_i and S_u are subgraphs of some switch of G, they
are acyclic. So, we left to prove that S_i and S_u are connected.
Let B, C be two formulas in H_i. As M and N are proof nets, there are
a path φ_M of S_M from B to C and a path φ_N of S_N from B to C. But,
φ_M and φ_N are paths of S_u and, as S_u is acyclic, $\varphi_M = \varphi_N$. Therefore,
as $\varphi_M = \varphi_N$ is a path of S_i too, we can conclude that S_i is connected.
Let B, C be two formulas in H_u. The only relevant case is when B is in
M and C is in N. As M and N are proof nets, there are a path φ_M of
S_M from B to A and a path φ_N of S_N from C to A. For φ_M and φ_N
are paths of S_u also, by composition of φ_M and φ_N, there is a path of
S_u from B to C. Therefore, S_u is connected. $\square$

The previous lemma allows to conclude that $\text{PN}_G(A)$ has a minimum
and a maximum w.r.t. to inclusion.

Definition 2.15 *The* empire *of A is the largest $H \in \mathsf{PN}_G(A)$. The* kingdom *of A is the least $H \in \mathsf{PN}_G(A)$.*

2.2.1.7 Algorithmic definition of empires

Given a proof net G, we present now a linear time algorithm for the construction of the empire $e_G(A)$ of a formula A in G.

Definition 2.16 *Let $V_G^e(A)$ be the least set of formulas in $G \in \mathsf{PN}$ s.t.:*

(i) $A \in V_G^e(A)$;

(ii) *if $B \in V_G^e(A)$ and $B \preccurlyeq C$, then $C \in V_G^e(A)$;*

(iii) *if $B \in V_G^e(A)$ and either $\overset{ax}{\rhd} B, B^\perp \in G$ or $\overset{ax}{\rhd} B^\perp, B \in G$, then $B^\perp \in V_G^e(A)$;*

(iv) *if $B \neq A$ and $B \in V_G^e(A)$ and either $B, B^\perp \overset{cut}{\rhd} \in G$ or $B^\perp, B \overset{cut}{\rhd} \in G$, with $A \neq B^\perp$, then $B^\perp \in V_G^e(A)$;*

(v) *if $B \neq A$ and $B \in V_G^e(A)$ and, for some $C \neq A$, either $B, C \overset{\otimes}{\rhd} D \in G$ or $C, B \overset{\otimes}{\rhd} D \in G$, then $D \in V_G^e(A)$;*

(vi) *if $B, C \neq A$ and $B, C \in V_G^e(A)$ and $B, C \overset{\pi}{\rhd} D \in G$, then $D \in V_G^e(A)$.*

We want to prove that the proof structure

$$\varepsilon_G(A) = \{\Gamma \rhd \Delta \in G \mid \Gamma, \Delta \in V_G^e(A)\}$$

is the empire of A (by construction, it is readily seen that $\varepsilon_G(A) \in \mathsf{PS}$).

Lemma 2.17 *Given $G \in \mathsf{PN}$, let A be premise of some link of G with conclusion C. There is $S \in \mathsf{SW}(G)$ s.t. the path of S from A to B touches formulas in $V_G^e(A)$ only and does not traverse the edge $A \frown C$.*

Proof By induction of the length of the derivation of $B \in V_G^e(A)$. $\quad\square$

By the latter lemma, the formula C below A cannot be in $V_G^e(A)$, otherwise we might construct a switch with a cycle. Thus, $\varepsilon_G(A) \vdash \Gamma, A$ and, for every $B \in \Gamma$, one of the following cases holds:

(i) B is a conclusion of G;

(ii) $B, A \overset{\pi}{\rhd} C \in G$ or $A, B \overset{\pi}{\rhd} C \in G$;

(iii) $B, C \overset{\pi}{\rhd} D \in G$ or $C, B \overset{\pi}{\rhd} D \in G$ for some $C \notin V_G^e(A)$.

This allows to say that $S \in \mathsf{SW}(G)$ is a *principal switch* of A when: for every $B \in \Gamma$, if B is premise of a par link with conclusion C, then $B \frown C$.

Lemma 2.18 *Given $G \in \mathsf{PN}$, let A be premise of some link of G with conclusion C and S be any principal switch of A. Then, $B \in V_G^e(A)$ iff the path from A to B in S does not traverse the edge $A \frown C$.*

Proof Let φ be the path of S from A to B. Let $D \frown E$ be an edge of φ s.t. $D \in V_G^e(A)$ and $E \notin V_G^e(A)$. By the definition of principal switch, the only edge of S with that property is $A \frown C$. Therefore, it is readily seen that, if φ does not traverses $A \frown C$, then $B \in V_G^e(A)$, otherwise $B \notin V_G^e(A)$. $\square$

Summing up, we can conclude that $V_G^e(A)$ is the set of formulas in the empire of A.

Proposition 2.19 *Let A be any formula in $G \in \mathsf{PN}$. $\mathsf{e}_G(A) = \varepsilon_G(A)$*

Proof W.l.o.g., let us assume that G is cut free (see Remark 2.6) and that A is premise of a link with conclusion C (otherwise, add an axiom $\overset{\text{ax}}{\triangleright} B, B^\perp$ and a tensor $A, B \overset{\otimes}{\triangleright} C$).
By hypothesis, every switch of $\varepsilon_G(A)$ is acyclic; by Lemma 2.18, every switch of $\varepsilon_G(A)$ is connected. Therefore, $\varepsilon_G(A) \in \mathsf{PN}_G(A)$ and, by the definition of empire, $\varepsilon_G(A) \subseteq \mathsf{e}_G(A)$.
Now, let B be any formula of G s.t. $B \notin V_G^e(A)$. By Lemma 2.18, the only path from A to B in a principal switch S traverses $A \frown C$. Therefore, as C is not a formula of $\mathsf{e}_G(A)$, the path that connects A to B is not a path of $\mathsf{e}_G(A)$ and B is not a formula in $\mathsf{e}_G(A)$. Then, every formula of $\mathsf{e}_G(A)$ is a formula of $\varepsilon_G(A)$, that is, $\mathsf{e}_G(A) \subseteq \varepsilon_G(A)$. $\square$

2.2.1.8 Sequentialization via empires

The algorithm that constructs $\mathsf{e}_G(A)$ can be used to give another proof of the sequentialization theorem. Indeed, this is the original one given by Girard, see [15] or [7].

Lemma 2.20 *Let A and B be distinct formulas s.t. $B \notin V_G^e(A)$. Then*

(i) *$A \in V_G^e(B)$ implies $\mathsf{e}_G(A) \subset \mathsf{e}_G(B)$;*
(ii) *$A \notin V_G^e(B)$ implies $\mathsf{e}_G(A) \cap \mathsf{e}_G(B) = \emptyset$.*

Proof Given $C \in V_G^e(A)$, let φ be the path from A to C in a principal switch S of B. Every formula in φ is in $V_G^e(A)$ (by the fact that $\mathsf{e}_G(A) \in \mathsf{PN}$). Moreover, if $D \frown E$ is an edge crossed by φ, the only case in which we can have $D \in V_G^e(B)$ and $E \notin V_G^e(B)$ is when $D = B$ and E is the conclusion of the link below B, if any (remind that, we are assuming that S is a principal switch of B), but, as $B \notin V_G^e(A)$ by hypothesis, this is not the case. As a consequence, if $A \in V_G^e(B)$, all the path φ is in $e_G(B)$ and, in particular, $C \in V_G^e(B)$; so, $V_G^e(A) \subset V_G^e(B)$. Otherwise, when $A \notin V_G^e(B)$, all the path φ is outside $e_G(B)$ and, in particular, $C \notin V_G^e(B)$; so, $V_G^e(A) \cap V_G^e(B) = \emptyset$ $\qquad\qquad\square$

The previous nesting property of empires induces a sequentialization procedure. As usual, let us assume w.l.o.g. that G is cut free. If $G \vdash \Gamma, A$ contains at least a par or tensor link and A is not conclusion of an axiom (that is, $G = G'$; $B, C \triangleright A$), we have the following two possibilities:

(i) $B, C \overset{\text{⅋}}{\triangleright} A$. In this case, it is readily seen that $G' \in \mathsf{PN}$ (remind that we are assuming $G \in \mathsf{PN}$). Therefore, given a sequentialization of G' (by the induction hypothesis), we can get a sequentialization of G by appending a par rule to it.

(ii) $B, C \overset{\otimes}{\triangleright} A$. In this case, by Lemma 2.20, $e_G(B) \cap e_G(C) = \emptyset$, and $e_G(B), e_G(C) \subset e_G(A)$. Now, let $H = e_G(B)$; $e_G(C)$; $B, C \overset{\otimes}{\triangleright} A$; the tensor link of A is splitting when $G = H$. By the previous item, we can always reduce to the case in which no conclusion of G is conclusion of a par link. Therefore, G is sequentializable only if one of the tensors above its conclusion is splitting.

Remark 2.21 *We have just seen that, when the link above the conclusion A is a par, we can get a sequentialization of the proof net in which that par corresponds to the last rule of the sequentialized proof. As a consequence, we say that the par rule is* invertible, *namely, a par rule that introduces a par formula in the final sequent of a proof can always be postponed to the end of the derivation. This is not the case for the tensor link. For instance, let $G_1 \vdash A, B$ and $G_2 \vdash C, D$ be two proof nets. It is readily seen that in any sequentialization of $G = G_1$; G_2; $A, C \overset{\otimes}{\triangleright} A \otimes C$; $B, D \overset{\text{⅋}}{\triangleright} B \,⅋\, D$, the par rule must be the last rule. However, even if we assume that every conclusion of the proof net is conclusion of a tensor link, not every concluding tensor is necessarily splitting (even if, there must be one). For instance, in the proof net in Figure 2.7 (where $G_1, G_2, G_3 \in \mathsf{PN}$), the tensor above $C \otimes B$ is not splitting, while the one above $A \otimes (B \,⅋\, C)$ it is.*

Summing up, for a cut free proof net G that does not contain any concluding par link (we assume that G has at least a concluding tensor link, otherwise G is an axiom), the problem of finding a sequentialization

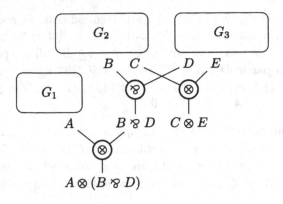

Fig. 2.7. Splitting tensor

can be reduced to that of finding a splitting tensor. So, let us choose any conclusion A_0 of G s.t. $B_0, C_0 \overset{\otimes}{\triangleright} A_0 \in G$. If A_0 is not splitting, there is a conclusion D of $e_G(B_0)$ or $e_G(C_0)$ that is not a conclusion of G. Let A_1 be the conclusion of G s.t. $A_1 \preccurlyeq_1 B_1 \preccurlyeq D$, where B_1 is one of the premises of the tensor link above A_1, we see that $A_1 \notin V_G^e(B_0) \cup V_G^e(C_0)$ and $A_0 \in V_G^e(B_1)$. Now, if A_1 is splitting, we have done; otherwise, let us iterate the procedure. After k steps we get a sequence of distinct conclusions $A_0, A_1, \ldots, A_k$ s.t. $A_0, \ldots, A_{k-1} \in V_G^e(B_k)$, where B_k is one of the premises of the tensor above A_k. Therefore, we must eventually terminate finding a splitting tensor (and the two proof nets in which the tensor splits G).

$$\frac{}{A \vdash A}\ \text{ax} \qquad\qquad \frac{\Gamma \vdash A \qquad \Delta, A \vdash B}{\Gamma, \Delta \vdash B}\ \text{cut}$$

$$\frac{\Gamma \vdash A \qquad \Delta, B \vdash C}{\Gamma, \Delta, A \multimap B \vdash C}\ \multimap\text{L} \qquad\qquad \frac{\Gamma, A \vdash B}{\Gamma \vdash A \multimap B}\ \multimap\text{R}$$

$$\frac{\Gamma, A, B \vdash C}{\Gamma, A \otimes B \vdash C}\ \otimes\text{L} \qquad\qquad \frac{\Gamma \vdash A \qquad \Delta \vdash B}{\Gamma, \Delta \vdash A \otimes B}\ \otimes\text{R}$$

Fig. 2.8. IMLL.

2.2.2 IMLL

The sequent calculus for Intuitionistic MLL (IMLL) is given in Figure 2.8. IMLL is a subsystem of MLL. In fact, let us take the *polarized* version of MLL, say pMLL, defined by:

1. the formulas of pMLL are pairs (A, x), also written A^x, with $x \in \{+, -\}$ and s.t. one of the following cases holds:

   (a) A is an atom;
   (b) $x = +$ and either $A = B^+ \otimes C^-$ or $A = B^- \otimes C^+$;
   (c) $x = +$ and $A = B^+ \otimes C^+$;
   (d) $x = -$ and either $A = B^+ \otimes C^-$ or $A = B^- \otimes C^+$
   (e) $x = -$ and $A = B^- \otimes C^-$.

2. $(A^+)^\perp = \overline{A}^-$ and $(A^-)^\perp = \overline{A}^+$ where $\overline{p} = p^\perp$ and

$$\overline{B^x \otimes C^y} = (B^x)^\perp \otimes (C^y)^\perp \qquad \overline{B^x \otimes C^y} = (B^x)^\perp \otimes (C^y)^\perp$$

3. the rules of pMLL are those of MLL where MLL formulas are replaced by pMLL formulas. This implies that:

   (a) the principal formulas of the axiom and of the cut rule have distinct polarities.
   (b) every formula in the proof must be a valid pMLL formula; therefore, the two premises of a par rule cannot be both positive, while the two premises of a tensor rule cannot be both negative.
   (c) every sequent derivable in pMLL has the shape $\vdash \Gamma, A^+$ where Γ is a set of negative formulas.

Assuming that $[A^+]_| = [A]_|$, the following rules define a translation function from positive pMLL formulas to IMLL formulas

$$[p]_| = p \qquad [A^+ \otimes B^+]_| = [A]_| \otimes [B]_|$$

$$[A^- \otimes B^+]_| = [B^+ \otimes A^-]_| = [(A^-)^\perp]_| \multimap [B]_|$$

The translation of a pMLL sequent into an IMLL sequent is defined by

$$[\vdash \Gamma, A^+]_| = [\Gamma^\perp]_| \vdash [A^+]_|$$

where $\Gamma^\perp$ is obtained by dualizing every formula in Γ and $[\Gamma^\perp]_|$ is the translation of every formula in $\Gamma^\perp$.

It is readily seen that the function that replaces every pMLL sequent $\vdash \Delta$ in a proof with the IMLL sequent $[\vdash \Delta]_|$ maps any pMLL proof of $\vdash \Gamma, A^+$ into an IMLL proof of the sequent $[\Gamma^\perp]_| \vdash [A^+]_|$.

By the way, there is an inverse translation from IMLL to pMLL. In fact, let us define the positive translation of an IMLL formula:

$$[p]_{\mathsf{p}} = p^+ \quad [A \multimap B]_{\mathsf{p}} = ([A]_{\mathsf{p}}^{\perp} \,\mathfrak{N}\, [B]_{\mathsf{p}})^+ \quad [A \otimes B]_{\mathsf{p}} = ([A]_{\mathsf{p}} \otimes [B]_{\mathsf{p}})^+$$

The translation of an IMLL sequent is

$$[\Gamma \vdash A]_{\mathsf{p}} = \vdash [\Gamma]_{\mathsf{p}}^{\perp}, [A]_{\mathsf{p}}$$

Again, the function that replaces every IMLL sequent $\Delta \vdash B$ in a proof with the pMLL sequent $[\Delta \vdash B]_{\mathsf{p}}$ maps any IMLL proof of $\Gamma \vdash A$ into a pMLL proof of the sequent $\vdash [\Gamma]_{\mathsf{p}}^{\perp}, [A]_{\mathsf{p}}$.

Remark 2.22 *The latter correspondence is not one to one for very bureaucratic reasons, namely, because of $[A^- \,\mathfrak{N}\, B^+]_{\mathsf{I}} = [B^+ \,\mathfrak{N}\, A^-]_{\mathsf{I}} = [(A^-)^{\perp}]_{\mathsf{I}} \multimap [B^+]_{\mathsf{I}}$. We might get a bijection by adding the connective $B \mathbin{\circ\!\!-} A$ (to get its rules, just replace $B \mathbin{\circ\!\!-} A$ in the rules for $\multimap$) with $[B^+ \,\mathfrak{N}\, A^-]_{\mathsf{I}} = [B^+]_{\mathsf{I}} \mathbin{\circ\!\!-} [(A^-)^{\perp}]_{\mathsf{I}}$, and $[B \mathbin{\circ\!\!-} A]_{\mathsf{p}} = [B]_{\mathsf{p}} \,\mathfrak{N}\, [A]_{\mathsf{p}}^{\perp}$. But, as we are in the commutative case, there is no real need for this.*

2.2.2.1 Oriented proof nets

Polarity may be extended to proof nets. Namely, let us say that an *oriented proof structure* is a pair (G, π) where G is a proof structure and π is a map from the formulas of G to $\{+, -\}$ that respects the restrictions in Figure 2.9 (where only the polarities of the formulas have been drawn).

Oriented proof structures can be used to represent pMLL and then IMLL proofs also. In fact, let us assume w.l.o.g. that all the conclusions of the axioms in G are atomic; the formulas in G can be replaced in a natural way by polarized formulas. Moreover, as MLL sequentialization Π of G is a valid derivation in pMLL, Π corresponds to a sequentialization in IMLL also.

Every switch S of an MLL proof net G defines a set of valid orientations for G, one for every choice of a conclusion of G as the only positive conclusion of the net. Therefore, fixed a conclusion of G, every $S \in \mathbb{SW}(G)$ gives a valid IMLL proof of $[\Gamma]_{\mathsf{I}}^{\perp} \vdash [A]_{\mathsf{I}}$ (see [7]).

Definition 2.23 *Let A be a conclusion of the proof net G and S be a switch of G. The orientation $\pi_{A,S}$ induced by the pair A, S is defined by the following rules:*

(i) $\pi_{A,S}(A) = +;$

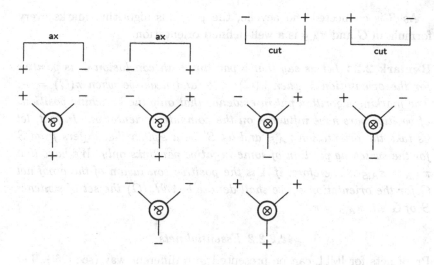

Fig. 2.9. Oriented links.

(ii) *if* $\pi_{A,S}(B) = +$ *and* $B \frown_S C$ *and either* $C, D \overset{\mathfrak{N}}{\rhd} B \in G$ *or* $D, C \overset{\mathfrak{N}}{\rhd} B \in G$, *then* $\pi_{A,S}(C) = +$;

(iii) *if* $\pi_{A,S}(B) = +$ *and* $C, D \overset{\otimes}{\rhd} B \in G$, *then* $\pi_{A,S}(C) = \pi_{A,S}(D) = +$;

(iv) *if* $\pi_{A,S}(B) = +$ *and either* $\overset{ax}{\rhd} B, B^{\perp} \in G$ *or* $\overset{ax}{\rhd} B^{\perp}, B \in G$, *then* $\pi_{A,S}(B^{\perp}) = -$;

(v) *if* $\pi_{A,S}(B) = -$ *and* $B \frown_S D$ *and either* $B, C \overset{\mathfrak{N}}{\rhd} D \in G$ *or* $C, B \overset{\mathfrak{N}}{\rhd} D \in G$, *then* $\pi_{A,S}(D) = -$;

(vi) *if* $\pi_{A,S}(B) = -$ *and either* $B, C \overset{\otimes}{\rhd} D \in G$ *or* $C, B \overset{\otimes}{\rhd} D \in G$, *then* $\pi_{A,S}(D) = -$ *and* $\pi_{A,S}(C) = +$;

(vii) *if* $\pi_{A,S}(B) = -$ *and either* $B, B^{\perp} \overset{cut}{\rhd} \in G$ *or* $B^{\perp}, B \overset{cut}{\rhd} \in G$, *then* $\pi_{A,S}(B^{\perp}) = +$.

The rules of the marking algorithm in Definition 2.23 correspond to the algorithm: visit the tree of S rooted at the conclusion A and mark every formula B according to the rules

(i) if B is the starting formula, then mark it with $+$;

(ii) if B is reached from the conclusion of the link below it or from the other premise of a cut, then mark B with $+$;

(iii) if B is reached from the premise of the link above it or from the other premise of an axiom, then mark B with $-$.

As S is connected and acyclic, the previous algorithm marks every formula in G and $\pi_{A,S}$ is a well-defined orientation.

Remark 2.24 *Let us say that a par link with conclusion C is positive for the orientation π when $\pi(C) = +$ and negative when $\pi(C) = -$. The marking algorithm put in evidence that only the switching positions of positive pars have influence on the computed orientation. In fact, let us take the orientation $\pi_{A,S}$ and let S' be a switch that differs from S for the switching position of some negative par links only. We have that $\pi_{A,S} = \pi_{A,S'}$. Therefore, if A is the positive conclusion of the proof net G for the orientation π, we shall denote by $\mathbb{SW}_\pi(G)$ the set of switches S of G s.t. $\pi_{A,S} = \pi$.*

2.2.2.2 Essential nets

Proof nets for IMLL can be presented in a different way (see [26]). Let (G, π) be the oriented proof net corresponding to the translation in pMLL of the IMLL proof Π. The *essential net E* corresponding to the orientation π of G is the least directed graph s.t. there is a directed edge from B to C, say $B \curvearrowright_E C$, when (see the thick lines in Figure 2.10):

  (i) $\pi(B) = +$ and B, C are the conclusions of an axiom;
 (ii) $\pi(C) = +$ and C is premise of the link with conclusion B;
(iii) $\pi(B) = -$ and B, C are the premises of a cut;
 (iv) $\pi(B) = \pi(C) = -$ and B is premise of a link with conclusion C.

The positive conclusion A of (G, π) is the only root (or source) of E. Then, let S_π be the directed switch obtained by orienting the edges of a switch of G according to the rules in Figure 2.10. If $S \in \mathbb{SW}_\pi(G)$, then S_π is a directed tree rooted at A, and every S_π s.t. $S \in \mathbb{SW}_\pi(G)$ (say $S_\pi \in \mathbb{SW}_\pi(G)$ for short) is a subtree of E obtained by removing, for every negative par link, one of the edges entering into its conclusion. Indeed, E is the superposition of all the $S_\pi \in \mathbb{SW}_\pi(G)$; so, in E, there is a path—not a unique one, in the general case—from A to every formula B. Nevertheless, E is acyclic. In fact, a path φ of E is not a path of some directed switch S_π of E iff it contains both the edges entering into the conclusion of some negative par; therefore, let $C \curvearrowright_E B \varphi D \curvearrowright_E B$ be the shortest path of E that contains both the premises C, D of a negative par with conclusion B; by hypothesis, $B \varphi D \curvearrowright_E B$ is a cyclic path of some directed switch of E; but, as every directed switch of E is a tree, such a path φ cannot exist. Summing up, E is a directed and acyclic graph rooted at A.

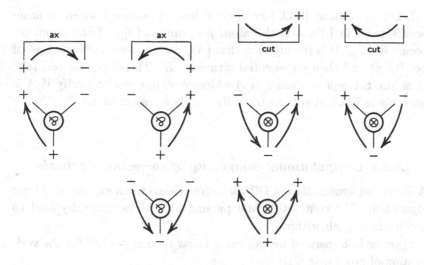

Fig. 2.10. Essential-net edges.

Now, given a positive par with conclusion B, let C, D be the premises of the par with $\pi(C) = +$ and $\pi(D) = -$. As D is a conclusion of the essential net $e_E(C)$ associated to $e_G(C)$, in every directed switch of $e_E(C)$, and then in every directed switch of E, there is a path form C to D. Moreover, as in every directed switch of E there is a path from A to B, we can conclude that every path of E from A to D splits into a path from A to B and a path from C to D connected by the directed edge $B \curvearrowright_E C$.

Summing up, let us say that a directed graph E is an *essential structure* when it has only one root, its vertices are polarized MLL formulas, and its directed edges respect the orientations in Figure 2.10. The proof structure G corresponding to E is a proof net only if E is acyclic and satisfies the following property:

EN: every path from the root of E to the negative premise of a positive par link passes through the conclusion of the par.

Proposition 2.25 *The essential structure E associated to an oriented proof structure (G, π) is an essential net, that is, G is a proof net iff E is acyclic and condition* **EN** *holds.*

Proof The only if direction has been already proved. For the if direction see [26]. □

Let us say that an IMLL formula A is linearly balanced when no atom occurs twice in $[A]_p$ and the atom p occurs in $[A]_p$ iff the atom $p^\perp$ occurs in $[A]_p$. It is readily seen that $[A]_p$ determines an oriented proof net (G, π) and then an essential structure E. Therefore, we can take E as the net representation of A. Moreover, in order to verify if A is provable in IMLL, it suffices to verify if E is an essential net.

2.2.3 Computational complexity of correctness criteria

A direct implementation of DR-correctness leads to an exponential time algorithm. The contraction and parsing algorithms naturally lead to quadratic time algorithms.

Starting from parsing we can get a linear time algorithm for the verification of proof nets [21].

Another linear time algorithm can be achieved exploiting the properties of essential nets [31].

2.2.4 Cut-elimination

The cut-elimination procedure for MLL proof nets is defined by the following rules (see Figure 2.11 also)

$$\overset{\text{ax}}{\rhd} A, X \; ; \; X, X^\perp \overset{\text{cut}}{\rhd} \; ; \; \overset{\text{ax}}{\rhd} X^\perp, A^\perp \qquad\qquad (\overset{\text{ax}}{\rhd}/\overset{\text{cut}}{\rhd})$$

$$\overset{\text{cut}}{\longrightarrow} \overset{\text{ax}}{\rhd} A, A^\perp$$

$$A, B \overset{\mathfrak{N}}{\rhd} A \mathfrak{N} B \; ; \; A \mathfrak{N} B, A^\perp \otimes B^\perp \overset{\text{cut}}{\rhd} \; ; \; A^\perp, B^\perp \overset{\otimes}{\rhd} A^\perp \otimes B^\perp \qquad (\overset{\mathfrak{N}}{\rhd}/\overset{\otimes}{\rhd})$$

$$\overset{\text{cut}}{\longrightarrow} A, A^\perp \overset{\text{cut}}{\rhd} \; ; \; B, B^\perp \overset{\text{cut}}{\rhd}$$

where X and $X^\perp$ in the first rule are occurrences of A and $A^\perp$, respectively.

The cut-elimination rules preserve DR-correctness. Therefore, if $G \in$ PN and $G \overset{\text{cut}}{\longrightarrow} G'$, then $G' \in$ PN. Moreover, they are confluent and, as every rule decreases the number of links in the structure, cut-elimination is terminating and the unique normal form of any proof net is cut free.

Theorem 2.26 *If* $\vdash \Gamma$ *is provable in* MLL, *there is a cut free* MLL *proof of* $\vdash \Gamma$.

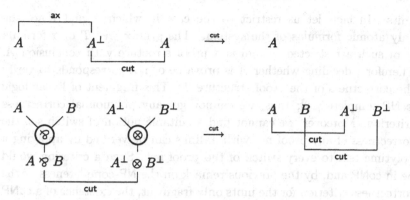

Fig. 2.11. MLL cut-elimination rules.

2.2.5 Units

The sequent calculus rules for the tensor unit 1 and for the par unit $\perp = 1^\perp$ are

$$\frac{}{\vdash 1} 1 \qquad \frac{\vdash \Gamma}{\vdash \Gamma, \perp} \perp$$

It is readily seen that $(A \otimes 1) \sim (1 \otimes A) \sim A$ and that $(A \,\%\, \perp) \sim (\perp \,\%\, A) \sim A$.

The natural way to extend proof nets to MLL plus units is by introducing two new links with no premises and one conclusion: $\triangleright \overset{1}{} 1$ and $\triangleright \overset{\perp}{} \perp$. The parsing algorithm for the system is that in Figure 2.6 plus the rules:

$$\triangleright \overset{1}{} 1 \overset{p}{\to} \overset{*}{\triangleright} 1 \tag{$\triangleright$}$$

$$\overset{*}{\triangleright} \Gamma; \overset{}{\triangleright} \overset{\perp}{} \perp \overset{p}{\to} \overset{*}{\triangleright} \Gamma, \perp \tag{$\triangleright$}$$

By the way, if $G \vdash \Gamma$ is a proof structure that contains unit links, it is still true that when $G \overset{p^*}{\to} \overset{*}{\triangleright} \Gamma$, then G is the image of a sequent proof $\Pi \vdash \Gamma$; vice versa, although when G is the image of some proof there is a parsing reduction from G to $\overset{*}{\triangleright} \Gamma$, it is no longer true that this is the only parsing normal form of G. The problem is the representation of $\perp$. In fact, we can easily construct a proof net that reduces to $G = \overset{*}{\triangleright} A; \overset{*}{\triangleright} \perp; \overset{*}{\triangleright} B; A, \perp \overset{\%}{\triangleright} C; B, C \overset{\otimes}{\triangleright} D$; then, by reducing $\overset{*}{\triangleright} A; \overset{}{\triangleright} \overset{\perp}{} \perp$, we get $G \overset{p^*}{\to} \overset{*}{\triangleright} D$; while, by reducing $\overset{*}{\triangleright} B; \overset{}{\triangleright} \overset{\perp}{} \perp$, the parsing reduction immediately ends with another normal form.

The key problem with the multiplicative units is that they cannot be characterized by any criterion similar to DR-correctness for MLL without units. In fact, let us restrict to the case in which $\bot$ and 1 are the only atomic formulas of the system. The syntax tree T of a formula A of such a restricted system is a proof structure with conclusion A. Therefore, deciding whether A is provable or not corresponds to verify the correctness of the proof structure T. This fragment of linear logic is NP-complete [29]; thus, we cannot get any polynomial correctness criterion. Moreover, we cannot find a suitable notion of switch s.t. the correctness of any proof net with $\bot$-links can be verified by applying a polytime test to every switch of the proof net. Such a criterion would be in coNP and, by the previous remark on the NP-completeness of the correctness criterion for the units only fragment, the existence of a coNP criterion would imply coNP = NP, that is very unlikely.

Another possible representation for $\bot$ would be to explicit its connections with the formulas in the context that justifies its introduction (*e.g.*, by linking any formula in the context Γ to the rule $\bot$). An easy representation of such a connection might be an edge from every $\bot$-link to the formula A justifying it, a so-called *jump*. Otherwise, we might decide to allow the introduction of $\bot$ at the level of axioms only (it is readily seen that this system has the same power of the general one where $\bot$ can be introduced at every point of the derivation). However, none of that solutions is completely satisfactory. In fact, first of all, there is no canonical way to link a $\bot$-link to a formula or to an axiom that justifies it. Moreover, both that solutions give problems when we try to eliminate a cut below the formula of the axiom linked to the $\bot$-link.

Another approach to the treatment of units—indeed, the original one given by Girard (see [13])—would be to link every $\bot$-link to the whole proof net corresponding to the proof above its introduction. This corresponds to what we shall do for the representation of the so-called exponentials; for every $\bot$-link, we should introduce a box (see Definition 2.29) with its $\bot$ as principal door and with no restriction on the shape of the auxiliary door. Nevertheless, this amounts to loose, at least for $\bot$, some of the main advantages of proof nets, as we would distinguish two proof nets that simply differ for the permutation of a $\bot$-rule with the rule above it.

2.2.5.1 A relevant case

There are cases in which units can be represented as distinct links and correctness can be formulated by a suitable extension of Definition 2.4. In his thesis, Regnier [33] has shown that the formulas $\bot \otimes A$ or $A \otimes \bot$— and $1 \mathbin{⅋} A$ and $A \mathbin{⅋} 1$ by duality—are the problematic cases. In fact, let us say that A is a $\perp$-formula when A is a par combination of $\perp$-atoms (that is, $A = \perp$ or $A = A_1 \,\mathbin{\rotatebox[origin=c]{180}{\&}}\, A_2$ where A_1, A_2 are $\perp$-formulas). It is readily seen that $A \sim \perp$, for every $\perp$-formula A. By duality, we define the 1-formulas as those obtained by tensor combination of 1-atoms.

Let us say that a formula A is *well-formed* when:

1. A is an atom, A is a $\perp$-formula, or A is a 1-formula;
2. $A = B \,\mathbin{\rotatebox[origin=c]{180}{\&}}\, C$ and neither B nor C is a 1-formula, or $A = B \otimes C$ and neither B nor C is a $\perp$-formula.

A proof structure G with units is well-formed when every formula in G is well-formed.

Definition 2.27 *A well-formed proof structure G with k $\perp$-links is* DR-*correct, G is a well-formed proof net, when every switch of G is acyclic and has $k+1$ connected components.*

The rules for parsing well-formed proof nets are those in Figure 2.6 plus the rules

$$\overset{1}{\rhd}\, A;\; \overset{1}{\rhd}\, B;\; A, B \overset{\otimes}{\rhd} A \otimes B \overset{p}{\to}\; \overset{1}{\rhd}\, A \otimes B \qquad (\overset{1}{\rhd})$$

$$\overset{\perp}{\rhd}\, A;\; \overset{\perp}{\rhd}\, B;\; A, B \overset{\rotatebox[origin=c]{180}{\&}}{\rhd} A \,\mathbin{\rotatebox[origin=c]{180}{\&}}\, B \overset{p}{\to}\; \overset{\perp}{\rhd}\, A \,\mathbin{\rotatebox[origin=c]{180}{\&}}\, B \qquad (\overset{\perp}{\rhd})$$

$$\overset{\perp}{\dot{\rhd}}\, \Gamma, A;\; \overset{\perp}{\rhd}\, B;\; A, B \overset{\rotatebox[origin=c]{180}{\&}}{\rhd} A \,\mathbin{\rotatebox[origin=c]{180}{\&}}\, B \overset{p}{\to}\; \overset{\perp}{\dot{\rhd}}\, \Gamma, A \,\mathbin{\rotatebox[origin=c]{180}{\&}}\, B \qquad (\perp\overset{\rotatebox[origin=c]{180}{\&}}{\rhd}_l)$$

$$\overset{\perp}{\dot{\rhd}}\, \Gamma, B;\; \overset{\perp}{\rhd}\, A;\; A, B \overset{\rotatebox[origin=c]{180}{\&}}{\rhd} A \,\mathbin{\rotatebox[origin=c]{180}{\&}}\, B \overset{p}{\to}\; \overset{\perp}{\dot{\rhd}}\, \Gamma, A \,\mathbin{\rotatebox[origin=c]{180}{\&}}\, B \qquad (\perp\overset{\rotatebox[origin=c]{180}{\&}}{\rhd}_r)$$

$$\overset{\perp}{\dot{\rhd}}\, \Gamma, A;\; \overset{1}{\rhd}\, B;\; A, B \overset{\otimes}{\rhd} A \otimes B \overset{p}{\to}\; \overset{\perp}{\dot{\rhd}}\, \Gamma, A \otimes B \qquad (1\overset{\otimes}{\rhd}_l)$$

$$\overset{\perp}{\dot{\rhd}}\, \Gamma, B;\; \overset{1}{\rhd}\, A;\; A, B \overset{\otimes}{\rhd} A \otimes B \overset{p}{\to}\; \overset{\perp}{\dot{\rhd}}\, \Gamma, A \otimes B \qquad (1\overset{\otimes}{\rhd}_r)$$

Any parsing structure obtained by the reduction of a well-formed proof structure is well-formed. Moreover, in a well-formed parsing structure the conclusion of a 1-link ($\perp$-link) can be any 1-formula ($\perp$-formula).

Proposition 2.28 *Let $G \vdash \Gamma, A_1, \ldots, A_n$ be a proof structure s.t. the conclusions $A_1, \ldots, A_n$ are the only $\perp$-formulas among the conclusions of G. Let $N = \overset{1}{\rhd}\, A;\; \overset{\perp}{\rhd}\, A_1;\; \ldots;\; \overset{\perp}{\rhd}\, A_n$, when $\Gamma = A$ for some 1-formula A, or $N = \overset{}{\dot{\rhd}}\, \Gamma;\; \overset{\perp}{\rhd}\, A_1;\; \ldots;\; \overset{\perp}{\rhd}\, A_n$, otherwise. Then, G is well-formed and* DR-*correct iff $G \overset{p}{\to}{}^* N$; moreover, if G is well-formed and* DR-*correct, N is its only $\overset{p}{\to}$-normal form.*

Proof First of all, let us remark that $G \overset{p}{\to}{}^* N$ only if G is well-formed.

Then, let us remark that DR-correctness is invariant under $\overset{p}{\to}$-reduction; that is, if $G\overset{p}{\to}G'$, then G is DR-correct iff G' is DR-correct. As a consequence, we immediately get the if direction.

Let us assume that G is well-formed. By Remark 2.6, every cut whose principal formula is not a $\perp$-formula can be replaced by a tensor, as this transformation preserves well-formedness. Otherwise, if the principal formula is a $\perp$-formula, it is readily seen that we have a substructure $G_0 = G_1\,;\; G_\perp\,;\; A_1, A_\perp \overset{\text{cut}}{\triangleright}$ s.t. $G_0 \overset{p}{\to}{}^* \emptyset$, where $G_1 \vdash A_1$ and $G_\perp \vdash A_\perp$, and $A_\perp = A_1^\perp$ is a $\perp$-formula (A_1 is a 1-formula). Therefore, w.l.o.g. we can restrict to the cut free case. Now, let $G \overset{p}{\to}{}^* G'$ be any parsing reduction such that G' does not contain any $\overset{\perp}{\triangleright}$-redex or any $\overset{1}{\triangleright}$-redex. It is readily seen that $G' = H\,;\; \overset{\perp}{\triangleright} A_1\,;\; \ldots\,;\; \overset{\perp}{\triangleright} A_n$, for some well-formed proof structure H s.t.: (i) H is DR-correct iff G' is DR-correct iff G is DR-correct (by the initial remark on the invariance of DR-correctness); (ii) every $\perp$-formula in H is one of the premises of a par link whose other premise (and its conclusion) is not a $\perp$-link; (iii) if $\Gamma = A$ for some 1-formula A, then $H = \overset{1}{\triangleright} A$. By (iii) and (i), we see that we left to show that $H \overset{p}{\to}{}^* \overset{*}{\triangleright} \Gamma$ when H is DR-correct and Γ is not a 1-formula.

Let us proceed by induction on the number of $\perp$-links in H. Let B be a $\perp$-formula in H and let C and D be the other premise and the conclusion of the par link below B. We see that the new structure obtained by removing the $\perp$-link above B and the par link below B and by replacing C for the occurrence of D in every formula $X \preccurlyeq C$ leads to a DR-correct structure $H' \vdash \Gamma'$ (if $\Gamma = \Delta, A$ with $A \preccurlyeq B$, then $\Gamma' = \Delta, A'$ where A' is obtained by replacing C for D in A); moreover, $H \overset{p}{\to}{}^* \overset{*}{\triangleright} \Gamma$ iff $H' \overset{p}{\to}{}^* \overset{*}{\triangleright} \Gamma'$. Therefore, by induction hypothesis, we conclude. Now, let us assume that H does not contain any $\perp$-link. Let H' be the structure obtained from H by replacing every 1-link in H with a $*$-link with the same conclusion. We see that H' is DR-correct iff H is DR-correct and that $H' \overset{p}{\to}{}^* \overset{*}{\triangleright} \Gamma$ iff $H \overset{p}{\to}{}^* \overset{*}{\triangleright} \Gamma$. Therefore, by Proposition 2.12, we conclude. $\qquad\square$

2.2.5.2 Cut-elimination

When units are introduced by means of links without connections to other formulas or links in the net, as in the relevant case considered above, the cut-elimination rule is

$$\overset{1}{\triangleright} 1\,;\; 1, \perp \overset{\text{cut}}{\triangleright}\,;\; \overset{\perp}{\triangleright} \perp \overset{\text{cut}}{\longrightarrow} \qquad\qquad (\overset{1}{\triangleright}/\overset{\perp}{\triangleright})$$

It is readily seen that adding that rule to those in section 2.2.4 preserves strong normalization and confluence.

2.3 The exponentials: MELL

MLL is too weak: as everything is linear, one can code linear λ-calculus only. In order to get λ-calculus we have to add *structural rules*: contraction and weakening. However, adding the rules in an uncontrolled way would destroy the system—we would just get a redundant system for classical logic. Because of this, in order to control the use of the structural rules, let us introduce two dual unary connectives ?, named *why-not*, and !, named *of-course*, with $(!A)^{\perp} = ?A^{\perp}$, and the rules in Figure 2.12, where $?\Gamma$ stands for a set of ?-formulas $?A_1, \ldots, ?A_k$.

$$\frac{\vdash \Gamma, A}{\vdash \Gamma, ?A} \text{ dereliction} \qquad \frac{\vdash \Gamma, ?A, ?A}{\vdash \Gamma, ?A} \text{ contraction} \qquad \frac{\vdash \Gamma}{\vdash \Gamma, ?A} \text{ weakening}$$

$$\frac{\vdash ?\Gamma, A}{\vdash ?\Gamma, !A} \text{ promotion}$$

Fig. 2.12. Exponential rules.

The connectives ? and ! are the *exponentials* and the fragment (without constants) of linear logic formed of the rules of MLL (Figure 2.1) plus the rules in Figure 2.12 is the so-called Multiplicative Exponential fragment of Linear Logic, MELL for short.

In MELL the structural rules are allowed on ?-formulas only and the promotion rule (the introduction rule for !) requires that the formulas in the context, the auxiliary conclusions, are ?-formulas.

Let us remark the constraint in the promotion rule. The elimination of a cut between a pair of exponential formulas requires the duplication or the erasing of the proof ending with the !-formula, but, in order to preserve the conclusions of the resulting proof, after a duplication, we have to contract the auxiliary conclusions of the duplicated proof, after an erasing, we have to introduce a weakening for every auxiliary conclusion of the erased proof. In both cases, the operation is sound only if all the auxiliary conclusions of the proof ending with the promotion rule are ?-formulas. (We omit the cut-elimination rules of MELL sequent calculus; we shall define cut-elimination directly for exponential proof nets, see 2.3.4).

2.3.1 Boxes

The representation of exponential proof nets requires the introduction of the links corresponding to the exponential rules (see Figure 2.13); moreover, because of the side-condition on the promotion rule, we need a new object named *box*.

Definition 2.29 (box) *A box B is a proof structure whose conclusions are ?-formulas but one, its* principal door, *which is the conclusion of an !-link. The ?-conclusions of B are its* auxiliary doors.

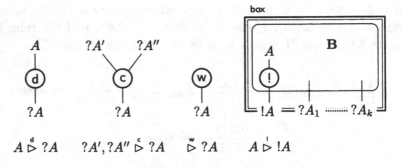

$$A \overset{d}{\rhd} ?A \qquad ?A', ?A'' \overset{c}{\rhd} ?A \qquad \overset{w}{\rhd} ?A \qquad A \overset{!}{\rhd} !A$$

Fig. 2.13. Exponential links.

Definition 2.30 (exponential proof structure) *An exponential proof structure $\mathbf{G}$ with conclusions Γ (written $\mathbf{G} \vdash \Gamma$) is a pair $(G, \mathbb{B})$ s.t.:*

   (i) $G \vdash \Gamma$ a proof structure;

   (ii) $\mathbb{B}$ is a set of boxes, one for each !-formula in G which is conclusion of an !-link, that satisfies the so-called box nesting condition: *for every pair $B_1, B_2 \in \mathbb{B}$, either $B_1 \subseteq B_2$ or $B_2 \subseteq B_1$, or $B_1 \cap B_2 = \emptyset$.*

According to the box nesting condition, two distinct boxes may nest but not partially overlap. Moreover, two boxes cannot have the same principal door, even if, they may share one or more auxiliary doors.

If $\mathbf{G} = (G, \mathbb{B})$ is an exponential proof structure, let us define $|\mathbf{G}| = G$ and $\mathbf{G}^{\square} = \mathbb{B}$. The inclusion relation between proof structures naturally extends to exponential proof structures. Namely, $\mathbf{H} \subseteq \mathbf{G}$ if $|\mathbf{H}| \subseteq |\mathbf{G}|$ and $\mathbf{H}^{\square} \subseteq \mathbf{G}^{\square}$.

Let $B \in \mathbf{G}^{\square}$. If $B = \overset{o}{B}; A \overset{!}{\rhd} !A$ is a box with principal door $!A$, the proof structure $\overset{o}{B}$ is the *opened box* of $!A$. The (opened) *proof box* of B is the exponential proof structure $\mathbf{B} \subseteq \mathbf{G}$ s.t. $|\mathbf{B}| = B$ ($\overset{\circ}{\mathbf{B}} \subseteq \mathbf{G}$ s.t $|\overset{\circ}{\mathbf{B}}| = \overset{\circ}{B}$). In the following, given the proof box $\mathbf{B} \vdash \,?\Gamma, !A$ (and its opened proof box $\overset{\circ}{\mathbf{B}} \vdash \,?\Gamma, A$), we shall use the notation

$$\mathbf{B} = [\,\overset{\circ}{\mathbf{B}}; \; A \overset{!}{\triangleright} !A\,]$$

or $\mathbf{B} = [\,\overset{\circ}{\mathbf{B}}; \; A \overset{!}{\triangleright} !A\,]^{!A}_{?\Gamma}$ when we need to explicit the doors of the box.

2.3.2 Parsing

The rules for parsing exponential proof nets are those in Figure 2.6 plus the rules

$$\overset{\bullet}{\triangleright} \Gamma, A; \; A \overset{d}{\triangleright} ?A \overset{p}{\to} \overset{d}{\triangleright} \Gamma, ?A \qquad (\overset{d}{\triangleright})$$

$$\overset{\bullet}{\triangleright} \Gamma, ?A', ?A''; \; ?A', ?A'' \overset{c}{\triangleright} ?A \overset{p}{\to} \overset{c}{\triangleright} \Gamma, ?A \qquad (\overset{c}{\triangleright})$$

$$\overset{\bullet}{\triangleright} \Gamma; \; \overset{w}{\triangleright} ?A \overset{p}{\to} \overset{w}{\triangleright} \Gamma, ?A \qquad (\overset{w}{\triangleright})$$

$$[\overset{\bullet}{\triangleright} \,?\Gamma, A; \; A \overset{!}{\triangleright} !A] \overset{p}{\to} \overset{!}{\triangleright} \,?\Gamma, !A \qquad (\overset{!}{\triangleright})$$

that applies to contexts in which the redex does not cross the border of any box (*i.e.*, $[\,\overset{\bullet}{\triangleright} \,?A', ?A'', A; \; A \overset{!}{\triangleright} !A\,]^{!A}_{?A', ?A''}$; $?A', ?A'' \overset{c}{\triangleright} ?A$ contains an $\overset{!}{\triangleright}$-redex but not a $\overset{c}{\triangleright}$-redex).

An exponential proof structure $\mathbf{G} \vdash \Gamma$ is sequentializable when $\mathbf{G} \overset{p}{\to}^{*} \overset{\bullet}{\triangleright} \Gamma$. Remarkably, $\mathbf{G}$ is sequentializable only if every box in $\mathbf{G}^{\square}$ is sequentializable.

2.3.3 Weakening and DR-correctness

Exponential proof structures does not have a satisfactory definition of correctness in the style of the DR-criterion. If we compare weakening and $\bot$, we see that their rules have the same shape—as a matter of fact, the $\bot$-rule is a particular case of weakening. Therefore, the issues pointed out in 2.2.5 applies to weakening as well. By restricting the use of $\bot$ (see 2.2.5.1), we have seen how to get a subsystem for which correctness can be expressed by means of an extension of the DR-criterion. We might try to do the same for weakening; for instance, by imposing that no weakening formula is one of the premises of a tensor link (a weakening formula is the conclusion of a weakening link or the contraction or the par of two weakening formulas). However, by analysis of the cut-elimination rules in 2.3.4, we see that such a constraint cannot be preserved by cut-elimination. For instance,

$$\mathbf{G};\ C,?B \overset{\otimes}{\rhd} C \otimes ?B;\ [\mathbf{G}]^{!A}_{?B};\ !A,?A^{\perp} \overset{cut}{\rhd};\ \overset{w}{\rhd} ?A^{\perp}$$

$$\overset{cut}{\longrightarrow} \mathbf{G};\ C,?B \overset{\otimes}{\rhd} C{\otimes}?B;\ \overset{w}{\rhd} ?B$$

We might argue that this happens as the weakening formula is the premise of a cut, which morally is a tensor. But we cannot force the weakening to not be the premise of a cut—that would force to forbid the cut of any conclusion containing a weakening formula as a subformula as well—for that would correspond to forbid the weakening at all. If we want to obtain a subsystem of MELL with a good definition of correctness, we have to require that no ?-formula might be the premise of a tensor—by duality, that restriction extends to the par combination of ?-formulas as well. Remarkably, the subsystem of MELL in which we can encode the λ-calculus has that property (see 2.4.3). For that subsystem, as for the well-formed proof structures with units (2.2.5.1), one has a good formulation of DR-correctness (see [33]).

2.3.4 Cut-elimination

The cut-elimination rules of MELL are those of MLL (see section 2.2.4) plus the rules (see Figure 2.14 also)

$$A \overset{d}{\rhd} ?A;\ ?A,!A^{\perp} \overset{cut}{\rhd};\ [\mathbf{G};\ A^{\perp} \overset{!}{\rhd} !A^{\perp}] \qquad (\overset{d}{\rhd}/\overset{!}{\rhd})$$

$$\overset{cut}{\longrightarrow} A,A^{\perp} \overset{cut}{\rhd};\ \mathbf{G}$$

$$?A',?A'' \overset{c}{\rhd} ?A;\ ?A,!A^{\perp} \overset{cut}{\rhd};\ [\mathbf{G};\ A^{\perp} \overset{!}{\rhd} !A^{\perp}]^{!A^{\perp}}_{?A_1,\dots,?A_k} \qquad (\overset{c}{\rhd}/\overset{!}{\rhd})$$

$$\overset{cut}{\longrightarrow} ?A',!A'^{\perp} \overset{cut}{\rhd};\ [\mathbf{G}';\ A'^{\perp} \overset{!}{\rhd} !A'^{\perp}]^{!A'^{\perp}}_{?A'_1,\dots,?A'_k};$$

$$?A'',!A''^{\perp} \overset{cut}{\rhd};\ [\mathbf{G}'';\ A''^{\perp} \overset{!}{\rhd} !A''^{\perp}]^{!A''^{\perp}}_{?A''_1,\dots,?A''_k};$$

$$?A'_1,?A''_1 \overset{c}{\rhd}?A_1;\ \dots?A'_k,?A''_k \overset{c}{\rhd}?A_k$$

$$\overset{w}{\rhd} ?A;\ ?A,!A^{\perp} \overset{cut}{\rhd};\ [\mathbf{G};\ A^{\perp} \overset{!}{\rhd} !A^{\perp}]^{!A^{\perp}}_{?A_1,\dots,?A_k} \qquad (\overset{w}{\rhd}/\overset{!}{\rhd})$$

$$\overset{cut}{\longrightarrow} \overset{w}{\rhd} ?A_1;\ \dots \overset{w}{\rhd} ?A_k$$

$$[\mathbf{G}';\ B \overset{!}{\rhd} !B]^{!B}_{\Gamma,?A};\ ?A,!A^{\perp} \overset{cut}{\rhd};\ [\mathbf{G};\ A^{\perp} \overset{!}{\rhd} !A^{\perp}]^{!A^{\perp}}_{\Delta} \qquad ([\ ]/[\ ])$$

$$\overset{cut}{\longrightarrow} [\mathbf{G}';\ ?A,!A^{\perp} \overset{cut}{\rhd};\ [\mathbf{G};\ A^{\perp} \overset{!}{\rhd} !A^{\perp}]^{!A^{\perp}}_{\Delta};\ B \overset{!}{\rhd} !B]^{!B}_{\Gamma,\Delta}$$

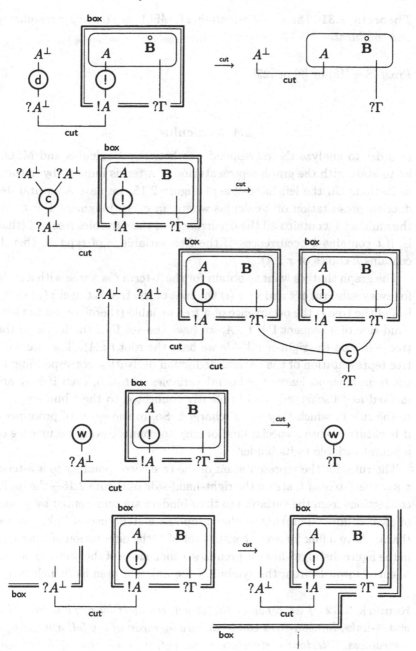

Fig. 2.14. Exponential cut-elimination rules.

Theorem 2.31 *The cut-elimination of* MELL *is strongly normalizing and confluent.*

Proof See [13] or [9] or [23]. □

2.4 λ-calculus

In order to analyze the correspondence between λ-calculus and MELL, let us start with the graph representation of λ-terms induced by natural deduction. On the left-hand-side of Figure 2.15, we have a natural deduction presentation of λ-calculus where, in every judgment $\Gamma \vdash t : A$, the multiset Γ contains all the occurrences of the variables free in t (that is, if t contains k-occurrences of the free variable x of type A, then Γ contains k copies of $x : A$).

The graph that we want to obtain for the λ-term t is a tree with a node for every subterm of t and s.t.: (i) the root of the tree is t itself; (ii) every leaf of the tree is the occurrence of a free variable (therefore, on the left-hand-side of a sequent $\Gamma \vdash t : A$, we have the set Γ of the leaves of the tree, while on the right-hand-side we have the root $t : A$). The standard tree representation of the natural deduction derivation corresponding to the term t has as leaves the bound variables of t also, even if they are marked as "discharged" and implicitly connected to their binders, *i.e.*, to the rule in which they are discharged. So, in the spirit of proof nets, it is natural to use a special kind of edge to connect every occurrence of a bound variable to its binder.

The rules for the representation of the tree corresponding to a λ-term t, say the *λ-tree* of t, are on the right-hand-side of Figure 2.15—the back-connections from the variables to their binders are represented by means of dashed lines. Thinking at that graph as a structure of links, we see that we have a link for every constructor of t (the orientation of the edges in the figure distinguishes the premises/conclusions of the links: an arrow oriented towards/from the symbol of the link is a premise/conclusion).

Remark 2.32 *If we consider the structures built by combination of* λ *and* @-links, *lambda-trees can be defined by mean of the following simple correctness criterion: a structure is correct, it is a λ-tree, iff by erasing the dashed edges we obtain a tree T and, for every dashed edge that connects the node* x *to the* λ-link *with conclusion* u, *the node* u *is an ancestor of* x *in the tree T.*

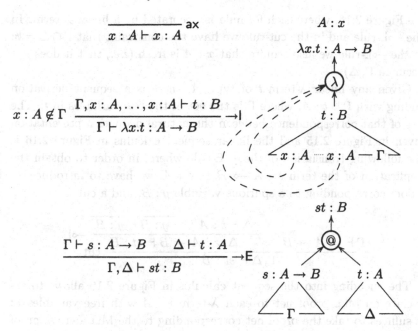

$$\frac{}{x : A \vdash x : A} \text{ ax}$$

$$x : A \notin \Gamma \quad \frac{\Gamma, x : A, \dots, x : A \vdash t : B}{\Gamma \vdash \lambda x.t : A \to B} \to \text{I}$$

$$\frac{\Gamma \vdash s : A \to B \quad \Delta \vdash t : A}{\Gamma, \Delta \vdash st : B} \to \text{E}$$

Fig. 2.15. λ-terms and λ-trees.

2.4.1 Linear λ-calculus

In order to see the correspondence between λ-trees and proof nets, let us start with the linear case. A λ-term t is linear when every free variable of t occurs exactly once and every proper subterm of t is linear. In linear λ-calculus types the linear implication $\multimap$ replaces the intuitionistic implication $\to$. The natural deduction presentation of linear λ-calculus is the same as that in Figure 2.15, with the restriction that, in the $\multimap$E-rule, $\Gamma \cap \Delta = \emptyset$, and that in the $\multimap$I-rule, $x : A$ must occur on the left-hand-side of the sequent—this means that it occurs exactly once also.

$$\frac{}{x : A \vdash x : A} \text{ ax} \qquad \frac{\Gamma \vdash s : A \quad \Delta, x : A \vdash t : B}{\Gamma, \Delta \vdash t[s/x] : B} \text{ cut}$$

$$\frac{\Gamma \vdash s : A \quad \Delta, y : B \vdash t : C}{\Gamma, \Delta, x : A \multimap B \vdash t[xs/y] : C} \multimap \text{L} \qquad \frac{\Gamma, x : A A \vdash t : B}{\Gamma \vdash \lambda x.t : A \multimap B} \multimap \text{R}$$

Fig. 2.16. Linear λ-calculus.

The linear λ-calculus corresponds to the implicative fragment of IMLL, see Figure 2.16, where each formula is decorated by a linear λ-term. In the $\multimap$L-rule and in the cut-rule we have the constraint that $\Gamma \cap \Delta = \emptyset$; in the $\multimap$L-rule we also require that $x : A$ is fresh (*i.e.*, that it does not occur in Γ, Δ).

Given any linear λ-term t of type A, there is a sequent derivation ending with $\Gamma \vdash t : A$, where Γ is the set of the free variables in t. The key of that correspondence between the natural deduction presentation given in Figure 2.15 and the linear sequent calculus in Figure 2.16 is the following translation of the $\multimap$E-rule where, in order to obtain the application of the term $s : A \multimap B$ to $t : A$, we have to introduce an axiom corresponding to a spurious variable $y : B$, and a cut:

$$\cfrac{\Gamma \vdash s : A \multimap B \qquad \cfrac{\Delta \vdash t : A \qquad \cfrac{}{y : B \vdash y : B}\ \text{ax}}{\Delta, z : A \multimap B \vdash zt : B}\ \multimap\text{L}}{\Gamma, \Delta \vdash st : B}\ \text{cut}$$

The encoding into the sequent calculus in Figure 2.16 allows to associate an MLL proof net to each λ-term $t : A$ with free variables Γ: it suffices to take the proof net corresponding to the MLL derivation of $\Gamma \vdash t : A$. In Figure 2.17, that correspondence is defined by means of the rules that transforms a λ-tree T into an oriented proof net N. The resulting net has one positive conclusion only, and a negative conclusion for every free variable in T.

Definition 2.33 (linear λ-nets) *A linear λ-net is an oriented proof net in which there are no positive tensors or negative pars.*

The translation in Figure 2.17 transforms a linear λ-tree into a linear λ-net. Vice versa, every linear λ-net represents a linear λ-tree, and therefore a linear λ-term. In fact, let us assume w.l.o.g. that: (i) every conclusion of a tensor in the linear λ-net N is the negative premise of a cut and that every negative premise of a tensor is the negative conclusion of an axiom; (ii) it is not the case that both the conclusions of an axiom are premises of cut links or that both the premises of a cut are conclusions of axiom links. (The previous conditions can always be obtained by the insertion or the elimination of identity cut redexes). By condition (i), we can reverse the transformation rules in Figure 2.17 and replace an @-link for every tensor and a λ-link for every $\invamp$-link; by condition (ii), the previous reverse transformation does not leave any spurious axiom or cut. The graph T obtained in this way, apart for the orientation of the edges that are reversed, is isomorphic to the essential

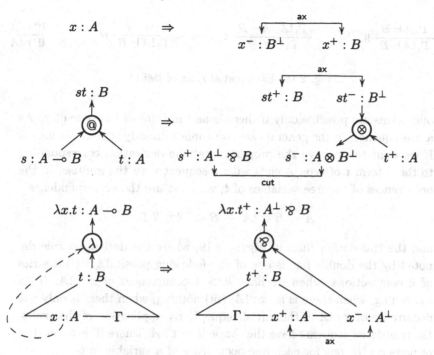

Fig. 2.17. From linear λ-trees to MLL proof nets.

net of N (2.2.2.2). Moreover, by condition **EN**, for every λ-link, the path that connects the variable bound by a λ-link to the root of the tree crosses that λ-link, that is, T is a linear λ-tree (see Remark 2.32).

2.4.2 IMELL *encoding of λ-calculus*

The encoding of full λ-calculus requires the exponentials for the treatment of non-linear variables. Reminding that the two-sided sequent $\Gamma \vdash \Delta$ corresponds to the one-sided sequent $\vdash \Gamma^{\perp}, \Delta$, the exponential fragment of intuitionistic **MELL**, say **IMELL**, are those in Figure 2.18. We see that the !R-rule corresponds to a promotion, that the !L corresponds to a dereliction, and that contraction and weakening (c-rule and w-rule, respectively) are allowed on the left-hand-side only.

For the encoding of λ-calculus into **IMELL**, the key case is the →I-rule. In fact, in order to encode it using ⊸, we need to contract several occurrences of $x : A$ into a single one (when x occurs more than once in t), or to introduce an $x : A$ (when x does not occur in t); both that

$$\frac{\Gamma, A \vdash B}{\Gamma, !A \vdash B} \text{ !L} \qquad \frac{\Gamma, !A, !A \vdash B}{\Gamma, !A \vdash B} \text{ c} \qquad \frac{\Gamma \vdash B}{\Gamma, !A \vdash B} \text{ w} \qquad \frac{!\Gamma \vdash A}{!\Gamma \vdash !A} \text{ !R}$$

Fig. 2.18. Exponential rules of IMELL.

operations are possible only if there is an ! in front of the type of x. As a consequence, in the general case, we cannot directly replace $\multimap$ for $\rightarrow$. If we want to preserve the property that the derivation corresponding to the λ-term t of type A ends with a sequent with the multiset of the occurrences of the free variables of t, we must use the correspondence

$$A \rightarrow B \equiv !A \multimap B \equiv ?A^\perp \parr B$$

and the translation rules in Figure 2.19, where the derivation rule denoted by the double line is one of the following possibility: (i) a series of k contractions, when we have $k > 1$ occurrences of $x : !A$; (ii) a weakening, when there is no $x : !A$; (iii) nothing, when there is only one occurrence of $x : !A$ in the initial sequent. By application of that rules, the translated sequents have the shape $!\Gamma \vdash t : A$, where $!\Gamma$ is a multiset of pairs $x : !B$, one for each free occurrence of a variable in t.

$$\frac{\frac{}{x : A \vdash x : A} \text{ ax}}{x : !A \vdash x : A} \text{ !L} \qquad \frac{\dfrac{!\Gamma, x : !A, \ldots, x : !A \vdash t : B}{!\Gamma, x : !A \vdash t : B} \text{ c/w}}{!\Gamma \vdash \lambda x.t : !A \multimap B} \multimap\text{R}$$

$$\frac{!\Gamma \vdash s : !A \multimap B \qquad \dfrac{\dfrac{!\Delta \vdash t : A}{!\Delta \vdash t : !A} \text{ !R} \qquad \dfrac{}{y : B \vdash y : B} \text{ ax}}{!\Delta, z : !A \multimap B \vdash zt : B} \multimap\text{L}}{!\Gamma, !\Delta \vdash st : B} \text{ cut}$$

Fig. 2.19. IMELL encoding of λ-calculus.

2.4.2.1 *Other encodings*

The translation of λ-calculus corresponding to the encoding $A \rightarrow B \equiv !A \multimap B$ is not the only possibility. Even if we can imagine many exotic encodings, this translation and that corresponding to

$$A \rightarrow B \equiv !(A \multimap B)$$

are the most used ones–both translations can be found in [13].

The encodings $!A \multimap B$ and $!(A \multimap B)$ are usually referred to as *call-by-name* and *call-by-value*, respectively. In fact, let us assume that a box freezes the reduction of the cuts inside it, that is, let us assume that a boxed net is a value. In the $!(A \multimap B)$ encoding, a function is a value and, during a computation, we can reduce inside the body of a function only after the application of the function to a value. Therefore, $!(A \multimap B)$ corresponds to call-by-value. This is not the case in the $!A \multimap B$ encoding, where a box surrounds every argument of a function; in this case, the argument of a function can be evaluated only after its replacement inside the body of the function. Therefore, $!A \multimap B$ corresponds to call-by-name. For more details see [32].

2.4.3 λ-nets

In the the previous sections we have seen how to translate λ-trees into λ-nets. In 2.3.3 (and in 2.2.5), we have seen that (in the general case) weakening links (and ⊥-links) does not allow to have a characterization of MELL proof nets in the Danos-Regnier style. However, as in the relevant case analyzed for ⊥-links (2.2.5.1), the weakening links are not a problem in the proof nets resulting from the translation of λ-terms.

Definition 2.34 (λ-structure) *Let us say that a* well-formed λ-formula *is any* MELL *formula in the following grammar:*

$$T = p \mid ?T_1^\perp \,\mathbin{⅋}\, T_2 \mid !T_1 \otimes T_2^\perp$$

where p ranges over atomic formulas. A λ-structure is an exponential proof structure that contains well-formed λ-formulas only. A λ-structure is closed *when it has only one conclusion of type $?A^\perp \,\mathbin{⅋}\, B \equiv !A \multimap B$, for some A and B.*

In particular, let us remark that $A_? \otimes B$, where $A_?$ is a par combination of ?-formulas, is not well-formed and that the translation of a λ-term contains well-formed λ-formulas only.

Definition 2.35 (exponential switch) *A* switch *of a λ-structure is a graph obtained in the following way:*

(i) *replace any number of boxes with a ∗-link whose conclusions are the doors of the box;*

(ii) *apply the usual switching rules for the multiplicative links;*

(iii) *apply the switching rule of pars to contractions and add an edge for every !-link and every dereliction.*

The switching edges corresponding to the exponential links are given in Figure 2.20. We see that, for every !-link, we can either choose to replace its box with the switch of the corresponding *-link, or to connect the premise to the conclusion of the link; for the d-link, the premise is directly connected to the conclusion of the link; for the contraction link, we have two possibilities as in the case of the par. For all the other links, we have the same switchings of MLL.

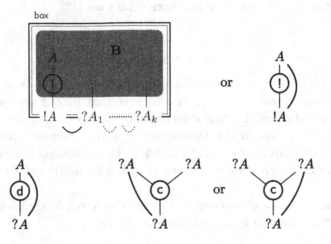

Fig. 2.20. Exponential switch edges.

Definition 2.36 (λ-net) *A λ-structure* **G** *is correct, say that* **G** *is a λ-net, when every switch of* **G** *that contains k weakening links is acyclic and has k + 1 connected components.*

The $!A \multimap B$ encoding ensures that the image of a closed λ-term is a closed λ-net. However, there are many closed λ-nets that are not images (at least, according to the given translation) of a λ-term, but that differ from the image of a λ-term for irrelevant details in the order of some exponential rules. In particular, for the order in which the occurrences of a variable are contracted, or for the place where that occurrences are contracted. Moreover, in the translation in Figure 2.19, we have not specified the shape of the tree of contraction links that merges the occurrences of a variable into a unique node, as a consequence, we have many possibilities for the same term. On the other side, given a bound variable, we have chosen to contract all its occurrences immediately before the introduction of its binder—outside all the boxes surrounding the occurrences of the variable—but other choices might be done. For instance, we might decide to always keep only one occurrence of the same variable in the context. According to this, the translation of an application should be followed by a sequence of contractions merging the variables that occur in both the sides of the application—in this way, it is no longer true that a tree of contractions does not crosses the border of any box.

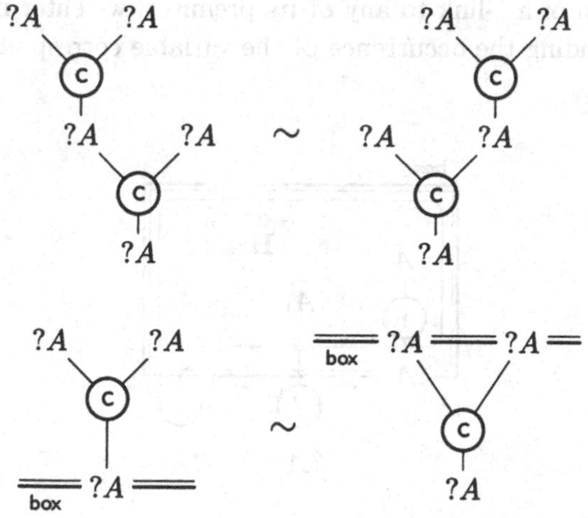

Fig. 2.21. Congruences.

In order to equate all the previous possibilities, a first approach is to consider the congruences in Figure 2.21. That congruences allow to study the connections between λ-calculus with explicit substitutions (see [8]). A more compact approach is to collapse dereliction, contraction and weakening links. The idea is to replace a cluster of that nodes by a unique ?-link with k premises of the same type, say A, and one conclusion with type $?A$ (see Figure 2.22). By the way, for $k = 0$, that link corresponds to a weakening; for $k = 1$, it is a dereliction; for $k > 1$, it corresponds to a tree of contractions with k derelictions at its leaves.

In a switch, every ?-link with $k > 0$ premises is replaced by a unique edge that connects the conclusion of the link to one of its premises (see Figure 2.22). The introduction of the ?-link forces a slight change in the definition of boxes: the proviso that every auxiliary door of a box

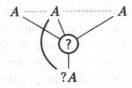

Fig. 2.22. ?-link.

is a ?-formula becomes that every auxiliary door of a box must be the
premise of a ?-link (see Figure 2.23). We also remark that, moving from
the conclusion of a ?-link to any of its premises, we enter inside all the
boxes surrounding the occurrence of the variable corresponding to that
premise.

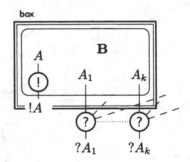

Fig. 2.23. Reformulation of boxes in the presence of ?-links.

The introduction of ?-links and the corresponding modification of
boxes does not change the definition of λ-nets. Definition 2.35 is still
valid if one takes the proper definition of switching for the ?-links; in
Definition 2.36 it suffices to assume that a weakening is a ?-link without
premises.

A recapitulation of the rules that transforms a λ-tree into a λ-net with
?-links is given in Figure 2.24. That translation is well-defined for closed
λ-terms only. The result of the translation of a closed λ-term is a *closed*
λ-*net* with only one conclusion of type $!A \multimap B$, for some A and B.

We also remark that the introduction of the ?-link collapses into a
unique rule the four exponential cut-elimination rules in Figure 2.14.
We omit to draw the rule, we simply remark that the elimination of a
cut between a k-ary ?-link and a box $\mathbf{B}$ creates k copies of $\overset{\circ}{\mathbf{B}}$, one for

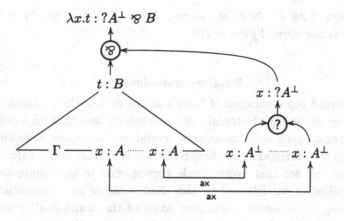

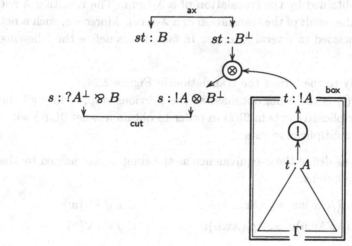

Fig. 2.24. Translation of λ-terms.

each premise of the ?-link, and that the i-th copy of $\overset{\circ}{\mathbf{B}}$ moves inside all the boxes that have the i-th premise of the ?-link as auxiliary door.

Theorem 2.37 *Let $\mathfrak{N}(t)$ be the proof net corresponding to the term t. If $t \xrightarrow{\beta}{}^{*} s$, then $\mathfrak{N}(t) \xrightarrow{cut}{}^{*} \mathfrak{N}(s)$.*

As a consequence of the previous theorem and the fact that $\xrightarrow{cut}$ is confluent and strongly normalizing, we see that λ-term normalization and λ-net normalization coincide.

Corollary 2.38 *Let* **N** *be the normal form of* $\mathfrak{N}(t)$. *Then* $\mathbf{N} = \mathfrak{N}(s)$, *where* s *is the normal form of* t.

2.4.4 σ-equivalence

The compact representation of ?-links solves part of the problems only. In fact, by analysis of the translation, it is readily seen that no λ-net that is image of a λ-term contains an exponential cut. Moreover, if we analyze in details the correspondence between λ-net reduction and λ-calculus β-reduction, we see that every β-rule corresponds to the elimination of a multiplicative cut followed by the elimination of an exponential cut. Therefore, let us assume to reduce some of the multiplicative cuts in the λ-net obtained by the translation of a λ-term. The resulting λ-net cannot be the result of the translation of a λ-term. Moreover, such a net can be associated to several λ-terms. In fact, let us define the following translation:

(i) apply to the term t the translation in Figure 2.24;
(ii) if $\mathfrak{N}(t)$ is the λ-net obtained at the previous step, reduce all the multiplicative cuts in $\mathfrak{N}(t)$ in order to obtain a λ-net $\mathfrak{N}_\sigma(t)$ without multiplicative cuts.

Now, let us define the σ-equivalence as the congruence defined by the following equations:

$$(\lambda x.u)vw =_\sigma (\lambda x.uw)v \qquad \text{if } x \notin \mathsf{FV}(w)$$
$$(\lambda x.\lambda y.u)v =_\sigma \lambda y.(\lambda x.u)v \qquad \text{if } y \notin \mathsf{FV}(v)$$

We have that

$$\mathfrak{N}_\sigma(u) = \mathfrak{N}_\sigma(v) \quad \text{iff} \quad u =_\sigma v$$

For more details on the σ-equivalence and for the proof of the previous assertion we refer the reader to [34] or [33].

2.4.5 Pure proof nets

In the previous sections we have seen how to translate a typed λ-term into a MELL proof net. Such a translation can be extended to the pure λ-calculus. As usual, the key step in that extension is the introduction of the isomorphism $(O \to O) \simeq O$, that in MELL becomes $(!O \multimap O) \simeq O$.

In the *pure λ-nets* corresponding to pure λ-terms, the only positive types are O and $!O$, while the only negative types are I and $?I$, with the obvious dualities $O^{\perp} = I$ and $(!O)^{\perp} = ?I$. Correspondingly, the links that can appear in a pure λ-net are those represented in Figure 2.25, where $X \in \{I, ?I\}$.

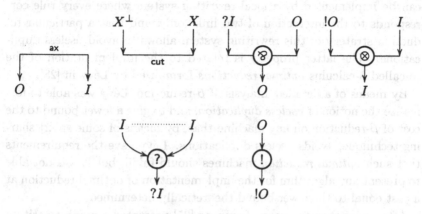

Fig. 2.25. Pure proof net links.

We remark that there are no exponential axioms. In fact, in the translation of λ-terms, every variable corresponds to an axiom of type O; moreover, because of the collapse of contraction and dereliction into the ?-link, we would not be able to contract the $?I$ conclusion of an axiom. All the results on the correctness of λ-nets and the simulation of β-reduction hold for pure λ-nets too. However, in the case of pure proof nets $\xrightarrow{\text{cut}}$ is confluent but not strongly normalizing (by the way, this is coherent with Theorem 2.37). As a consequence, in the pure case, we cannot have a correspondence between normal forms.

Theorem 2.39 *If* $\mathfrak{N}(t) \xrightarrow{\text{cut}*} N$, *there is* s *s.t.* $t \xrightarrow{\beta}{}^* s$ *and* $N \xrightarrow{\text{cut}*} \mathfrak{N}(s)$.

For more details on pure λ-nets, we refer the reader to [33].

2.5 Sharing graphs

The cut-elimination rule for the exponential cut is a global rule. In fact, the reduction of an exponential cut involving a k-ary ?-link requires the creation of k copies of the box associated to the !-link in the redex. Thus, the cost of an exponential cut reduction cannot be bound by any constant. Moreover, the duplication of a box creates k-copies of all the redexes already present in the box, or that will appear in it during the reduction; as a consequence, in the general case, the duplication of a box may cause the duplication of part of the work required to reach the normal form.

In this section, we shall see that cut-elimination, and then β-reduction, can be implemented by a local rewriting system where every rule corresponds to the interaction of two links only, and that a particular reduction strategy of this rewriting system allows to avoid useless duplications. The latter property is related to the implementation of the so-called λ-calculus *optimal reductions* formalized by Lévy in [28].

By means of a detailed analysis of β-reduction, Lévy was able to formalize the notion of *useless duplication*, and to give a lower bound to the cost of β-reduction on any machine that, by means of some smart sharing technique, avoids useless duplications. Lévy gave the requirements that such *optimal reduction* machines should fulfill, but it was not able to present any algorithm for the implementation of optimal reduction at a cost equal to the lower-bound theoretically determined.

After more than ten years, Lamping [25] presented a graph algorithm for optimal reduction that reached Lévy's lower-bound in terms of the number of β-reductions executed, but that, as later discovered by Asperti and Mairson [5], has a non-polynomial overhead w.r.t. the number of β-rules executed, because of some bookkeeping operations that makes impossible to reach Lévy's lower-bound. Indeed, Asperti and Mairson showed that this is not a problem of Lamping's algorithm, but of any algorithm that tries to implement Lévy's optimal reduction. In [12], Gonthier et al. showed that Lamping's algorithm could have been related to Girard's GOI (Geometry of Interaction [14]) and reformulated in terms of Linear Logic proof nets. For a first approach to β-reduction in terms of GOI see [11, 34]. For more details on the relations between GOI and optimal reductions see [2]. For a complete account of the results on optimal reductions see [3].

In the following, we shall give the basic ideas that allow to obtain a local rewriting system for the implementation of proof net cut-elimination, and then of λ-calculus β-reduction, by means of the so-called *sharing graphs*. Even if we tackle the problem starting from the idea of a local implementation of cut-elimination, it is possible to see that optimal reductions can be obtained as a reduction strategy of the rewriting system that we shall introduce. More details on the results that we shall see in the following can be found in [20]; for a more general approach to sharing graphs, see [22].

Another solution to the local implementation of cut-elimination can be found in [30].

2.5.1 λ-nets with levels

The first step for the local implementation of cut-elimination is to find a local representation for boxes. The idea is to use an implicit representation of boxes based on indexes. In fact, as boxes are a sort of brackets, we can associate to each node of the net a nesting depth *level*. According to this, every formula in the net has a level, and the only links where there is a difference between the levels of the premises and the levels of the conclusions are the !-link and the ?-link (see Figure 2.26, where at the nodes we have omitted the formulas, reporting the corresponding levels only). In particular, given an !-link surrounded by n boxes, its conclusion $!A$ is at level n, while its premise A is at level $n + 1$, that is, $!A$ is outside the box of the !-link, while A is inside that box. Instead, given a ?-link surrounded by n boxes, its conclusion $?A$ is at level n, while any of its premises A is at level $n + p$, where $p \geq 0$ is the number of box auxiliary doors crossed while moving from $?A$ to A.

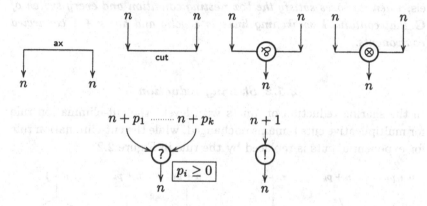

Fig. 2.26. Levels.

Let us say that a λ-structure with levels is a λ-structure in which every formula has a level and s.t. the restrictions on levels in Figure 2.26 hold.

Definition 2.40 (box) *In a λ-structure with levels, the box of an !-link whose conclusion has level n is the largest connected subnet B that contains the premise of the !-link and s.t. all the formulas in B have a level greater than n.*

Remark 2.41 *The previous definition of box cannot be directly extended* *to the whole case of* MELL *proof nets with levels, because of the presence of weakening. In the general case, weakening links may introduce disconnected parts on the net, causing the loss of the connectedness of boxes (for more details see [19]).*

A λ-structure with levels can be transformed into a λ-structure by associating to each !-link the box computed according to Definition 2.40. If the boxes assigned in this way satisfy the box nesting condition—in particular, every box has one principal port of !-type only—the correctness of the λ-structure can be verified by applying the criterion in Definition 2.36.

Definition 2.42 (λ-net with levels) *The switches of a λ-structure with levels can be obtained by application of Definition 2.35, provided that the boxes of the structure are computed according to Definition 2.40. A λ-structure with levels* G *is correct, say that* G *is a λ-net with levels, when its boxes satisfy the box nesting condition and every switch of* G *that contains k weakening links is acyclic and has $k + 1$ connected components.*

2.5.2 Sharing reduction

In the sharing reduction of λ-nets with levels, the cut-elimination rule for multiplicative cuts remains unchanged, while the cut-elimination rule for exponential cuts is replaced by the rule in Figure 2.27.

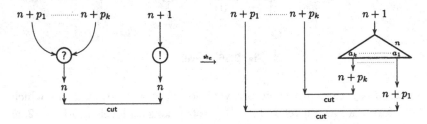

Fig. 2.27. Sharing reduction: exponential cut rule.

Instead of performing the duplication of the box B of the !-link in the redex, the cut is replaced by a new kind of link named *mux* (or multiplexer) with one premise and k conclusions, and an index n, named the *threshold* of the mux, equal to the level of the cut formulas. In this way, the inside of the box $\overset{\circ}{B}$ is not duplicated but shared. In the following, we shall see that there is a complementary mux too, with k premises and one conclusion. Because of this, we say that the premises and the conclusions of a mux are its doors and that every mux has one principal door only. The doors of a mux are connected to the ports of the mux. Each auxiliary port of a mux has a name, that is different from the name of any other port of the same mux. The difference between the level of any auxiliary door and the level of the principal door of a mux is a constant proper of the corresponding port named the *offset* of the port, and is always greater or equal than -1.

In order to complete the reduction, the sharing introduced by the elimination of an exponential cut must be unfolded. According to this, a mux is a link that dynamically performs a step-by-step duplication of the links in the interior of some box, according to the rule in Figure 2.28. Such a rule can be applied whichever is the type α of the duplicating link and whichever is the formula of the α-link at the principal door of the mux, provided that the side-condition in Figure 2.27 holds. If the α-link has $h + 1$ premises/conclusions, after the application of the rule, the mux splits in h copies; these copies of the mux proceed duplicating in different directions. We remark that the offsets of the ports does not change and that, when the α-link is a cut or an axiom, the doors of the mux change their orientations—if they were premises, they become conclusions, and vice versa.

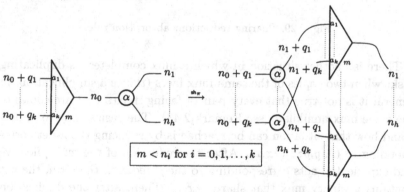

$$m < n_i \text{ for } i = 0, 1, \ldots, k$$

Fig. 2.28. Sharing reduction: propagation rule.

The threshold m of the mux plays a key role in the duplication rule in Figure 2.28, because of the side-condition that the level of any formula connected to the duplicating link is greater than m. In fact, according to the definition of box in Definition 2.40, a mux must duplicate all and only the links that it can reach and whose formulas have a level greater than m, that is, which are inside the box left shared by the cut-elimination step that has introduced the mux.

According to the conditions on the formulas of a link (remind Figure 2.26), and to the fact that a box has one principal door only, we see that a mux has completed its duplication task when it reaches the auxiliary door of the duplicating box. Again, the threshold of the mux plays a key role in recognizing such a situation: a mux with threshold m has reached the border of its box when it is at the premise of a ?-link whose conclusion has level $n \leq m$. In this case, all the premises of the mux become auxiliary doors of the new box, that is, they are added to the premises of the ?-link pointed by the mux. The corresponding rule is represented in Figure 2.29. The rule is named *absorption*, for we may think that the mux is absorbed by the ?-link.

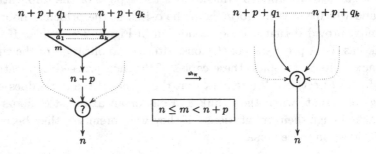

Fig. 2.29. Sharing reduction: absorption rule.

There is another situation in which a mux completes its duplicating task: when two copies of the same mux faces (this is a simplification, in general it is not true that every pair of facing muxes that are copies of the same mux annihilate, see Remark 2.43). The easiest way to understand how that situation can be reached is by reducing the λ-net corresponding to $(\lambda zy.yzz)(\lambda x.x)$. After the reduction of the multiplicative and exponential cuts corresponding to the β-redex in the term, the net contains a binary mux that shares $\lambda x.x$. Then, after the duplication of the $\mathfrak{N}$-link corresponding to the abstraction in $\lambda x.x$, we get a graph with two complementary muxes that faces. At that point, in order to complete the computation, the muxes must be replaced by direct connections between the matching doors of the facing muxes, obtaining the λ-net of $\lambda y.y(\lambda x.x)(\lambda x.x)$. The exact definition of the mux *annihilation* rule that allows to remove two complementary muxes that face is given in Figure 2.30. We see that the rule can be applied only when the two facing muxes have exactly the same shape, *i.e.*, the same threshold, the same number of doors, and the same offsets. After the annihilation of the muxes, the formulas connected to any pair of matching doors collapse into a unique formula with the same level (in all the rules we have omitted the formulas at the nodes, however, it is not difficult to see that after the reduction every node keeps the correct type).

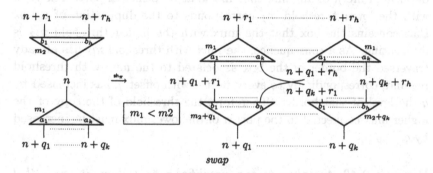

Fig. 2.30. Mux rules.

Let t be a λ-term that contains a redex s.t. $t \xrightarrow{\beta} s$. It is an easy exercise to see that, after the reduction $\mathfrak{N}(t) \xrightarrow{sh_m} \xrightarrow{sh_e} \mathbf{N}$ of the multiplicative and of the exponential cut of the redex, we get a sharing graph s.t. $\mathbf{N} \xrightarrow{sh_\pi}{}^* \mathfrak{N}(s)$. Then, a sound and complete implementation of the usual λ-net reduction can be obtained by application of the following reduction strategy: after the sharing reduction of any exponential cut, applies a sequence of $\xrightarrow{sh_\pi}$-rules until all the muxes in the graph disappear.

However, we can do much better. In fact, what does it happen if, instead of completing the $\xrightarrow{sh_\pi}$-reduction that leads to a new λ-net, we reduce a multiplicative or an exponential cut? The elimination of a multiplicative cut does not pose any problem. Instead, the elimination of an exponential cut introduces into the net a new mux before that the old ones have disappeared from the graph. In this way, we get two different kinds of muxes in the net that correspond to distinct duplication tasks. Therefore, we need a way to distinguish the two kinds of muxes in order to avoid that their corresponding duplication tasks might interfere. After realizing that such duplications may interfere only if the duplicating boxes nest, we can exploit the fact that two boxes that nest cannot be at the same level; therefore, we may use the thresholds of the muxes to recognize when they come from different cuts.

In addition to the mux annihilation rule, we have a mux *swap* rule that deals with the case in which the thresholds of two facing muxes differ, e.g., $m_1 < m_2$ (see Figure 2.30). We remark that the highest threshold m_2 is lifted by the offset of the ports that causes its duplication. The reason should be intuitively clear. During its propagation, for each of its auxiliary doors, a mux creates a new copy, lifted by the offset q of the port (that is, q is added to the level of all the formulas at the duplicated link), of any link that meets at its principal port. The mux with the lower threshold m_1 corresponds to the duplication of a box that contains the box that the mux with the higher threshold m_2 is duplicating. As a consequence, the mux with threshold m_1 has already traversed the border of the box associated to the mux with threshold m_2 and, correspondingly to every port a with offset q, has increased by q the level of that border. Therefore, the threshold of the copy of the higher mux connected to the port a of the lower mux must be increased by q.

Remark 2.43 *According to our simplified presentation, it seems that two muxes annihilate when they are copies of the same mux, or swap when they come from distinct muxes. In the general case, this is not completely true. Two complementary muxes that face and annihilate have always the same progenitor, but, two muxes that come from the same progenitor may face with different thresholds. This situation corresponds to involved cases in which one of the copies of the duplicating box enters inside another copy of the same box. According to the definition of the rules in Figure 2.30, in that case, the two muxes swap.*

Finally, we have the complete rewriting system for sharing reduction. The case without restrictions on the order in which the rules can be applied is not as direct as the simplified case from which we started, and the proofs of soundness and completeness of the sharing rewriting system are particularly involved (see [22] or [20]). Apart for the technical difficulties in the proof, we can get a result similar to that of Theorem 2.37 and Theorem 2.39.

Theorem 2.44 *The sharing reduction of a λ-net with levels is confluent, sound and complete. In particular, if $t \xrightarrow{\beta} s$, then $\mathfrak{N}(t) \xrightarrow{sh}{}^* \mathfrak{N}(s)$. Vice versa, if $\mathfrak{N}(t) \xrightarrow{sh}{}^* N$, there is $t \xrightarrow{\beta}{}^* s$ s.t. $N \xrightarrow{sh_\pi}{}^* \mathfrak{N}(s)$.*

The previous theorem holds for λ-nets and pure λ-nets with levels. In the typed case, we have strong normalization also.

2.5.3 Optimal reduction

If one tries to think at which is the best sharing reduction strategy for the normalization of λ-nets, one immediately realizes that the target is to avoid useless duplications of redexes. In particular, we see that axioms and cuts are not a problem—in our presentation their only purpose is to preserve typing and orientation, so they might simply be erased—and that for the other kind of links one must avoid to duplicate the link when a mux arrives at one of its premises. More precisely, from the point of view of its dynamics, a proof net can be seen as an interaction net, see [24], that is, a graph in which every link has a principal door and the interactions between links take place when two links are connected through their principal doors. The rewriting rules definable in this way suffice to perform the normalization of λ-nets, as it can be seen that the rules that change the λ-term associated to the net are the multiplicative and exponential rewriting rules only, and that, if some muxes interpose between the principal doors of a $\mathcal{V}/\otimes$ or of a ?/! pair that has to interact, after the application of a sequence of mux annihilation and swap rules, that muxes move away leaving an interacting pair of $\mathcal{V}/\otimes$ or of ?/! links. Actually, the existence of such a reduction sequence is ensured by the fact that, in the sharing graph resulting from the reduction of a λ-net, we cannot have pairs of deadlocked muxes (*i.e.*, muxes that are connected through their principal doors and have the same thresholds, but that differ for the number, the names or the offsets of their ports). The proof of that key result is not trivial, moreover, it is one of the basic steps in the proof of Theorem 2.44, but, if one assume as proved that theorem, the fact that we cannot have deadlocks become a trivial corollary.

Summing up, one can define the following *optimal reduction strategy*: the propagation rule can be applied only when the α-link in the redex is an axiom or a cut, or when the principal door of the mux is the conclusion of the α-link in the redex.

The number of β-rules (pairs of multiplicative and exponential cuts) executed by application of the optimal strategy in the reduction of $\mathfrak{N}(t)$ is equal to the number of Lévy's families in the λ-term t. Thus, the optimal sharing reduction strategy implements Lévy's optimal reduction (see [3]).

Bibliography

[1] V. M. Abrusci. Noncommutative proof nets. In J.-Y. Girard, Y. Lafont, and L. Regnier, editors, *Advances in Linear Logic*, pages 271–296. Cambridge University Press, 1995. London Mathematical Society Lecture Note Series 222, Proceedings of the 1993 Workshop on Linear Logic, Cornell Univesity, Ithaca.

[2] Andrea Asperti, Vincent Danos, Cosimo Laneve, and Laurent Regnier. Paths in the lambda-calculus: three years of communications without understanding. In *9th Annual IEEE Symposium on Logic in Computer Science (LICS'94)*, pages 426–436, Paris, France, July 1994. IEEE.

[3] Andrea Asperti and Stefano Guerrini. *The Optimal Implementation of Functional Programming Languages*. Number 45 in Cambridges Tracts in Theoretical Computer Science. Cambridge University Press, 1998.

[4] Andrea Asperti and Cosimo Laneve. Paths, computations and labels in the λ-calculus. In C. Kirchner, editor, *Rewriting Techniques and Applications, Proceedings of the 5th International Conference, RTA 93*, volume 690 of *Lecture Notes in Computer Science*, pages 152–167, Montreal, Canada, June 1993. Springer Verlag.

[5] Andrea Asperti and Harry G. Mairson. Parallel beta reduction is not elementary recursive. In *Proceedings of the 25th ACM SIGPLAN-SIGACT Symposium on Principles of Programming Languages, POPL '98*, pages 303–315, San Diego, CA, 1998. ACM Press.

[6] V. M. Abrusci and P. Ruet. Non-commutative logic I: the multiplicative fragment. *Annals Pure Appl. Logic*, 101(1):29–64, 2000.

[7] G. Bellin and J. van de Wiele. Empires and kingdoms in MLL⁻. In J.-Y. Girard, Y. Lafont, and L. Regnier, editors, *Advances in Linear Logic*, pages 249–270. Cambridge University Press, 1995. London Mathematical Society Lecture Note Series 222, Proceedings of the 1993 Workshop on Linear Logic, Cornell Univesity, Ithaca.

[8] R. Di Cosmo and D. Kesner. Strong normalization of explicit substitutions via cut elimination in proof nets. In *12th Annual IEEE Symposium on Logic in Computer Science (LICS'97)*, pages 35–47. IEEE, June 1997.

[9] Vincent Danos. *Une Application de la Logique Linéaire à l'Ètude des Processus de Normalisation (principalement du λ-calcul)*. PhD Thesis, Université Paris 7, Paris, June 1990.

[10] V. Danos and L. Regnier. The structure of multiplicatives. *Archive for Mathematical Logic*, 28:181–203, 1989.

[11] Vincent Danos and Laurent Regnier. Local and asyncrhonous beta-reduction. In IEEE, editor, *8th Annual IEEE Symposium on Logic*

in *Computer Science (LICS '93)*, pages 296–306, Montreal, Canada, 1993.

[12] Georges Gonthier, Martín Abadi, and Jean-Jacques Lévy. Linear logic without boxes. In *Proceedings, Seventh Annual IEEE Symposium on Logic in Computer Science*, pages 223–234, Santa Cruz, California, 22–25 June 1992. IEEE Computer Society Press.

[13] Jean-Yves Girard. Linear logic. *Theoretical Computer Science*, 50(1):1–102, 1987.

[14] Jean-Yves Girard. Geometry of interaction 1: Interpretation of system F. In R. Ferro, C. Bonotto, S. Valentini, and A. Zanardo, editors, *Logic Colloqium '88*, pages 221–260. Elsevier (North-Holland), 1989.

[15] J.-Y. Girard. Quantifiers in linear logic II. Prépublications 19, Équipe de Logique Mathématique, Univeristé Paris VII, jan 1991.

[16] J.-Y. Girard. Linear logic: Its syntax and semantics. In J.-Y. Girard, Y. Lafont, and L. Regnier, editors, *Advances in Linear Logic*, pages 1–42. Cambridge University Press, 1995. Proceedings of the Workshop on Linear Logic, Ithaca, New York, June 1993.

[17] Jean-Yves Girard. Proof-nets: The parallel syntax for proof-theory. In P. Agliano and A. Ursini, editors, *Logic and Algebra*. Marcel Dekker, New York, 1996.

[18] Stefano Guerrini and Andrea Masini. Parsing MELL Proof Nets. *Theoretical Computer Science*, 254(1–2):317–335, March 2001.

[19] Stefano Guerrini, Simone Martini, and Andrea Masini. Proof nets, garbage, and computations. *Theoretical Computer Science*, 253(2):185–237, February 2001.

[20] Stefano Guerrini, Simone Martini, and Andrea Masini. Coherence for sharing proof nets. *Theoretical Computer Science*, 294(3):379–409, February 2003.

[21] Stefano Guerrini. Correctness of multiplicative proof nets is linear. In *14th Annual IEEE Symposium on Logic in Computer Science (LICS '99)*, pages 454–463, Trento, Italy, July 1999. IEEE Computer Society.

[22] Stefano Guerrini. A general theory of sharing graphs. *Theoretical Computer Science*, 227(1–2):99–151, 1999.

[23] J.-B. Joinet. *Etude de la normalisation du calcul des séquents classique à travers la logique linéaire*. PhD thesis, Université Paris VII, January 1993.

[24] Y. Lafont. From proof nets to interaction nets. In J.-Y. Girard, Y. Lafont, and L. Regnier, editors, *Advances in Linear Logic*, pages 225–247. Cambridge University Press, 1995. Proceedings of the Workshop on Linear Logic, Ithaca, New York, June 1993.

[25] John Lamping. An algorithm for optimal lambda calculus reduction. In *Conference Record of the Seventeenth Annual ACM Symposium on Principles of Programming Languages*, pages 16–30, San Francisco, California, January 1990.

[26] Franccois Lamarche. Proof nets for intuitionistic linear logic I: Essential nets. Preprint available at http://hypatia.dcs.qmw.ac.uk/data/L/LamarcheF/prfnet1.ps.gz, 1994.

[27] Olivier Laurent. Polarized proof-nets and λµ-calculus. *Theoretical Computer Science*, 2002. To appear.

[28] Jean-Jacques Lévy. *Réductions Correctes et Optimales dans le*

lambda-calcul. PhD Thesis, Université Paris VII, Paris, 1978.

[29] P. Lincoln and T. Winkler. Constant-only multiplicative linear logic is np-complete. *Theoretical Computer Science*, 135:155–169, 1994.

[30] Ian Mackie. Linear logic with boxes. In IEEE, editor, *13th Annual IEEE Symposium on Logic in Computer Science (LICS '98)*, pages 309–320, Indianapolis, Indiana, 1998.

[31] A. S. Murawski and C.-H. L. Ong. Dominator trees and fast verification of proof nets. In *15th Annual IEEE Symposium on Logic in Computer Science (LICS 2000)*, Santa Barbara, CA, June 2000. IEEE Computer Society.

[32] Maraist, Odersky, Turner, and Wadler. Call-by-name, call-by-value, call-by-need and the linear lambda calculus. *TCS: Theoretical Computer Science*, 228, 1999.

[33] Laurent Regnier. *Lambda-Calcul et Reseaux.* Phd Thesis, Université Paris 7, Paris, January 1992.

[34] Laurent Regnier. Une équivalence sur les lambda-termes. *Theoretical Computer Science*, 126(2):281–292, April 1994. Note.

[35] Lorenzo Tortora de Falco. *Réseaux, cohérence et expériences obsessionelles.* Phd thesis, Université Paris 7, Paris, January 2000.

[36] Lorenzo Tortora de Falco. The additive multiboxes. *Annals of Pure and Applied Logic*, 2002.

3

An Overview of Linear Logic Programming

Dale Miller

INRIA - Futurs & Laboratoire d'Informatique LIX, École Polytechnique

Abstract

Logic programming can be given a foundation in sequent calculus by viewing computation as the process of building a cut-free sequent proof bottom-up. The first accounts of logic programming as *proof search* were given in classical and intuitionistic logic. Given that linear logic allows richer sequents and richer dynamics in the rewriting of sequents during proof search, it was inevitable that linear logic would be used to design new and more expressive logic programming languages. We overview how linear logic has been used to design such new languages and describe briefly some applications and implementation issues for them.

3.1 Introduction

It is now commonplace to recognize the important role of logic in the foundations of computer science. When a major new advance is made in our understanding of logic, we can thus expect to see that advance ripple into many areas of computer science. Such rippling has been observed during the years since the introduction of linear logic by Girard in 1987 [35]. Since linear logic embraces computational themes directly in its design, it often allows direct and declarative approaches to computational and resource sensitive specifications. Linear logic also provides new insights into the many computational systems based on classical and intuitionistic logics since it refines and extends these logics.

There are two broad approaches by which logic, via the theory of proofs, is used to describe computation [69]. One approach is the *proof reduction* paradigm, which can be seen as a foundation for *functional pro-*

gramming. Here, programs are viewed as natural deduction or sequent calculus proofs and computation is modeled using proof normalization. Sequents are used to type a functional program: a program fragment is associated with the *single-conclusion* sequent $\Delta \longrightarrow G$ if the code has the type declared in G when all its free variables have types declared for them in the set of type judgments Δ. Abramsky [2] has extended this interpretation of computation to *multiple-conclusion* sequents of linear logic, $\Delta \longrightarrow \Gamma$, where Δ and Γ are both multisets of typing judgments. In that setting, cut-elimination can be seen as specifying concurrent computations. See also [17, 60, 61] for related uses of concurrency in proof normalization in linear logic. The more expressive types made possible by linear logic have also been used to provide static analysis of run-time garbage, aliases, reference counters, and single-threadedness [34, 83, 91, 111].

Another approach to using proof theory to specify computation is the *proof search* paradigm, which can be seen as a foundation for *logic programming*. In this paper (which is an update to [71]), we first provide an overview of the proof search paradigm and then outline the impact that linear logic has made to the design and expressivity of new logic programming languages.

3.2 Goal-directed proof search

When logic programming is considered abstractly, sequents directly encode the state of a computation and the changes that occur to sequents during bottom-up search for cut-free proofs encode the dynamics of computation. In particular, following the framework described in [80], a logic programming language consists of two kinds of formulas: *program clauses* describe the meaning of non-logical constants and *goals* are the possible consequences considered from collections of program clauses. A single-conclusion sequent $\Delta \longrightarrow G$ represents the state of an idealized logic programming interpreter in which the current logic program is Δ (a set or multiset of formulas) and the goal is G. These two classes of formulas are duals of each other in the sense that a negative subformula of a goal is a program clause and a negative subformula of a program clause is a goal formula.

3.2.1 Uniform proofs

The constants that appear in logical formulas are of two kinds: logical constants (connectives and quantifiers) and non-logical constants (predicates and function symbols). The "search semantics" of the former is fixed and independent of context: for example, the search for the proof of a disjunction or universal quantifier should be the same no matter what program is contained in the sequent for which a proof is required. On the other hand, the instructions for proving a formula with a non-logical constant head (that is, an atomic formula) are provided by the logic program in the sequent.

This separation of constants into logical and non-logical yields two different phases in proof search for a sequent. One phase is that of *goal reduction*, in which the search for a proof of a non-atomic formula uses the introduction rule for its top-level logical constant. The other phase is *backchaining*, in which the meaning of an atomic formula is extracted from the logic program part of the sequent.

The technical notion of *uniform proofs* is used to capture the notion of *goal-directed search*. When sequents are single-conclusion, a *uniform proof* is a cut-free proof in which every sequent with a non-atomic right-hand side is the conclusion of a right-introduction rule [80]. An interpreter attempting to find a uniform proof of a sequent would directly reflect the logical structure of the right-hand side (the goal) into the proof being constructed. As we shall see, left-introduction rules are used only when the goal formula is atomic and as part of the backchaining phase.

A specific notion of goal formula and program clause along with a proof system is called an *abstract logic programming language* if a sequent has a proof if and only if it has a uniform proof. As we shall illustrate below, first-order and higher-order variants of Horn clauses paired with classical provability [89] and hereditary Harrop formulas paired with intuitionistic provability [80] are two examples of abstract logic programming languages.

While backchaining is not part of the definition of uniform proofs, the structure of backchaining is consistent across several abstract logic programming languages. In particular, when proving an atomic goal, applications of left-introduction rules can be used in a coordinated decomposition of a program clause that yields not only a matching atomic formula occurrence to the atomic goal but also possibly new goals formulas for which additional proofs must be attempted [89, 80, 47].

3.2.2 *Logic programming in classical and intuitionistic logics*

In the beginning of the logic programming literature, there was one example of logic programming, namely, the first-order classical theory of Horn clauses, which was the logic basis of the popular programming language Prolog. However, no general framework existed for connecting logic and logic programming. The operational semantics of logic programs was presented as resolution [11], an inference rule optimized for classical reasoning. Miller and Nadathur [78, 65, 89] were probably the first to use the sequent calculus to examine design and correctness issues for logic programming languages. Moving to the sequent calculus made it nature to consider logic programming in settings other than just classical logic.

We first consider the design of logic programming languages within classical and intuitionistic logic, where the logical constants are taken to be *true*, $\wedge$, $\vee$, $\supset$, $\forall$, and $\exists$ (false and negation are not part of the first logic programs we consider).

Horn clauses can be defined simply as those formulas built from *true*, $\wedge$, $\supset$, and $\forall$ with the proviso that no implication or universal quantifier is to be left of an implication. A goal in this setting would then be any negative subformula of a Horn clause: more specifically, they would be either *true* or a conjunction of atomic formulas. It is shown in [89] that a proof system similar to the one in Figure 3.1 is complete for the classical logic theory of Horn clauses and their associated goal formulas. It then follows immediately that Horn clauses are an abstract logic programming language. (The syntactic variable A in Figure 3.1 denotes atomic formulas.) Notice that sequents in this and other proof systems contain a *signature* Σ as its first element: this signature contains type declarations for all the non-logical constants in the sequent. Notice also that there are two different kinds of sequent judgments: one with and one without a formula on top of the sequent arrow. The sequent $\Sigma : \Delta \xrightarrow{D} A$ denotes the sequent $\Sigma : \Delta, D \longrightarrow A$ but with the D formula being distinguished (that is, marked for backchaining).

Inference rules in Figure 3.1, and those that we shall show in subsequent proof systems, can be divided into four categories. The right-introduction rules (goal-reduction) are those using the unlabeled sequent arrow and in which the goal formula is non-atomic. The left-introduction rules (backchaining) are those with sequent arrows labeled with a formula and it is on that formula that introduction rules are applied. The initial rule forms the third category and is the only rule with a repeated

$$\overline{\Sigma : \Delta \longrightarrow true}$$

$$\frac{\Sigma : \Delta \longrightarrow G_1 \quad \Sigma : \Delta \longrightarrow G_2}{\Sigma : \Delta \longrightarrow G_1 \wedge G_2}$$

$$\frac{\Sigma : \Delta \xrightarrow{D} A}{\Sigma : \Delta \longrightarrow A} \ decide$$

$$\overline{\Sigma : \Delta \xrightarrow{A} A} \ initial$$

$$\frac{\Sigma : \Delta \xrightarrow{D_i} A}{\Sigma : \Delta \xrightarrow{D_1 \wedge D_2} A}$$

$$\frac{\Sigma : \Delta \longrightarrow G \quad \Sigma : \Delta \xrightarrow{D} A}{\Sigma : \Delta \xrightarrow{G \supset D} A}$$

$$\frac{\Sigma : \Delta \xrightarrow{D[t/x]} A}{\Sigma : \Delta \xrightarrow{\forall_\tau x.D} A}$$

Fig. 3.1. In the decide rule, $D \in \Delta$; in the left rule for $\wedge$, $i \in \{1,2\}$; and in the left rule for $\forall$, t is a Σ-term of type τ.

$$\frac{\Sigma : \Delta, D \longrightarrow G}{\Sigma : \Delta \longrightarrow D \supset G}$$

$$\frac{\Sigma, c : \tau : \Delta \longrightarrow G[c/x]}{\Sigma : \Delta \longrightarrow \forall_\tau x.G}$$

Fig. 3.2. The rule for universal quantification has the proviso that c is not declared in Σ.

occurrence of a schema variable in its conclusion. The decide rule forms the forth and final category: this rule is responsible for moving a formula from the logic program to above the sequent arrow.

In this proof system, left-introductions are now applied only on the formula annotating the sequent arrow. The usual notion of backchaining can be seen as an instance of a decide rule, which places a formula from the program (the left-hand context) on top of the sequent arrow, and then a sequence of left-introductions work on that distinguished formula. Backchaining ultimately performs a linking between a goal formula and a program clause via the repeated schema variable in the initial rule. In Figure 3.1, there is one decide rule and one initial rule: in a subsequent inference system, there are more of each category. Also, proofs in this system involving Horn clauses have a simple structure: all sequents in a given proof have identical left hand sides: signatures and programs are fixed and global during the search for a proof. If changes in sequents are meant to be used to encode dynamics of computation, then Horn clauses provide a weak start: the only dynamics are changes in goals which relegates such dynamics entirely to the non-logical domain of atomic formulas. As we illustrate with an example in Section 3.6, if one can use a logic programming language where sequents have more dynamics, then one can reason about some aspects of logic programs directly using logical tools.

Hereditary Harrop formulas can be presented simply as those formulas built from *true*, $\wedge$, $\supset$, and $\forall$ with no restrictions. Goal formulas, *i.e.*, negative subformulas of such formulas, would thus have the same structure. It is shown in [80] that a proof system similar to the one formed by adding to the inference rules in Figure 3.1 the rules in Figure 3.2 is complete for the intuitionistic logic theory of hereditary Harrop formulas and their associated goal formulas. It then follows immediately that hereditary Harrop formulas are an abstract logic programming language. The classical logic theory of hereditary Harrop formulas is not, however, an abstract logic programming language: Peirce's formula $((p \supset q) \supset p) \supset p$, for example, is classically provable but has no uniform proof. The original definition of hereditary Harrop formulas permitted disjunctions and existential quantifiers at the top-level of goal formulas. Such an extension makes little change to the logic's proof theory properties but does help to justify its name since all positive subformulas of program clauses are then Harrop formulas [43].

Notice that sequents in this new proof system have a slightly greater ability to change during proof search: in particular, both signatures and programs can increase as proof search moves upward. Thus, not all constants and program clauses need to be available at the beginning of a computation: instead they can be made available as search continues. For this reason, the hereditary Harrop formulas have been used to provide logic programming with approaches to modular programming [67] and abstract datatypes [66, 87].

3.2.3 Higher-order quantification and proof search

The impact of using higher-order quantification in proof search was systematically studied in the contexts of Horn clauses [78, 86, 89] and hereditary Harrop formulas [80, 90]. The higher-order setting for these studies was done using the subset of Church's Simple Theory of Types [22] in which the "mathematical axioms" of extensionality, infinity, choice, etc, are not assumed.

Allowing quantification of variables of functional types only (that is, not at predicate type) is not a challenge for the high-level treatment of proof search. Such an extension to the first-order setting does make logic programming much more expressive and more challenging to implement. In particular, the presence of quantification at function types and of simply typed λ-terms [22] endowed logic programming with the encoding technique called *higher-order abstract syntax* [79, 45, 94]. It was, in fact, the λProlog programming language [88] in which this style programming was first supported.

Allowing quantification at predicate type does provide some significant challenges to the proof theoretical analysis of proof search: we illustrate two such issues here.

One issue with predicate quantification is that during proof search, the careful restriction to having program clauses on the left of the sequent arrow and goal formulas on the right might be broken via higher-order instantiation with terms containing logical connectives. For example, consider a logic program containing the following two clauses:

$$\forall P[P\ a \supset q\ b] \quad \text{and} \quad \forall x[q\ x \supset r].$$

Here, the first clause is a higher-order Horn clause following the definition in [89]. If we take an instance of this logic program in which P in the first clause is instantiated by $\lambda w.\neg q\ w$, we have clauses logically equivalent to

$$[q\ a \vee q\ b] \quad \text{and} \quad \forall x[q\ x \supset r].$$

Notice that with respect to this second logic program the atomic goal r has a classical logic proof but does not have a uniform proof. Thus, the instance of a higher-order Horn clause does not necessarily result in another higher-order Horn clause. Fortunately, for both the theory of higher-order Horn clauses and higher-order hereditary Harrop formulas, it is possible to prove that the only higher-order instances that are required during proof search are those that preserve the invariance of the initial syntactic restriction to Horn clauses or hereditary Harrop formulas [80, 89].

A second issue is more related the operational reading of clauses: program clauses are generally seen as contributing meaning to specific predicates, such as those that, for example, define the relations of concatenation or sorting of lists. These predicate constants have occurrences at strictly positive positions within program clauses: such a positive occurrence is called a *head* of a clause. If one allows predicate variables instead of constants in such head positions, then in a sense, such program clauses would be contributing meaning to any predicate. For this reason, head symbols are generally restricted to be constants. If all head symbols in a logic program are constant, it is also easy to show that that logic program is consistent (that is, some formulas are not deducible from it).

If certain mild restrictions are placed on the occurrences of logical connectives within the scope of non-logical constants (that is, within atoms), then higher-order variants of Horn clauses and hereditary Harrop formulas are known to be abstract logic programming languages [89, 80]. Higher-order quantification of predicates can provide logic programming specifications with direct and natural ways to do higher-order programming, as is popular in functional programming languages, as well as providing a means of lexically scoping and hiding predicates [66].

Allowing higher-order head positions does have, at least, a theoretical interest. Full first-order intuitionistic logic is not an abstract logic programming language since both $\vee$ and $\exists$ can cause incompleteness of uniform proofs. For example, both

$$p \vee q \longrightarrow q \vee p \quad \text{and} \quad \exists x.B \longrightarrow \exists x.B$$

have intuitionistic proofs but neither sequent has a uniform proof. As we have seen above, eliminating disjunction and existential quantification yields immediately abstract logic programming languages (at least within intuitionistic logic). As is well known, higher-order quantification allows one to define the intuitionistic disjunction $B \vee C$ as $\forall p((B \supset p) \supset (C \supset p) \supset p)$ and the existential quantifier $\exists x.Bx$ as $\forall p((\forall x.Bx \supset p) \supset p)$. Both of these formulas have the predicate variable p in a head position. Notice that if the two sequents displayed above are rewritten using these two definitions, the resulting sequents would have uniform proofs. Felty has shown that higher-order intuitionistic logic based on *true*, $\wedge$, $\supset$, and $\forall$ for all higher-order types (with no restriction of predicate variable occurrences) is an abstract logic programming language [29].

3.2.4 Uniform proofs with multiple conclusion sequents

In the multiple-conclusion setting, goal-reduction should continue to be independent not only from the logic program but also from other goals, *i.e.*, multiple goals should be reducible simultaneously. Although the sequent calculus does not directly allow for simultaneous rule application, it can be simulated easily by referring to permutations of inference rules [56]. In particular, we can require that if two or more right-introduction rules can be used to derive a given sequent, then all possible orders of applying those right-introduction rules can be obtained from any other order simply by permuting right-introduction inferences. It is easy to see that the following definition of uniform proofs for multiple-conclusion sequents generalizes that for single-conclusion sequents: a cut-free, sequent proof Ξ is *uniform* if for every subproof Ψ of Ξ and for every non-atomic formula occurrence B in the right-hand side of the end-sequent of Ψ, there is a proof Ψ' that is equal to Ψ up to permutation of inference rules and is such that the last inference rule in Ψ' introduces the top-level logical connective occurring in B [69, 72]. The notion of abstract logic programming language can be extended to the case where this extended notion of uniform proof is complete. As evidence of the usefulness of this definition, Miller in [69] used it to specify a π-calculus-like process calculus in linear logic and showed that it was an abstract logic programming language in this new sense.

Given this definition of uniform proofs for multiple conclusion sequent calculus, an interesting next step would be to turn to linear logic and start to identify subsets for which goal directed search is complete and to identify backchaining rules. Fortunately and surprisingly, the work of Andreoli in his PhD thesis [5] on focused proof search for linear logic provides a complete analysis along these lines for all of linear logic.

3.3 Linear logic and focused proofs

As we have seen, the goal-directed proof search analysis of logic programming in classical and intuitionistic logic revealed three general observations: (1) Two sets of formulas can be identified for use as goals and as program clauses. (2) These two classes are duals of each other, at least in the sense that a negative subformula of a formula in one class is a formula in the other class. (3) Goal formulas are processed immediately by a sequence of invertible right-rules and program clauses are used via a focused application of left-rules know as backchaining.

Andreoli analyzed the structure of proof search in linear logic using the notion of *focused proof* [5, 6]. His analysis made it possible to extend the above three observations to all of linear logic and to provide a deep and elegant explanation for why they hold. Andreoli classified the logical connectives into two sets of connectives. *Asynchronous* connectives are those whose right-introduction rule is invertible and *synchronous* connectives are those whose right-introduction is not invertible; that is, the success of applying a right-introduction rule for a synchronous connective required information from the context. (We say that a formula is asynchronous or synchronous depending on the top-level connective of the formula.) He also observed that these two classes of connectives are de Morgan duals of each other.

Given these distinctions between formulas, Andreoli showed that a complete bottom-up proof search procedure for cut-free proofs in linear logic (using one-sided sequents) can be described roughly as follows: first decompose all asynchronous formulas and when none remain, pick some synchronous formula, introduce its top-level connective and then continue decomposing all synchronous subformulas that might arise. Thus interleaving between asynchronous reductions and synchronous reductions yields a highly normalized proof search mechanism. Proofs built in this fashion are called *focused proofs*.

A consequence of this completeness of focused proofs is that all of linear logic can be seen as logic programming, at least once we choose the proper presentation of linear logic. In such a presentation, focused proofs capture the notion of uniform proofs and backchaining at the same time. Since all of linear logic can be seen as logic programming, we delay presenting more details about focused proofs until the next section where we present several linear logic programming languages.

3.4 Linear logic programming languages

We now present the designs of some linear logic programming languages. Our first language, Forum, provides a basis for considering all of linear logic as logic programming. We shall also look at certain subsets of Forum since they will allow us to focus on particular structural features of proof search and particular application areas.

3.4.1 The Forum presentation of linear logic

The logic programming languages based on classical and intuitionistic logics considered earlier used the connectives *true*, $\land$, $\supset$, and $\forall$. We shall now consider a presentation of linear logic using the corresponding connectives, namely, $\top$, $\&$, $\Rightarrow$, and $\forall$, along with the distinctly linear connectives $\multimap$, $\bot$, $\bindnasrepma$, and $?$. Together, this collection of connectives yields a presentation of all of linear logic since the missing connectives are directly definable using the following logical equivalences.

$$B^\perp \equiv B \multimap \bot \quad 0 \equiv \top \multimap \bot \quad 1 \equiv \bot \multimap \bot \quad \exists x.B \equiv (\forall x.B^\perp)^\perp$$
$$!B \equiv (B \Rightarrow \bot) \multimap \bot \quad B \oplus C \equiv (B^\perp \& C^\perp)^\perp \quad B \otimes C \equiv (B^\perp \bindnasrepma C^\perp)^\perp$$

This collection of connectives is not minimal: for example, $?$ and $\bindnasrepma$, can be defined in terms of the remaining connectives

$$? B \equiv (B \multimap \perp) \Rightarrow \perp \quad \text{and} \quad B \mathbin{\rotatebox[origin=c]{180}{\&}} C \equiv (B \multimap \perp) \multimap C.$$

Unlike many treatments of linear logic, we shall treat $B \Rightarrow C$ as a logical connective (which corresponds to $! B \multimap C$). From the proof search point-of-view, the four intuitionistic connectives *true*, $\wedge$, $\supset$, and $\forall$ correspond naturally with the four linear logic connectives $\top$, $\&$, $\Rightarrow$, and $\forall$ (in fact, the correspondence is so strong for the quantifiers that we write them the same in both settings). We shall call this particular presentation of linear logic the *Forum* presentation of linear logic or simply *Forum*.

Notice that all the logical connectives used in Forum are asynchronous: that is, their right-introduction rules are invertible. Since we are using two sided sequents, asynchronous formulas have a synchronous behavior when they are introduced on the left of the sequent arrow. Thus, goal reduction correspondences to the reduction of asynchronous connectives and backchaining correspondences to the focused decomposition of synchronous connectives (via left-introduction rules).

The proof systems in Figures 3.1 and 3.2 that describe logic programming in classical and intuitionistic logic used two styles of sequents: $\Sigma : \Delta \longrightarrow G$ and $\Sigma : \Delta \xrightarrow{D} A$, where Δ is a set of formulas. These sequent judgments are generalized here to $\Sigma : \Psi; \Delta \longrightarrow \Gamma; \Upsilon$ (for goal-reduction) and $\Sigma : \Psi; \Delta \xrightarrow{D} \mathcal{A}; \Upsilon$ (for backchaining), where Ψ and Υ are sets of formulas (classical maintenance), Δ and Γ are multisets of formulas (linear maintenance), $\mathcal{A}$ is a multiset of atomic formulas, and D is a formula. Notice that placement of the linear context next to the sequent arrow and classical context away from the arrow is standard notation in the literature of linear logic programming, but is the opposite convention used by Girard in his LU proof system [37].

The focusing result of Andreoli [6] can be formulated [72] as the completeness of the proof system for linear logic using the proof system in Figure 3.3. This proof system appears rather complicated at first glance, so it is worth noting the following organization of these inference rules: there are 8 right-introduction rules, 7 left-introduction rules, 2 initial rules, and 3 decide rules. Notice that 2 of the decide rules place a formula on the sequent arrow while the third copies of formula from the classically maintained right context to the linear maintained right context. This third decide rule is a combination of contraction and dereliction rule for ? and is used to "decide" on a new goal formula on which to do reductions.

$$\frac{}{\Sigma : \Psi; \Delta \longrightarrow \top, \Gamma; \Upsilon} \qquad \frac{\Sigma : \Psi; \Delta \longrightarrow B, \Gamma; \Upsilon \quad \Sigma : \Psi; \Delta \longrightarrow C, \Gamma; \Upsilon}{\Sigma : \Psi; \Delta \longrightarrow B \& C, \Gamma; \Upsilon}$$

$$\frac{\Sigma : \Psi; \Delta \longrightarrow \Gamma; \Upsilon}{\Sigma : \Psi; \Delta \longrightarrow \bot, \Gamma; \Upsilon} \qquad \frac{\Sigma : \Psi; \Delta \longrightarrow B, C, \Gamma; \Upsilon}{\Sigma : \Psi; \Delta \longrightarrow B \parr C, \Gamma; \Upsilon}$$

$$\frac{\Sigma : \Psi; B, \Delta \longrightarrow C, \Gamma; \Upsilon}{\Sigma : \Psi; \Delta \longrightarrow B \multimap C, \Gamma; \Upsilon} \qquad \frac{\Sigma : B, \Psi; \Delta \longrightarrow C, \Gamma; \Upsilon}{\Sigma : \Psi; \Delta \longrightarrow B \Rightarrow C, \Gamma; \Upsilon}$$

$$\frac{y : \tau, \Sigma : \Psi; \Delta \longrightarrow B[y/x], \Gamma; \Upsilon}{\Sigma : \Psi; \Delta \longrightarrow \forall_\tau x.B, \Gamma; \Upsilon} \qquad \frac{\Sigma : \Psi; \Delta \longrightarrow \Gamma; B, \Upsilon}{\Sigma : \Psi; \Delta \longrightarrow ?B, \Gamma; \Upsilon}$$

$$\frac{\Sigma : B, \Psi; \Delta \xrightarrow{B} A; \Upsilon}{\Sigma : B, \Psi; \Delta \longrightarrow A; \Upsilon} \qquad \frac{\Sigma : \Psi; \Delta \xrightarrow{B} A; \Upsilon}{\Sigma : \Psi; B, \Delta \longrightarrow A; \Upsilon} \qquad \frac{\Sigma : \Psi; \Delta \longrightarrow A, B; B, \Upsilon}{\Sigma : \Psi; \Delta \longrightarrow A; B, \Upsilon}$$

$$\frac{}{\Sigma : \Psi; \cdot \xrightarrow{A} A; \Upsilon} \qquad \frac{}{\Sigma : \Psi; \cdot \xrightarrow{A} \cdot; A, \Upsilon}$$

$$\frac{}{\Sigma : \Psi; \cdot \xrightarrow{\bot} \cdot; \Upsilon} \qquad \frac{\Sigma : \Psi; \Delta \xrightarrow{G_i} A; \Upsilon}{\Sigma : \Psi; \Delta \xrightarrow{G_1 \& G_2} A; \Upsilon} \qquad \frac{\Sigma : \Psi; B \longrightarrow \cdot; \Upsilon}{\Sigma : \Psi; \cdot \xrightarrow{?B} \cdot; \Upsilon}$$

$$\frac{\Sigma : \Psi; \Delta_1 \xrightarrow{B} A_1; \Upsilon \quad \Sigma : \Psi; \Delta_2 \xrightarrow{C} A_2; \Upsilon}{\Sigma : \Psi; \Delta_1, \Delta_2 \xrightarrow{B \parr C} A_1, A_2; \Upsilon} \qquad \frac{\Sigma : \Psi; \Delta \xrightarrow{B[t/x]} A; \Upsilon}{\Sigma : \Psi; \Delta \xrightarrow{\forall_\tau x.B} A; \Upsilon}$$

$$\frac{\Sigma : \Psi; \Delta_1 \longrightarrow A_1, B; \Upsilon \quad \Sigma : \Psi; \Delta_2 \xrightarrow{C} A_2; \Upsilon}{\Sigma : \Psi; \Delta_1, \Delta_2 \xrightarrow{B \multimap C} A_1, A_2; \Upsilon}$$

$$\frac{\Sigma : \Psi; \cdot \longrightarrow B; \Upsilon \quad \Sigma : \Psi; \Delta \xrightarrow{C} A; \Upsilon}{\Sigma : \Psi; \Delta \xrightarrow{B \Rightarrow C} A; \Upsilon}$$

Fig. 3.3. A proof system for the Forum presentation of linear logic. The right-introduction rule for $\forall$ has the proviso that y is not declared in the signature Σ, and the left-introduction rule for $\forall$ has the proviso that t is a Σ-term of type τ. In left-introduction rule for $\&$, $i \in \{1, 2\}$.

Because linear logic can be seen as the logic behind classical and intuitionistic logic, it is possible to see both Horn clauses and hereditary Harrop formulas as subsets of Forum. It is a simple matter to see that the proof systems in Figures 3.1 and 3.2 result from restricting the proof system for Forum in Figure 3.3 to Horn clauses and to hereditary Harrop formulas. Below we overview various other subsets of linear logic that have been proposed as specification languages and as abstract logic programming languages.

$$\frac{}{\Sigma : \Psi; \Delta \longrightarrow \top} \qquad \frac{\Sigma : \Psi; \Delta \longrightarrow G_1 \qquad \Sigma : \Psi; \Delta \longrightarrow G_2}{\Sigma : \Psi; \Delta \longrightarrow G_1 \,\&\, G_2}$$

$$\frac{\Sigma : \Psi, G_1; \Delta \longrightarrow G_2}{\Sigma : \Psi; \Delta \longrightarrow G_1 \Rightarrow G_2} \qquad \frac{\Sigma : \Psi; \Delta, G_1 \longrightarrow G_2}{\Sigma : \Psi; \Delta \longrightarrow G_1 \multimap G_2} \qquad \frac{y : \tau, \Sigma : \Psi; \Delta \longrightarrow B[y/x]}{\Sigma : \Psi; \Delta \longrightarrow \forall_\tau x.B}$$

$$\frac{\Sigma : \Psi, D; \Delta \xrightarrow{D} A}{\Sigma : \Psi, D; \Delta \longrightarrow A} \qquad \frac{\Sigma : \Psi; \Delta \xrightarrow{D} A}{\Sigma : \Psi; \Delta, D \longrightarrow A} \qquad \frac{}{\Sigma : \Psi; \cdot \xrightarrow{A} A}$$

$$\frac{\Sigma : \Psi; \Delta \xrightarrow{D_i} A}{\Sigma : \Psi; \Delta \xrightarrow{D_1 \wedge D_2} A} \qquad \frac{\Sigma : \Psi; \cdot \longrightarrow G \qquad \Sigma : \Psi; \Delta \xrightarrow{D} A}{\Sigma : \Psi; \Delta \xrightarrow{G \Rightarrow D} A}$$

$$\frac{\Sigma : \Psi; \Delta_1 \longrightarrow G \qquad \Sigma : \Psi; \Delta_2 \xrightarrow{D} A}{\Sigma : \Psi; \Delta_1, \Delta_2 \xrightarrow{G \multimap D} A} \qquad \frac{\Sigma : \Psi; \Delta \xrightarrow{D[t/x]} A}{\Sigma : \Psi; \Delta \xrightarrow{\forall_\tau x.D} A}$$

Fig. 3.4. The proof system for Lolli. The rule for universal quantification has the proviso that y is not in Σ. In the $\forall$-left rule, t is a Σ-term of type τ.

3.4.2 Lolli

The connectives $\perp$, $\parr$, and ? force the genuinely classical feel of linear logic. (In fact, using the two linear logic equivalences ? $B \equiv (B \multimap \perp) \Rightarrow \perp$ and $B \parr C \equiv (B \multimap \perp) \multimap C$ we see that we only need to add $\perp$ to a system with the two implication $\multimap$ and $\Rightarrow$ to get full classical linear logic.) Without these three connectives, the multiple-conclusion sequent calculus given for Forum in Figure 3.3 can be replaced by one with only single-conclusion sequents.

The collection of connectives one gets from dropping these three connectives from Forum, namely $\top$, $\&$, $\Rightarrow$, $\multimap$, and $\forall$, form the *Lolli* logic programming language. Presenting a sequent calculus for Lolli is a simple matter. First, remove any inference rule in Figure 3.3 involving $\perp$, $\parr$, and ?. Second, abbreviate the sequents $\Sigma : \Psi; \Delta \longrightarrow G; \cdot$ and $\Sigma : \Psi; \Delta \xrightarrow{D} A; \cdot$ as $\Sigma : \Psi; \Delta \longrightarrow G$ and $\Sigma : \Psi; \Delta \xrightarrow{D} A$. The resulting proof system for Lolli is given in Figure 3.4. The completeness of this proof system for Lolli was given directly by Hodas and Miller in [48], although it follows directly from the completeness of focused proofs [6], at least once focused proofs are written as the Forum proof system.

3.4.3 Uncurrying program clauses

Frequently it is convenient to view a program clause, such as

$$\forall \bar{x}[G_1 \Rightarrow G_2 \multimap A],$$

which contains two goals, as a program clause containing one goal: the formula

$$\forall \bar{x}[(!\, G_1 \otimes G_2) \multimap A].$$

is logically equivalent to the formula above and brings the two goals into the one expression $!\, G_1 \otimes G_2$. Such a rewriting of a formula to a logically equivalent formula is essentially the *uncurrying* of the formula, where uncurrying is the rewriting of formulas using the following equivalences in the forward direction.

$$
\begin{array}{rcl}
H & \equiv & 1 \multimap H \\
B \multimap C \multimap H & \equiv & (B \otimes C) \multimap H \\
B \Rightarrow H & \equiv & !\, B \multimap H \\
(B \multimap H)\,\&\,(C \multimap H) & \equiv & (B \oplus C) \multimap H \\
\forall x.(B(x) \multimap H) & \equiv & (\exists x.B(x)) \multimap H
\end{array}
$$

(The last equivalence assumes that x is not free in H.) Allowing occurrences of 1, $\otimes$, $!$, $\oplus$, and $\exists$ into goals does not cause any problems with the completeness of uniform provability and some presentations of linear logic programming language [48, 72, 95] allow for such occurrences.

3.4.4 Other subsets of Forum

Although all of linear logic can be seen as abstract logic programming, it is still of interest to examine subsets of linear logic for use as specification languages or as programming languages. These subsets are often motivated by picking a small subset of linear logic that is expressive enough to specify problems of a certain application domain. Below we list some subsets of linear logic that have been identified in the literature.

If one maps *true* to $\top$, $\wedge$ to $\&$, and $\supset$ to $\Rightarrow$, then both Horn clauses and hereditary Harrop formulas can be identified with linear logic formulas. Proofs given for these two sets of formulas in Figures 3.1 and 3.2 are essentially the same as those for the corresponding proofs in Figure 3.4. Thus, viewing these two classes of formulas as being based on linear instead of intuitionistic logic does not change their expressiveness. In this sense, Lolli can be identified as being hereditary Harrop formulas extended with linear implication. When one is only interested in cut-free proofs, a second translation of Horn clauses and hereditary Harrop formulas into linear logic is possible. In particular, if negative occurrences of *true*, $\wedge$, and $\supset$ are translated to 1, $\otimes$, and $\multimap$, respectively, while positive occurrences of *true*, $\wedge$, and $\supset$ are translated to $\top$, $\&$, and $\Rightarrow$, respectively, then the resulting proofs in Figure 3.4 of the linear logic formulas yield proofs similar to those in Figures 3.1 and 3.2 [48]. (The notion here of positive and negative occurrences are with respect to occurrences within a cut-free proof: for example, a positive occurrence in a formula on the left of a sequent arrow is judged to be a negative occurrence for this translation.) Thus, if the formula

$$\forall \bar{x}[(A_1 \wedge (A_2 \supset A_3) \wedge A_4) \supset A_0]$$

appears on the left of the sequent arrow, it is translated as

$$\forall \bar{x}[(A_1 \otimes (A_2 \Rightarrow A_3) \otimes A_4) \multimap A_0]$$

and if it appears on the right of the sequent arrow, it is translated as

$$\forall \bar{x}[(A_1 \& (A_2 \multimap A_3) \& A_4) \Rightarrow A_0].$$

Historically speaking, the first proposal for a linear logic programming language was LO (Linear Objects) by Andreoli and Pareschi [8, 9]. LO is an extension to the Horn clause paradigm in which atomic formulas are generalized to multisets of atomic formulas connected by $\otimes$. In LO, backchaining is again multiset rewriting, which was used to specify object-oriented programming and the coordination of processes. LO is a subset of the LinLog [5, 6], where formulas are of the form

$$\forall \bar{y}(G_1 \multimap \cdots \multimap G_m \multimap (A_1 \otimes \cdots \otimes A_p)).$$

Here $p > 0$ and $m \geq 0$; occurrences of $\multimap$ are either occurrences of $\multimap$ or $\Rightarrow$; $G_1, \ldots G_m$ are built from $\bot$, $\otimes$, $?$, $\top$, $\&$, and $\forall$; and $A_1, \ldots A_m$ are atomic formulas. In other words, these are formula in Forum where the "head" of the formula is not empty (*i.e.*, $p > 0$) and where the goals $G_1, \ldots G_m$ do not contain implications. Andreoli argues that arbitrary linear logic formulas can be "skolemize" (by introducing new non-logical constants) to yield only LinLog formulas, such that proof search involving the original and the skolemize formulas are isomorphic. By applying uncurrying, the displayed formula above can be written in the form

$$\forall \bar{y}(G \multimap (A_1 \otimes \cdots \otimes A_p))$$

where G is composed of the top-level synchronous connectives and of the subformulas $G_1, \ldots, G_m$, which are all composed of asynchronous connectives. In LinLog, goal formulas have no synchronous connective in the scope an asynchronous connective.

3.4.5 Other language designs

Another linear logic programming language that has been proposed is the Lygon system of Harland and Pym [52]. They based their design on notions of goal-directed proof and multiple conclusion uniform proofs [95] that unfortunately differ from those presented here. The operational semantics for proof search that they developed is different and more complex than the alternation of asynchronous and synchronous search that is used for, say, Forum.

Let G and H be formulas composed of $\bot$, $⅋$, and $\forall$. Closed formulas of the form $\forall \bar{x}[G \multimap H]$ where H is not (logically equivalent to) $\bot$ have been called *process clauses* in [69] and are used there to encode a calculus similar to the π-calculus: the universal quantifier in goals are used to encode name restriction. These clauses written in the contrapositive (replacing, for example, $⅋$ with $\otimes$) have been called *linear Horn clauses* by Kanovich and have been used to model computation via multiset rewriting [55].

Various other specification logics have also been developed, often designed directly to deal with particular application areas. In particular, the language ACL by Kobayashi and Yonezawa [58, 59] captures simple notions of asynchronous communication by identifying the send and read primitives with two complementary linear logic connectives. Lincoln and Saraswat have developed a linear logic version of concurrent constraint programming [64, 107] and Fages, Ruet, and Soliman have analyzed similar extensions to the concurrent constraint paradigm [31, 104].

Some aspects of dependent typed λ-calculi overlap with notions of abstract logic programming languages. Within the setting of intuitionistic, single-side sequents, uniform proofs are similar to $\beta\eta$-long normal forms in natural deduction and typed λ-calculus. The LF logical framework [44] can be mapped naturally [28] into a higher-order extension of hereditary Harrop formulas [80]. Inspired such a connection and by the design of Lolli, Cervesato and Pfenning developed a linear extension to LF called Linear LF [23, 24].

3.5 Applications of linear logic programming

One theme that occurs often in applications of linear logic programming is that of *multiset rewriting*: a simple paradigm that has wide applications in computational specifications. To see how such rewriting can be captured in proof search, consider the rewriting rule

$$a, a, b \Rightarrow c, d, e,$$

which specifies that a multiset can be rewritten by first removing two occurrences of a and one occurrence of b and then have one occurrence each of c, d, and e added. Since the left-hand of sequents in Figure 3.4 and the left- and right-hand sides of sequents in Figure 3.3 have multisets of formulas, it is an easy matter to write clauses in linear logic which can rewrite multisets when they are used in backchaining.

To rewrite the right-hand multiset of a sequent using the rule above, simply backchain over the clause $c \,\aftperp\, d \,\aftperp\, e \multimap a \,\aftperp\, a \,\aftperp\, b$. To illustrate such rewriting directly via Forum, consider the sequent $\Sigma : \Psi; \Delta \longrightarrow a, a, b, \Gamma; \Upsilon$ where the above clause is a member of Ψ. A proof for this sequent can then look like the following (where the signature Σ is not displayed).

$$
\cfrac{
\cfrac{\Psi; \Delta \longrightarrow c, d, e, \Gamma; \Upsilon}{\Psi; \Delta \longrightarrow c \,\aftperp\, d \,\aftperp\, e, \Gamma; \Upsilon}
\qquad
\cfrac{
\cfrac{\Psi; \cdot \xrightarrow{a} a; \Upsilon \quad \Psi; \cdot \xrightarrow{a} a; \Upsilon \quad \Psi; \cdot \xrightarrow{b} b; \Upsilon}{\Psi; \cdot \xrightarrow{a \,\aftperp\, a \,\aftperp\, b} a, a, b; \Upsilon}
}{\Psi; \Delta \xrightarrow{c \,\aftperp\, d \,\aftperp\, e \,-\!\circ\, a \,\aftperp\, a \,\aftperp\, b} a, a, b, \Gamma; \Upsilon}
}{\Psi; \Delta \longrightarrow a, a, b, \Gamma; \Upsilon}
$$

We can interpret this fragment of a proof as a rewriting of the multiset a, a, b, Γ to the multiset c, d, e, Γ using the rule displayed above.

To rewrite the left-hand context instead, a clause such as

$$a \multimap a \multimap b \multimap (c \multimap d \multimap e \multimap A_1) \multimap A_0$$

or (using the uncurried form)

$$(a \otimes a \otimes b) \otimes ((c \otimes d \otimes e) \multimap A_1) \multimap A_0$$

can be used in backchaining. Operationally this clause means that to prove the atomic goal A_0, first remove two occurrence of a and one of b from the left-hand multiset, then add one occurrence each of c, d, and e, and then proceed to attempt a proof of A_1.

Of course, there are additional features of linear logic than can be used to enhance this primitive notion of multiset rewriting. For examples, the

? modal on the right and the ! modal on the left can be used to place items in multisets than cannot be deleted and the additive conjunction & can be used to copy multisets.

Listed below are some application areas where proof search and linear logic have been used. A few representative references for each area are listed.

Object-oriented programming Capturing inheritance was a goal of the early LO system [9] and modeling state encapsulation was a motivation [46] for the design of Lolli. State encapsulation was also addressed using Forum in [26, 70].

Concurrency Linear logic is often been considered a promising declarative foundation for concurrency primitives in specification languages and programming languages. Via reductions to multiset rewriting, several people have found encodings of Petri nets into linear logic [33, 3, 13, 27]. The specification logic ACL of Kobayashi and Yonezawa is an asynchronous calculus in which the send and read primitives are identified with two complementary linear logic connectives [58, 59]. Miller [69] described how features of the π-calculus [85] can be modeled in linear logic and Bruscoli and Guglielmi [14] showed how specifications in the Gamma language [15] can be related to linear logic.

Operational semantics Forum has been used to specify the operational semantics of the imperative features in Algol [70] and ML [20] and the concurrency features of Concurrent ML [72]. Forum was used by Chirimar to specify the operational semantics of a pipe-lined, RISC processor [20] and by Chakravarty to specify the operational semantics of a parallel programming language that combines functional and logic programming paradigms [19]. Linear logic has also been used to express and to reason about the operational semantics of security protocols [18, 73]. A similar approach to using linear logic was also applied to specifying real-time finite-state systems [57].

Object-logic proof systems Intuitionistic based systems, such as the LF dependent type system and hereditary Harrop formulas, are popular choices for the specification of natural deduction proof systems [44, 30]. The linear logic aspects of both Lolli and Linear LF have been used to specify natural deduction systems for a wider collection of object-logics than are possible with these non-linear logic frameworks [48, 23].

By admitting full linear logic and multiple conclusion sequents, Forum provides a natural setting for the specification of object-level sequent calculus proof systems. In classical linear logic, the duality between left and rule introduction rules and between the cut and initial rules is easily explained using the meta-level linear negation. Some examples of specifying object-level sequent proof systems in Forum are given in [70, 84, 105].

Natural language parsing Lambek's precursor to linear logic [62] was motivated in part to deal with natural language syntax. An early use of Lolli was to provide a simple and declarative approach to gap threading and island constraints within English relative clauses [48, 49] that built on an approach first proposed by Pareschi using intuitionistic logic [93, 96]. Researchers in natural language syntax are generally quick to look closely at most advances in proof theory, and linear logic has not been an exception: for a few additional references, see [25, 82, 81].

3.6 Examples of reasoning about a linear logic program

One of the reasons to use logic as the source code for a programming languages is that the actual artifact that is the program should be amenable to direct manipulation and analysis in ways that might be hard or impossible in more conventional programming languages. One method for reasoning directly on logic programming involves the cut rule (via modus ponens) and cut-elimination. We consider here two examples of how the meta-theory of linear logic can be used to prove properties of logic programs.

While much of the motivation for designing logic programming languages based on linear logic has been to add expressiveness to such languages, linear logic can also help shed some light on conventional programs. In this section we consider the linear logic specification for the reverse of lists and formally show it is symmetric.

Let the constants *nil* and $(\cdot :: \cdot)$ denote the two constructors for lists. Consider specifying the binary relation *reverse* that relates two lists if one is the reverse of the other. To specify the computation of reversing a list, consider making two piles on a table. Initialize one pile to the list you wish to reverse and initialize the other pile to be empty. Next, repeatedly move the top element from the first pile to the top of the second pile. When the first pile is empty, the second pile is the reverse of the original list. For example, the following is a trace of such a computation.

$(a :: b :: c :: nil)$	nil
$(b :: c :: nil)$	$(a :: nil)$
$(c :: nil)$	$(b :: a :: nil)$
nil	$(c :: b :: a :: nil)$

In more general terms: first pick a binary relation rv to denote the pairing of lists above (this predicate will be an auxiliary predicate to reverse). If we wish to reverse the list L to get K, then start with the atomic formula $(rv\ L\ nil)$ and do a series of backchaining over the clause

$$\forall X \forall P \forall Q (rv\ P\ (X :: Q) \multimap rv\ (X :: P)\ Q)$$

to get to the formula $(rv\ nil\ K)$. Once this is done, K is the result of reversing L. The entire specification of reverse can be written as the following single formula.

$$\forall L \forall K [\ \forall rv\ ((\forall X \forall P \forall Q (rv\ P\ (X :: Q) \multimap rv\ (X :: P)\ Q)) \Rightarrow$$
$$rv\ nil\ K \multimap rv\ L\ nil) \multimap reverse\ L\ K\ ]$$

Notice that the clause used for repeatedly moving the top elements of lists is to the left of an intuitionistic implication (so it can be used any number of times) while the formula representing the base case of the recursion, namely $(rv\ nil\ K)$, is to the left of a linear implication (thus it must be used exactly once).

Consider proving that reverse is symmetric: that is, if $(reverse\ L\ K)$ is proved from the above clause, then so is $(reverse\ K\ L)$. The informal proof of this is simple: in the trace table above, flip the rows and the columns. What is left is a correct computation of reversing again, but the start and final lists have exchanged roles. This informal proof is easily made formal by exploiting the meta-theory of linear logic. A more formal proof proceeds as follows. Assume that $(reverse\ L\ K)$ can be proved. There is only one way to prove this (backchaining on the above clause for *reverse*). Thus the formula

$$\forall rv((\forall X \forall P \forall Q(rv\ P\ (X :: Q) \multimap rv\ (X :: P)\ Q)) \Rightarrow rv\ nil\ K \multimap rv\ L\ nil)$$

is provable. Since we are using logic, we can instantiate this quantifier with any binary predicate expression and the result is still provable. So we choose to instantiate it with the λ-expression $\lambda x \lambda y (rv\ y\ x)^{\perp}$. The resulting formula

$$(\forall X \forall P \forall Q(rv\ (X :: Q)\ P)^{\perp} \multimap (rv\ Q\ (X :: P)^{\perp})) \Rightarrow$$

$$(rv\ K\ nil)^{\perp} \multimap (rv\ nil\ L)^{\perp}$$

can be simplified by using the contrapositive rule for negation and linear implication, and hence yields

$$(\forall X \forall P \forall Q (rv\ Q\ (X::P) \multimap rv\ (X::Q)\ P) \Rightarrow rv\ nil\ L \multimap rv\ K\ nil)$$

If we now universally generalize on rv we again have proved the body of the reverse clause, but this time with L and K switched. Notice that we have succeeded in proving this fact about reverse without explicit reference to induction.

For another example of using linear logic's meta-theory to reason directly on specifications, consider the problem of adding a global counter to a functional programming language that already has primitives for, say conditionals, application, abstraction, etc [72]. Now add *get* and *inc* expressions: evaluation of *get* causes the counter's value to be returned while evaluation of *inc* causes the counter's value to be incremented. Figure 3.5 contains three specifications, E_1, E_2, and E_3, of such a counter: all three specifications store the counter's value in an atomic formula as the argument of the predicate r. In these three specifications, the predicate r is existentially quantified over the specification in which it is used so that the atomic formula that stores the counter's value is itself local to the counter's specification (such existential quantification of predicates is a familiar technique for implementing abstract datatypes in logic programming [66]). The first two specifications store the counter's value on the right of the sequent arrow, and reading and incrementing the counter occurs via a synchronization between the *eval*-atom and the r-atom. In the third specification, the counter is stored as a linear assumption on the left of the sequent arrow, and synchronization is not used: instead, the linear assumption is "destructively" read and then rewritten in order to specify *get* and *inc* (counters such as these are described in [48]). Finally, in the first and third specifications, evaluating the *inc* symbol causes 1 to be added to the counter's value. In the second specification, evaluating the *inc* symbol causes 1 to be subtracted from the counter's value: to compensate for this unusual implementation of *inc*, reading a counter in the second specification returns the negative of the counter's value.

The use of $\otimes$, $!$, $\exists$, and negation in Figure 3.5 is for convenience in displaying these abstract datatypes. The curry/uncurry equivalence

$$\exists r (R_1^{\perp} \otimes !\,R_2 \otimes !\,R_3) \multimap G \equiv \forall r (R_2 \Rightarrow R_3 \Rightarrow G \,\mathop{\aparr}\, R_1)$$

$$E_1 = \exists r[\quad (r\ 0)^{\perp} \otimes$$
$$!\forall K \forall V (eval\ get\ V\ K\ \Im\ r\ V \circ\!\!- eval\ K\ \Im\ r\ V)) \otimes$$
$$!\forall K \forall V (eval\ inc\ V\ K\ \Im\ r\ V \circ\!\!- K\ \Im\ r\ (V+1))]$$

$$E_2 = \exists r[\quad (r\ 0)^{\perp} \otimes$$
$$!\forall K \forall V (eval\ get\ (-V)\ K\ \Im\ r\ V \circ\!\!- K\ \Im\ r\ V) \otimes$$
$$!\forall K \forall V (eval\ inc\ (-V)\ K\ \Im\ r\ V \circ\!\!- K\ \Im\ r\ (V-1))]$$

$$E_3 = \exists r[\quad (r\ 0) \otimes$$
$$!\forall K \forall V (eval\ get\ V\ K \circ\!\!- r\ V \otimes (r\ V \multimap K)) \otimes$$
$$!\forall K \forall V (eval\ inc\ V\ K \circ\!\!- r\ V \otimes (r\ (V+1) \multimap K))]$$

Fig. 3.5. Three specifications of a global counter.

directly converts a use of such a specification into a formula of Forum (given α-conversion, we may assume that r is not free in G).

Although these three specifications of a global counter are different, they should be equivalent in the sense that evaluation cannot tell them apart. Although there are several ways that the equivalence of such counters can be proved (for example, operational equivalence), the specifications of these counters are, in fact, *logically* equivalent. In particular, the three entailments $E_1 \vdash E_2$, $E_2 \vdash E_3$, and $E_3 \vdash E_1$ are provable in linear logic. The proof of each of these entailments proceeds (in a bottom-up fashion) by choosing an eigenvariable to instantiate the existential quantifier on the left-hand specification and then instantiating the right-hand existential quantifier with some term involving that eigenvariable. Assume that in all three cases, the eigenvariable selected is the predicate symbol s. Then the first entailment is proved by instantiating the right-hand existential with $\lambda x.s\ (-x)$; the second entailment is proved using the substitution $\lambda x.(s\ (-x))^{\perp}$; and the third entailment is proved using the substitution $\lambda x.(s\ x)^{\perp}$. The proof of the first two entailments must also use the equations

$$\{-0 = 0, -(x+1) = -x-1, -(x-1) = -x+1\}.$$

The proof of the third entailment requires no such equations.

Clearly, logical equivalence is a strong equivalence: it immediately implies that evaluation cannot tell the difference between any of these different specifications of a counter. For example, assume $E_1 \vdash eval\ M\ V\ \top$. Then by cut and the above entailments, we have $E_2 \vdash eval\ M\ V\ \top$.

3.7 Effective implementations of proof search

There are several challenges facing the implementers of linear logic programming languages. One problem is how to split multiset contexts when proving a tensor or backchaining over linear implications. If the multiset contexts of a sequent have $n \geq 0$ formulas in them, then can be as many as 2^n ways that a context can be partitioned into two multisets. Often, however, very few of these splits will lead to a successful proof. An obvious approach to address the problem of splitting context would be to do the split lazily. One approach to such lazy splitting was presented in [48] where proof search was seen to be a kind of input/output process. When proving one part of a tensor, all formulas are given to that attempt. If the proof process is successful, any formulas remaining would then be output from that attempt and handed to the remaining part of the tensor. A rather simple interpreter for such a model of resource consumption and its Prolog implementation is given in [48]. Experience with this interpreter showed that the presence of the additive connectives – ⊤ and & – caused significant problems with efficient interpretation. Several researchers have developed significant variations to the model of lazy splitting. See for example, [21, 50, 63]. Similar implementation issues concerning the Lygon logic programming language are described in [112]. More recent approaches to accounting for resource consumption in linear logic programming uses constraint solving to treat the different aspects of resources sharing and consumption in different parts of the search for a proof [7, 51].

Based on such approaches to lazy splitting, various interpreters of linear logic programming languages have been implemented. To date, however, only one compiling effort has been made. Tamura and Kaneda [110] have developed an extension to the Warren abstract machine (a commonly used machine model for logic programming) and a compiler for a subset of Lolli. This compiler was shown in [54] to perform surprisingly well for a certain theorem proving application where linear logic provided a particularly elegant specification.

3.8 Research in sequent calculus proof search

Since the majority of linear logic programming is described using sequent calculus proof systems, a great deal of the work in understanding and implementing these languages has focused on properties of the sequent calculus. Besides the work mentioned already concerning refinements to proof search, there is the related work of Galmiche, Boudinet, and Perrier [32, 40], Tammet [109], and Guglielmi [42], and Gabbay and Olivetti [39]. And, of course, there is the recent work of Girard on *Locus solum* [38].

Below is briefly described three areas certainly deserving additional consideration and which should significantly expand our understanding and application of proof search and logic programming.

3.8.1 Polarity and proof search.

Andreoli observed the critical role of polarity in proof search: the notion of asynchronous behavior (goal-reduction) and synchronous behavior (backchaining) are de Morgan duals of each other. There have been other uses of polarity in proof systems and proof search. In [37], Girard introduced the LU system in which classical, intuitionistic, and linear logics share a common proof system. Central to their abilities to live together is a notion of polarity: positive, negative, and neutral. As we have shown in this paper, linear logic enhances the expressiveness of logic programming languages presented in classical and intuitionistic logic, but this comparison is made after they have been *translated* into linear logic. It would be interesting to see if there is one logic programming language that contains, for example, a classical, intuitionistic, and linear implication.

3.8.2 Non-commutativity.

Having a non-commutative conjunction or disjunction within a logic programming language should significantly enhance the expressiveness of the language. Lambek's early calculus [62] was non-commutative but it was also weak in that it did not have modals and additive connectives. In recent years, a number of different proposals for non-commutative versions of linear logic have been considered. Abrusci [1] and later Ruet and Abrusci [10, 106] have developed one such approach. Remi Baudot [12] and Andreoli and Maieli [4] developed focusing strategies for this logic and have designed abstract logic programming languages based on the proposal of Abrusci and Ruet. Alessio Guglielmi has proposed a new approach to representing proofs via the *calculus of structures* and presents a non-commutative connective which is self-dual [41]. Paola Bruscoli has shown how that non-commutative connective can be used to code sequencing in the CCS specification language [16]. Christian

Retoré has also proposed a non-commutative, self dual connective within the context of proof nets [102, 103]. Finally, Pfenning and Polakow have developed a non-commutative version of intuitionistic linear logic with a sequential operator and have demonstrated its uses in several applications [97, 100, 98, 99]. Currently, non-commutativity has the appearance of being rather complicated and no single proposal seems to be canonical at this point.

3.8.3 Reasoning about specifications.

One of the reasons for using logic to make specifications in the first place must surely be that the meta-theory of logic should help in establishing properties of logic programs: cut and cut-elimination will have a central role here. While this was illustrated in Section 3.6, very little of this kind of reasoning has been done for logic programs written in logics programming languages more expressive than Horn clauses. The examples in Section 3.6 are also not typical: most reasoning about logic specifications will certainly involve induction. Also, many properties of computational specifications involve being able to reason about all paths that a computation may take: simulation and bisimulation are examples of such properties [68]. The proof theoretical notion of *fixpoint* [36] and of *definition* [53, 77, 108] has been used to help capture such notions. See, for example, the work on integrating inductions and definitions into intuitionistic logic [74, 75, 76]. Extending such work to incorporate co-induction and to embrace logics other than intuitionistic logic should certainly be considered.

Of course, there are many other avenues that work in proof search and logic programming design can take. For example, one can investigate rather different logics, for example, the logic of bunched implications [92, 101], for their suitability as logic programming languages. Model theoretic semantics suitable for reasoning about linear logic specification would certainly be desirable, especially if they can provide simple, natural, and compositional notions of meaning. Also, several application areas of linear logic programming seems convincing enough that work on improving the effectiveness of interpreters and compilers certainly seems appropriate.

Acknowledgments Parts of this overview where written while the author was at Penn State University and was supported in part by NSF grants CCR-9912387, CCR-9803971, INT-9815645, and INT-9815731. Alwen Tiu provided useful comments on a draft of this paper.

Bibliography

[1] V. Michele Abrusci. Phase semantics and sequent calculus for pure non-commutative classical linear propositional logic. *Journal of Symbolic Logic*, 56(4):1403–1451, December 1991.

[2] Samson Abramsky. Computational interpretations of linear logic. *Theoretical Computer Science*, 111:3–57, 1993.

[3] A. Asperti, G.-L. Ferrari, and R. Gorrieri. Implicative formulae in the 'proof as computations' analogy. In *Principles of Programming Languages (POPL'90)*, pages 59–71. ACM, January 1990.

[4] J.-M. Andreoli and R. Maieli. Focusing and proof nets in linear and noncommutative logic. In *International Conference on Logic for Programming and Automated Reasoning (LPAR)*, volume 1581 of *LNAI*. Springer, 1999.

[5] Jean-Marc Andreoli. *Proposal for a Synthesis of Logic and Object-Oriented Programming Paradigms*. PhD thesis, University of Paris VI, 1990.

[6] Jean-Marc Andreoli. Logic programming with focusing proofs in linear logic. *Journal of Logic and Computation*, 2(3):297–347, 1992.

[7] Jean-Marc Andreoli. Focussing and proof construction. *Annals of Pure and Applied Logic*, 107(1):131–163, 2001.

[8] J.-M. Andreoli and R. Pareschi. Communication as fair distribution of knowledge. In *Proceedings of OOPSLA 91*, pages 212–229, 1991.

[9] J.M. Andreoli and R. Pareschi. Linear objects: Logical processes with built-in inheritance. *New Generation Computing*, 9(3-4):445–473, 1991.

[10] V. Michele Abrusci and Paul Ruet. Non-commutative logic I: The multiplicative fragment. *Annals of Pure and Applied Logic*, 101(1):29–64, 2000.

[11] K. R. Apt and M. H. van Emden. Contributions to the theory of logic programming. *Journal of the ACM*, 29(3):841–862, 1982.

[12] Rémi Baudot. *Programmation Logique: Non commutativité et Polarisation*. PhD thesis, Université Paris 13, Laboratoire d'informatique de Paris Nord (L.I.P.N.), December 2000.

[13] C. Brown and D. Gurr. A categorical linear framework for petri nets. In *Logic in Comptuer Science (LICS'90)*, pages 208–219, Philadelphia, PA, June 1990. IEEE Computer Society Press.

[14] Paola Bruscoli and Alessio Guglielmi. A linear logic view of Gamma style computations as proof searches. In Jean-Marc Andreoli, Chris Hankin, and Daniel Le Métayer, editors, *Coordination Programming: Mechanisms, Models and Semantics*. Imperial College Press, 1996.

[15] Jean-Pierre Banâtre and Daniel Le Métayer. Gamma and the chemical reaction model: ten years after. In *Coordination programming: mechanisms, models and semantics*, pages 3–41. World Scientific Publishing, IC Press, 1996.

[16] Paola Bruscoli. A purely logical account of sequentiality in proof search. In Peter J. Stuckey, editor, *Logic Programming, 18th International Conference*, volume 2401 of *LNAI*, pages 302–316. Springer-Verlag, 2002.

[17] Gianluigi Bellin and Philip J. Scott. On the pi-calculus and linear logic. *Theoretical Computer Science*, 135:11–65, 1994.

[18] Iliano Cervesato, Nancy A. Durgin, Patrick D. Lincoln, John C. Mitchell, and Andre Scedrov. A meta-notation for protocol analysis. In R. Gorrieri, editor, *Proceedings of the 12th IEEE Computer Security Foundations Workshop — CSFW'99*, pages 55–69, Mordano, Italy, 28–30 June 1999. IEEE Computer Society Press.

[19] Manuel M. T. Chakravarty. *On the Massively Parallel Execution of Declarative Programs*. PhD thesis, Technische Universität Berlin, Fachbereich Informatik, February 1997.

[20] Jawahar Chirimar. *Proof Theoretic Approach to Specification Languages*. PhD thesis, University of Pennsylvania, February 1995.

[21] Iliano Cervesato, Joshua Hodas, and Frank Pfenning. Efficient resource management for linear logic proof search. In Roy Dyckhoff, Heinrich Herre, and Peter Schroeder-Heister, editors, *Proceedings of the 1996 Workshop on Extensions to Logic Programming*, pages 28–30, Leipzig, Germany, March 1996. Springer-Verlag LNAI.

[22] Alonzo Church. A formulation of the simple theory of types. *Journal of Symbolic Logic*, 5:56–68, 1940.

[23] Iliano Cervesato and Frank Pfenning. A linear logic framework. In *Proceedings, Eleventh Annual IEEE Symposium on Logic in Computer Science*, pages 264–275, New Brunswick, New Jersey, July 1996. IEEE Computer Society Press.

[24] Iliano Cervesato and Frank Pfenning. A Linear Logical Framework. *Information & Computation*, 179(1):19–75, November 2002.

[25] M. Dalrymple, J. Lamping, F. Pereira, and V. Saraswat. Linear logic for meaning assembly. In *Proceedings of the Workshop on Computational Logic for Natural Language Processing*, 1995.

[26] Giorgio Delzanno and Maurizio Martelli. Objects in Forum. In *Proceedings of the International Logic Programming Symposium*, 1995.

[27] U. Engberg and G. Winskel. Petri nets and models of linear logic. In A. Arnold, editor, *CAAP'90*, LNCS 431, pages 147–161. Springer Verlag, 1990.

[28] Amy Felty. Transforming specifications in a dependent-type lambda calculus to specifications in an intuitionistic logic. In Gérard Huet and Gordon D. Plotkin, editors, *Logical Frameworks*. Cambridge University Press, 1991.

[29] Amy Felty. Encoding the calculus of constructions in a higher-order logic. In M. Vardi, editor, *Eighth Annual Symposium on Logic in Computer Science*, pages 233–244. IEEE, June 1993.

[30] Amy Felty and Dale Miller. Specifying theorem provers in a higher-order logic programming language. In *Ninth International Conference on Automated Deduction*, pages 61–80, Argonne, IL, May 1988. Springer-Verlag.

[31] Franccois Fages, Paul Ruet, and Sylvain Soliman. Phase semantics and verification of concurrent constraint programs. In Vaughan Pratt, editor, *Symposium on Logic in Computer Science*. IEEE, July 1998.

[32] Didier Galmiche and E. Boudinet. Proof search for programming in intuitionistic linear logic. In D. Galmiche and L. Wallen, editors, *CADE-12 Workshop on Proof Search in Type-Theoretic Languages*, pages 24–30, Nancy, France, June 1994.

[33] Vijay Gehlot and Carl Gunter. Normal process representatives. In *Proceedings of the Fifth Annual Symposium on Logic in Computer Science*, pages 200–207, Philadelphia, Pennsylvania, June 1990. IEEE Computer Society Press.

[34] Juan C. Guzmán and Paul Hudak. Single-threaded polymorphic lambda calculus. In *Proceedings of the Fifth Annual Symposium on Logic in Computer Science*, pages 333–343, Philadelphia, Pennsylvania, June 1990. IEEE Computer Society Press.

[35] Jean-Yves Girard. Linear logic. *Theoretical Computer Science*, 50:1–102, 1987.

[36] Jean-Yves Girard. A fixpoint theorem in linear logic. Email to the linear@cs.stanford.edu mailing list, February 1992.

[37] Jean-Yves Girard. On the unity of logic. *Annals of Pure and Applied Logic*, 59:201–217, 1993.

[38] Jean-Yves Girard. Locus solum. *Mathematical Structures in Computer Science*, 11(3):301–506, June 2001.

[39] Dov M. Gabbay and Nicola Olivetti. *Goal-Directed Proof Theory*, volume 21 of *Applied Logic Series*. Kluwer Academic Publishers, August 2000.

[40] Didier Galmiche and Guy Perrier. Foundations of proof search strategies design in linear logic. In *Symposium on Logical Foundations of Computer Science*, pages 101–113, St. Petersburg, Russia, 1994. Springer-Verlag LNCS 813.

[41] Alessio Guglielmi and Lutz Straßburger. Non-commutativity and MELL in the calculus of structures. In L. Fribourg, editor, *CSL 2001*, volume 2142 of *LNCS*, pages 54–68, 2001.

[42] Alessio Guglielmi. *Abstract Logic Programming in Linear Logic—Independence and Causality in a First Order Calculus*. PhD thesis, Università di Pisa, 1996.

[43] R. Harrop. Concerning formulas of the types $A \rightarrow B \vee C$, $A \rightarrow (Ex)B(x)$ in intuitionistic formal systems. *Journal of Symbolic Logic*, pages 27–32, 1960.

[44] Robert Harper, Furio Honsell, and Gordon Plotkin. A framework for defining logics. *Journal of the ACM*, 40(1):143–184, 1993.

[45] John Hannan and Dale Miller. Uses of higher-order unification for implementing program transformers. In *Fifth International Logic Programming Conference*, pages 942–959, Seattle, Washington, August 1988. MIT Press.

[46] Joshua Hodas and Dale Miller. Representing objects in a logic programming language with scoping constructs. In David H. D. Warren and Peter Szeredi, editors, *1990 International Conference in Logic Programming*, pages 511–526. MIT Press, June 1990.

[47] Joshua Hodas and Dale Miller. Logic programming in a fragment of intuitionistic linear logic: Extended abstract. In G. Kahn, editor, *Sixth Annual Symposium on Logic in Computer Science*, pages 32–42, Amsterdam, July 1991.

[48] Joshua Hodas and Dale Miller. Logic programming in a fragment of intuitionistic linear logic. *Information and Computation*, 110(2):327–365, 1994.

[49] Joshua Hodas. Specifying filler-gap dependency parsers in a linear-logic programming language. In K. Apt, editor, *Proceedings of the Joint*

International Conference and Symposium on Logic Programming, pages 622–636, 1992.

[50] Joshua S. Hodas. *Logic Programming in Intuitionistic Linear Logic: Theory, Design, and Implementation*. PhD thesis, University of Pennsylvania, Department of Computer and Information Science, May 1994.

[51] James Harland and David Pym. Resource-distribution via boolean constraints. *ACM Transactional on Computational Logic*, 4(1):56–90, 2003.

[52] James Harland, David Pym, and Michael Winikoff. Programming in Lygon: An overview. In *Proceedings of the Fifth International Conference on Algebraic Methodology and Software Technology*, pages 391 – 405, July 1996.

[53] Lars Hallnäs and Peter Schroeder-Heister. A proof-theoretic approach to logic programming. II. Programs as definitions. *Journal of Logic and Computation*, 1(5):635–660, October 1991.

[54] Joshua S. Hodas and Naoyuki Tamura. lolliCop — A linear logic implementation of a lean connection-method theorem prover for first-order classical logic. In R. Goré, A. Leitsch, and T. Nipkow, editors, *Proceedings of IJCAR: International Joint Conference on Automated Reasoning*, number 2083 in LNCS, pages 670–684, 2001.

[55] Max Kanovich. The complexity of Horn fragments of linear logic. *Annals of Pure and Applied Logic*, 69:195–241, 1994.

[56] Stephen Cole Kleene. Permutabilities of inferences in Gentzen's calculi LK and LJ. *Memoirs of the American Mathematical Society*, 10:1–26, 1952.

[57] M.I. Kanovich, M. Okada, and A. Scedrov. Specifying real-time finite-state systems in linear logic. In *Second International Workshop on Constraint Programming for Time-Critical Applications and Multi-Agent Systems (COTIC)*, Nice, France, September 1998. Also appears in the *Electronic Notes in Theoretical Computer Science*, Volume 16 Issue 1 (1998) 15 pp.

[58] Naoki Kobayashi and Akinori Yonezawa. ACL - a concurrent linear logic programming paradigm. In Dale Miller, editor, *Logic Programming - Proceedings of the 1993 International Symposium*, pages 279–294. MIT Press, October 1993.

[59] Naoki Kobayashi and Akinori Yonezawa. Asynchronous communication model based on linear logic. *Formal Aspects of Computing*, 3:279–294, 1994.

[60] Yves Lafont. Functional programming and linear logic. Lecture notes for the Summer School on Functional Programming and Constructive Logic, Glasgow, United Kingdom, 1989.

[61] Yves Lafont. Interaction nets. In *Seventeenth Annual Symposium on Principles of Programming Languages*, pages 95–108, San Francisco, California, 1990. ACM Press.

[62] J. Lambek. The mathematics of sentence structure. *American Mathematical Monthly*, 65:154–169, 1958.

[63] Pablo López and Ernesto Pimentel. A lazy splitting system for Forum. In *AGP'97: Joint Conference on Declarative Programming*, 1997.

[64] P. Lincoln and V. Saraswat. Higher-order, linear, concurrent constraint programming. Available as

ftp://parcftp.xerox.com/pub/ccp/lcc/hlcc.dvi., January 1993.

[65] Dale Miller. A theory of modules for logic programming. In Robert M. Keller, editor, *Third Annual IEEE Symposium on Logic Programming*, pages 106–114, Salt Lake City, Utah, September 1986.

[66] Dale Miller. Lexical scoping as universal quantification. In *Sixth International Logic Programming Conference*, pages 268–283, Lisbon, Portugal, June 1989. MIT Press.

[67] Dale Miller. A logical analysis of modules in logic programming. *Journal of Logic Programming*, 6(1-2):79–108, January 1989.

[68] Robin Milner. *Communication and Concurrency*. Prentice-Hall International, 1989.

[69] Dale Miller. The π-calculus as a theory in linear logic: Preliminary results. In E. Lamma and P. Mello, editors, *Proceedings of the 1992 Workshop on Extensions to Logic Programming*, number 660 in LNCS, pages 242–265. Springer-Verlag, 1993.

[70] Dale Miller. A multiple-conclusion meta-logic. In S. Abramsky, editor, *Ninth Annual Symposium on Logic in Computer Science*, pages 272–281, Paris, July 1994. IEEE Computer Society Press.

[71] Dale Miller. A survey of linear logic programming. *Computational Logic: The Newsletter of the European Network in Computational Logic*, 2(2):63 – 67, December 1995.

[72] Dale Miller. Forum: A multiple-conclusion specification language. *Theoretical Computer Science*, 165(1):201–232, September 1996.

[73] Dale Miller. Encryption as an abstract data-type: An extended abstract. In Iliano Cervesato, editor, *Proceedings of FCS'03: Foundations of Computer Security*, pages 3–14, 2003.

[74] Raymond McDowell and Dale Miller. A logic for reasoning with higher-order abstract syntax. In Glynn Winskel, editor, *Proceedings, Twelfth Annual IEEE Symposium on Logic in Computer Science*, pages 434–445, Warsaw, Poland, July 1997. IEEE Computer Society Press.

[75] Raymond McDowell and Dale Miller. Cut-elimination for a logic with definitions and induction. *Theoretical Computer Science*, 232:91–119, 2000.

[76] Raymond McDowell and Dale Miller. Reasoning with higher-order abstract syntax in a logical framework. *ACM Transactions on Computational Logic*, 3(1):80–136, January 2002.

[77] Raymond McDowell, Dale Miller, and Catuscia Palamidessi. Encoding transition systems in sequent calculus. *TCS*, 294(3):411–437, 2003.

[78] Dale Miller and Gopalan Nadathur. Higher-order logic programming. In Ehud Shapiro, editor, *Proceedings of the Third International Logic Programming Conference*, pages 448–462, London, June 1986.

[79] Dale Miller and Gopalan Nadathur. A logic programming approach to manipulating formulas and programs. In Seif Haridi, editor, *IEEE Symposium on Logic Programming*, pages 379–388, San Francisco, September 1987.

[80] Dale Miller, Gopalan Nadathur, Frank Pfenning, and Andre Scedrov. Uniform proofs as a foundation for logic programming. *Annals of Pure and Applied Logic*, 51:125–157, 1991.

[81] Michael Moortgat. Categorial type logics. In Johan van Benthem and Alice ter Meulen, editors, *Handbook of Logic and Language*, pages 93–177. Elsevier, Amsterdam, 1996.

[82] Glyn Morrill. Higher-order linear logic programming of categorial deduction. In *7th Conference of the Association for Computational Linguistics*, pages 133–140, Dublin, Ireland, 1995.

[83] John Maraist, Martin Odersky, David N. Turner, and Philip Wadler. Call-by-name, call-by-value, call-by-need and the linear lambda calculus. *Theoretical Computer Science*, 228(1–2):175–210, 1999.

[84] Dale Miller and Elaine Pimentel. Using linear logic to reason about sequent systems. In Uwe Egly and Christian G. Fermüller, editors, *International Conference on Automated Reasoning with Analytic Tableaux and Related Methods*, volume 2381 of *Lecture Notes in Computer Science*, pages 2–23. Springer, 2002.

[85] Robin Milner, Joachim Parrow, and David Walker. A calculus of mobile processes, Part I. *Information and Computation*, pages 1–40, September 1992.

[86] Gopalan Nadathur. *A Higher-Order Logic as the Basis for Logic Programming*. PhD thesis, University of Pennsylvania, May 1987.

[87] Gopalan Nadathur, Bharat Jayaraman, and Keehang Kwon. Scoping constructs in logic programming: Implementation problems and their solution. *Journal of Logic Programming*, 25(2):119–161, November 1995.

[88] Gopalan Nadathur and Dale Miller. An Overview of λProlog. In *Fifth International Logic Programming Conference*, pages 810–827, Seattle, August 1988. MIT Press.

[89] Gopalan Nadathur and Dale Miller. Higher-order Horn clauses. *Journal of the ACM*, 37(4):777–814, October 1990.

[90] Gopalan Nadathur and Dale Miller. Higher-order logic programming. In Dov M. Gabbay, C. J. Hogger, and J. A. Robinson, editors, *Handbook of Logic in Artificial Intelligence and Logic Programming*, volume 5, pages 499 – 590. Clarendon Press, Oxford, 1998.

[91] P. W. O'Hearn. Linear logic and interference control: Preliminary report. In S. Abramsky, P.-L. Curien, A. M. Pitts, D. H. Pitt, A. Poigné, and D. E. Rydeheard, editors, *Proceedings of the Conference on Category Theory and Computer Science*, pages 74–93, Paris, France, 1991. Springer-Verlag LNCS 530.

[92] P. O'Hearn and D. Pym. The logic of bunched implications. *Bulletin of Symbolic Logic*, 5(2):215–244, June 1999.

[93] Remo Pareschi. *Type-driven Natural Language Analysis*. PhD thesis, University of Edinburgh, 1989.

[94] Frank Pfenning and Conal Elliott. Higher-order abstract syntax. In *Proceedings of the ACM-SIGPLAN Conference on Programming Language Design and Implementation*, pages 199–208. ACM Press, June 1988.

[95] David J. Pym and James A. Harland. The uniform proof-theoretic foundation of linear logic programming. *Journal of Logic and Computation*, 4(2):175 – 207, April 1994.

[96] Remo Pareschi and Dale Miller. Extending definite clause grammars with scoping constructs. In David H. D. Warren and Peter Szeredi, editors, *1990 International Conference in Logic Programming*, pages 373–389. MIT Press, June 1990.

[97] Jeff Polakow. *Ordered Linear Logic and Applications*. PhD thesis, Department of Computer Science, Caregnie Mellon, August 2001.

[98] Jeff Polakow and Frank Pfenning. Natural deduction for intuitionistic

non-commutative linear logic. In J.-Y. Girard, editor, *Proceedings of the 4th International Conference on Typed Lambda Calculi and Applications (TLCA'99)*, pages 295–309, L'Aquila, Italy, 1999. Springer-Verlag LNCS 1581.

[99] Jeff Polakow and Frank Pfenning. Relating natural deduction and sequent calculus for intuitionistic non-commutative linear logic. In Andre Scedrov and Achim Jung, editors, *Proceedings of the 15th Conference on Mathematical Foundations of Programming Semantics*, New Orleans, Louisiana, 1999.

[100] Jeff Polakow and Kwangkeun Yi. Proving syntactic properties of exceptions in an ordered logical framework. In *Fifth International Symposium on Functional and Logic Programming (FLOPS 2001)*, Tokyo, Japan, March 2001.

[101] David J. Pym. On bunched predicate logic. In G. Longo, editor, *Proceedings of LICS99: 14th Annual Symposium on Logic in Computer Science*, pages 183–192, Trento, Italy, 1999. IEEE Computer Society Press.

[102] Christian Retoré. Pomset logic: a non-commutative extension of classical linear logic. In *Proceedings of TLCA*, volume 1210, pages 300–318, 1997.

[103] Christian Retoré. Pomset logic as a calculus of directed cographs. In V. M. Abrusci and C. Casadio, editors, *Dynamic Perspectives in Logic and Linguistics: Proof Theoretical Dimensions of Communication Processes*, pages 221–247, 1999.

[104] P. Ruet and F. Fages. Concurrent constraint programming and non-commutative logic. In *Proceedings of the 11^{t}h Conference on Computer Science Logic*, Lecture Notes in Computer Science. Springer-Verlag, 1997.

[105] Giorgia Ricci. *On the expressive powers of a Logic Programming presentation of Linear Logic (FORUM)*. PhD thesis, Department of Mathematics, Siena University, December 1998.

[106] Paul Ruet. Non-commutative logic II: sequent calculus and phase semantics. *Mathematical Structures in Computer Science*, 10(2):277–312, 2000.

[107] V. Saraswat. A brief introduction to linear concurrent constraint programming. Available as ftp://parcftp.xerox.com/pub/ccp/lcc/lcc-intro.dvi., 1993.

[108] Peter Schroeder-Heister. Rules of definitional reflection. In M. Vardi, editor, *Eighth Annual Symposium on Logic in Computer Science*, pages 222–232. IEEE Computer Society Press, June 1993.

[109] T. Tammet. Proof strategies in linear logic. *Journal of Automated Reasoning*, 12:273–304, 1994.

[110] Naoyuki Tamura and Yukio Kaneda. Extension of wam for a linear logic programming language. In T. Ida, A. Ohori, and M. Takeichi, editors, *Second Fuji International Workshop on Functional and Logic Programming*, pages 33–50. World Scientific, November 1996.

[111] Philip Wadler. Linear types can change the world! In *Programming Concepts and Methods*, pages 561–581. North Holland, 1990.

[112] M. Winikoff and J. Harland. Implementing the Linear Logic Programming Language Lygon. In *Proceedings of the International Logic Programming Symposium*, pages 66–80, December 1995.

4

Linearity and Nonlinearity in Distributed Computation

Glynn Winskel
Cambridge University Computer Laboratory

Abstract

The copying of processes is limited in the context of distributed computation, either as a fact of life, often because remote networks are simply too complicated to have control over, or deliberately, as in the design of security protocols. Roughly, linearity is about how to manage without a presumed ability to copy. The meaning and mathematical consequences of linearity are studied for path-based models of processes which are also models of affine-linear logic. This connection yields an affine-linear language for processes in which processes are typed according to the kind of computation paths they can perform. One consequence is that the affine-linear language automatically respects open-map bisimulation. A range of process operations (from CCS, CCS with process-passing, mobile ambients, and dataflow) can be expressed within the affine-linear language showing the ubiquity of linearity. Of course, process code can be sent explicitly to be copied. Following the discipline of linear logic, suitable nonlinear maps are obtained as linear maps whose domain is under an exponential. Different ways to make assemblies of processes lead to different choices of exponential; the nonlinear maps of only some of which will respect bisimulation.

4.1 Introduction

In the film "Groundhog Day" the main character comes to relive and remember the same day repeatedly, until finally he gets it right (and the girl). Real life isn't like that. The world moves on and we cannot rehearse, repeat or reverse the effects of the more important decisions we take.

As computation becomes increasingly distributed and interactive the more it resembles life in this respect, and the more difficult or impossible it is for a state of computation to be frozen and copied, so that interaction is most often conducted in a single irreversible run. Mathematically, this amounts to a form of linearity, as it is understood within models of linear logic.

Although it can be hard for processes to copy processes, it is generally easy for processes to ignore other processes. For this reason distributed computation often involves affine-linear maps.

To get an idea of the nature of affine-linear maps, imagine a network of interacting processes with a single hole, into which a process, the input process, may be plugged to form a complete process, the output process. The network with the hole represents an affine-linear map; given an input process, one obtains an output process.

The input process cannot be copied, but can be ignored. Consequently, any computation path (or run) of the output process will have depended on at most one computation path (or run) of the input. This property of affine-linear maps rests on an understanding of the nature of the computation paths of a process. The property can be simplified provided we understand a computation path of the input processes to also admit the empty computation path—a computation path obtained even when the input process is ignored. Any output got while ignoring the input process, results from the empty computation path as input. An affine-linear map has the following property:

A computation path of the process arising from the application of an affine-linear map to an input process has resulted from a single computation path, possibly empty, of the input process.

As this suggests, an affine-linear map is determined by its action on single, possibly empty, computation paths.

This article presents models of processes based on computation paths and so can make precise the sense in which many operations on distributed systems are associated with affine-linear maps, investigates the consequences of linearity and affine linearity for the important equivalence of bisimulation, and delineates the boundaries of linearity with respect to one, fairly broad, mathematical model, in which non-deterministic processes are represented as presheaves.

Of course, sometimes code can be sent and copied, which can give rise to maps which are not affine-linear. The presheaf model exposes how different manners of copying lead to different kinds of nonlinear maps, some respecting bisimulation, others not.

4.2 Path-based models of processes

Consider processes, like those of CCS [24] and CSP [16], which can perform simple atomic actions, among which might be actions of synchronisations. An old idea is to represent the nondeterministic behaviour of a such a process as a "collection" of the computation paths it can perform. If we interpret this idea literally, and may assume that actions occur one at a time, we arrive at one of the early models of processes as *sets* of traces, where a trace is a finite sequence of actions that the process can perform [16]. It was realised very quickly that a problem with the trace model is that it is blind to deadlock; two processes may have the same trace sets and yet one may deadlock while the other does not (see for instance the discussion in [23], Ch.1). To detect possible deadlock, one way or another, one needs to keep track of nondeterministic branching in the representation of processes. An early proposal on how to do this is to represent a process as a tree, where branching stands for nondeterministic choice (*cf.* the synchronisation trees of [23]). The tree is still a sort of "collection" of the process's computation paths but one keeping track of where the paths overlap, through sharing a common subpath as history. Now the model is too concrete; many different trees represent what seems to be esentially the same behaviour, and this led to equivalences such as bisimulation on trees and transition systems [24].

The trace and tree models of processes are based on different ideas of what a "collection" of the computation paths means. A trace set of a process simply expresses whether or not a path of a certain shape is possible for the process. A tree expresses not only what paths are present but also how paths are subpaths, or restrictions, of others. This data, what paths are present and how they restrict to smaller paths, is precisely that caught in a presheaf over a category, a category in which the objects are path shapes and the maps express how one path shape can extend to another. In the category of all such presheaves we can view the tree as a *colimit* of its paths—another kind of "collection" of its paths.

To illustrate the idea, suppose that actions are drawn from some alphabet L, and consider processes whose computation paths have the shape of strings of actions, so members of L^*. A subpath will be associated with a substring. Regard L^* as a partial order where $s \leq t$ iff s is a substring of t. (So L^* is also a category where we view there as being an arrow from s to t precisely when $s \leq t$.)

A presheaf over L^* is a functor from the opposite category $(L^*)^{op}$, where all the arrows are reversed, to the category of sets and functions **Set**. Spelt out, a presheaf X over L^* is a function which to each string s gives a set $X(s)$ and which to any pair of strings (s, t), with $s \leq t$, gives a function $X(s, t) : X(t) \longrightarrow X(s)$ (note the reversal) in such a way that identities and composition are respected: $X(s, s) = 1_{X(s)}$ for each string s, and $X(s, u) = X(s, t) \circ X(t, u)$ whenever $s \leq t \leq u$.

When thinking of a presheaf X as representing a process, for a string s, the set $X(s)$ is the set of computation paths of shape s that the process can perform, and, when $s \leq t$, the function $X(s, t) : X(t) \longrightarrow X(s)$ tells how paths of shape t restrict to subpaths of shape s. For example, a tree whose branches consist of strings of actions in L is easily viewed as a presheaf X over L^*. The set $X(s)$ consists of all the branches of shape s. The function $X(s, t) : X(t) \longrightarrow X(s)$ restricts a branch of shape t to its sub-branch of shape s. The presheaf is rooted in the sense that the set $X(\epsilon)$ assigned to the empty string ϵ is a singleton—its only element is the root of the tree. Conversely, it is easy to see that a rooted presheaf over L^* determines a tree.

Suppose that we replace the category of sets used in the definition of presheaves by the simple subcategory **2**, consisting of two distinct elements, the empty set, $\emptyset$, and the singleton set, **1**, with the only non-identity map being $\emptyset \subseteq \mathbf{1}$. A functor X from $(L^*)^{op}$ to **2** is the same as a monotonic function from the reverse order $(L^*)^{op}$ to the order **2**, so that if $s \leq t$ then $X(t) \leq X(s)$. When thinking of X as representing a process, $X(s) = \mathbf{1}$ means that the process can perform a path of shape s while $X(s) = \emptyset$ means that it cannot. If $X(t) = \mathbf{1}$ and $s \leq t$, then $X(s) = \mathbf{1}$. The functor X is a characteristic function for a trace set.

So trees and trace sets arise as variants of a common idea, that of representing a process as a (generalised) characteristic function, in the form of a functor from path shapes to measures of the extent to which the path shapes can be realised by the process.

In what follows, we want to broaden computation paths to have more general shapes than sequences of atomic actions, to allow actions to occur concurrently in a computation path, and for individual actions to have a more complicated structure. Later on, processes will be allocated types; the type of a process will specify the shapes of computation path it might perform.

4.3 Processes as trace sets

To allow a broad understanding of the shape of computation paths, we take a path order to be a partial order $\mathbb{P}$ in which the elements are path shapes and the order $p \leq q$ means that p can be extended to q. We obtain a form of nondeterministic domain by imitating the definition of presheaf category, but replacing the use of the category of sets, **Set**, by its much simpler subcategory, **2**. The category **2** is essentially a very simple partial order consisting of $\emptyset$ and **1** ordered by $\emptyset \subseteq \mathbf{1}$. A partial order $\mathbb{P}$ can always be regarded as a category by taking the homset $\mathbb{P}(p, q)$ to be **1** if $p \leq q$, and $\emptyset$ otherwise. Functors between partial orders seen as categories correspond precisely to monotonic maps.

The functor category $\widehat{\mathbb{P}} = [\mathbb{P}^{op}, \mathbf{2}]$ consists of objects the functors from $\mathbb{P}^{op}$ to **2** and maps the natural transformations between them. A functor from $\mathbb{P}^{op}$ to **2** is essentially a monotonic function from $\mathbb{P}^{op}$ to **2**. It is not hard to see that an object X of $\widehat{\mathbb{P}}$ corresponds to a downwards-closed set given by $\{p \in \mathbb{P} \mid X(p) = \mathbf{1}\}$, and that a natural transformation from X to Y in $\widehat{\mathbb{P}}$ corresponds to the inclusion of $\{p \in \mathbb{P} \mid X(p) = \mathbf{1}\}$ in $\{p \in \mathbb{P} \mid Y(p) = \mathbf{1}\}$. So we can identify $\widehat{\mathbb{P}}$ with the partial order of downwards-closed subsets of $\mathbb{P}$, ordered by inclusion; the order $\widehat{\mathbb{P}}$ has joins got simply via unions with the empty join being the least element $\emptyset$. The partial orders obtained in this way are precisely the infinitely-distributive algebraic lattices (see *e.g.*, [27, 28]) and these are just the same as prime algebraic lattices [25], and free join completions of partial orders.

We are thinking of $\widehat{\mathbb{P}}$ as a nondeterministic domain [15, 14]. An object in $\widehat{\mathbb{P}}$ is thought of as a denotation of a nondeterministic process which can realise path shapes in $\mathbb{P}$. An object in $\widehat{\mathbb{P}}$ is a trace set, like those originally, in [16], but for general path shapes, standing for the set of computation paths a process can perform. The join operation on $\widehat{\mathbb{P}}$ is a form of nondeterministic sum.

This suggests that we take maps between nondeterministic domains to be join preserving functions, the choice dealt with in the next section. We will however be forced a little beyond this first mathematically obvious choice.

4.3.1 A linear category of domains

We mentioned that $\widehat{\mathbb{P}}$ is a free join completion of a partial order $\mathbb{P}$. We spell this out. There is a monotonic map $y_{\mathbb{P}} : \mathbb{P} \longrightarrow \widehat{\mathbb{P}}$ which on p yields

$y_{\mathbb{P}}(p) = \mathbb{P}(-, p)$; for $p' \in \mathbb{P}$,

$$y_{\mathbb{P}}(p)(p') = \mathbf{1} \text{ if } p' \le p, \text{ and } \emptyset \text{ otherwise.}$$

The map $y_{\mathbb{P}}$ satisfies the universal property that for any monotonic map $F : \mathbb{P} \to \mathcal{E}$, where $\mathcal{E}$ is a partial order with all joins, there is a unique join-preserving map $G : \widehat{\mathbb{P}} \to \mathcal{E}$ such that $F = G \circ y_{\mathbb{P}}$:

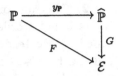

The proof of the universal property hinges on the fact that every object of $\widehat{\mathbb{P}}$ is the join of the "complete primes", objects $y_{\mathbb{P}}(p)$, below it. We will use an "inner product" notation and describe G above as taking $X \in \widehat{\mathbb{P}}$ to $F \cdot X$.

We can, in particular, instantiate $\mathcal{E}$ to a nondeterministic domain $\widehat{\mathbb{Q}}$, which certainly has all joins. By the universal property, monotonic maps $F : \mathbb{P} \longrightarrow \widehat{\mathbb{Q}}$ are in 1-1 correspondence with join-preserving maps $G : \widehat{\mathbb{P}} \longrightarrow \widehat{\mathbb{Q}}$. But monotonic maps $F : \mathbb{P} \to \widehat{\mathbb{Q}}$ are just the same as monotonic maps $F : \mathbb{P} \to [\mathbb{Q}^{op}, \mathbf{2}]$ and, uncurrying, these correspond to monotonic maps $H : \mathbb{P} \times \mathbb{Q}^{op} \to \mathbf{2}$ and so to objects of $\widehat{\mathbb{P}^{op} \times \mathbb{Q}} = [(\mathbb{P}^{op} \times \mathbb{Q})^{op}, \mathbf{2}]$.

So, on mathematical grounds it is natural to consider taking maps between nondeterministic domains as functions which preserve all joins. Such functions (often known as *additive* functions) compose as usual, have identities and give rise to a category rich in structure. Call this category $\mathbf{Lin_2}$; it consists of objects partial orders $\mathbb{P}$, $\mathbb{Q}$, $\cdots$, with maps $G : \mathbb{P} \to \mathbb{Q}$ the join-preserving functions from $\widehat{\mathbb{P}}$ to $\widehat{\mathbb{Q}}$.

As we have just seen, we can regard a map from $\mathbb{P}$ to $\mathbb{Q}$ in $\mathbf{Lin_2}$ in several ways and, in particular, as an object of $\widehat{\mathbb{P}^{op} \times \mathbb{Q}}$. Thus a map in $\mathbf{Lin_2}$ corresponds to a downward-closed subset of $\mathbb{P}^{op} \times \mathbb{Q}$ and so can be viewed as a relation between the partial orders $\mathbb{P}$ and $\mathbb{Q}$, with composition now given as the usual composition of relations.

This more symmetric, relational presentation exposes an involution central in understanding $\mathbf{Lin_2}$ as a categorical model of classical linear logic. The involution of linear logic, yielding $\mathbb{P}^{\perp}$ on an object $\mathbb{P}$, is given by $\mathbb{P}^{op}$; clearly downward-closed subsets of $\mathbb{P}^{op} \times \mathbb{Q}$ correspond to downward-closed subsets of $(\mathbb{Q}^{op})^{op} \times \mathbb{P}^{op}$, showing how maps $F : \mathbb{P} \longrightarrow \mathbb{Q}$ correspond to maps $F^{\perp} : \mathbb{Q}^{op} \longrightarrow \mathbb{P}^{op}$ in $\mathbf{Lin_2}$. The tensor product of $\mathbb{P}$ and $\mathbb{Q}$ is given by the product of partial orders $\mathbb{P} \times \mathbb{Q}$

and the function space from $\mathbb{P}$ to $\mathbb{Q}$ by $\mathbb{P}^{op} \times \mathbb{Q}$. On objects $\mathbb{P}$ and $\mathbb{Q}$, products and coproducts are both given by $\mathbb{P} + \mathbb{Q}$, the disjoint juxtaposition of $\mathbb{P}$ and $\mathbb{Q}$. One choice of interpretation of the exponential of linear logic is got by taking $!\mathbb{P}$, for a partial order $\mathbb{P}$, to be the partial order obtained as the restriction of $\widehat{\mathbb{P}}$ to its finite (or isolated) elements. (An element of a partial order is finite if whenever it is dominated by a directed join, it is dominated by an element of the join.) The partial order $\widehat{\mathbb{P}}$, with the inclusion $!\mathbb{P} \longrightarrow \widehat{\mathbb{P}}$, is the free closure of $!\mathbb{P}$ under directed joins (the "ideal completion" of $!\mathbb{P}$.) Consequently, there is 1-1 correspondence between linear maps from $!\mathbb{P}$ to $\mathbb{Q}$ in $\mathbf{Lin_2}$ and Scott continuous (*i.e.*, directed-join preserving) functions from $\widehat{\mathbb{P}}$ to $\widehat{\mathbb{Q}}$. In fact, $!(-)$ extends straightforwardly to a comonad on $\mathbf{Lin_2}$ whose coKleisli category is isomorphic to the category of Scott continuous functions between nondeterministic domains.

Linear maps preserve joins. The join of the empty set is $\emptyset$, to be thought of as a *nil* process, which is unable to perform any computation path. So linear maps always send the *nil* process to the *nil* process. Going back to the intuitions in the introduction, if a network context gave rise to a linear map, then plugging a dead process into the network would always be catastrophic, and lead to the whole network going dead. We could extend to maps from $!\mathbb{P}$ to $\mathbb{Q}$, for objects $\mathbb{P}$ and $\mathbb{Q}$ in $\mathbf{Lin_2}$, but by the properties of the exponential, this would allow arbitrary copying of the argument process. All we often need is to allow maps to ignore their arguments and this can be got much more cheaply, by moving to a model of affine linear logic.

4.3.2 An affine-linear category of domains

A common operation in process algebras is that of prefixing a process by an action, so that a computation path of the prefixed process consists of first performing the action and then resuming by following a computation path of the original process. To understand prefix operations, we first need to lift path shapes by an initial action.

The operation of *lifting* on a partial order $\mathbb{P}$ produces a partial order $\mathbb{P}_\perp$, got by adjoining a new element $\perp$ below a copy of $\mathbb{P}$. Denote by $\lfloor p \rfloor$ the copy in $\mathbb{P}_\perp$ of the original element p in $\mathbb{P}$—each $\lfloor p \rfloor$ is assumed distinct from $\perp$. The order of $\mathbb{P}_\perp$ is given by $\perp \leq \lfloor p \rfloor$, for any $p \in \mathbb{P}$, with $\lfloor p \rfloor \leq \lfloor p' \rfloor$ iff $p \leq p'$ initially in $\mathbb{P}$.

Prefixing operations on processes make essential use of an operation associated with lifting. The operation is the function

$$\lfloor - \rfloor : \widehat{\mathbb{P}} \to \widehat{\mathbb{P}_\perp}$$

such that $\lfloor X \rfloor(\perp) = \mathbf{1}$ and $\lfloor X \rfloor(\lfloor p \rfloor) = X(p)$ for $X \in \widehat{\mathbb{P}}$. The function $\lfloor - \rfloor$ is not a map from $\mathbb{P}$ to $\mathbb{P}_\perp$ in $\mathbf{Lin_2}$ as it clearly does not preserve joins of the empty set. It does however preserve all nonempty joins (*i.e.*, joins of nonempty sets). This provides a clue as to how to expand the maps of $\mathbf{Lin_2}$.

To accommodate the functions $\lfloor - \rfloor$ we move to a slightly broader category, though fortunately one that inherits a good many properties from $\mathbf{Lin_2}$. The category $\mathbf{Aff_2}$ has the same objects, partial orders, but its maps from $\mathbb{P}$ to $\mathbb{Q}$, written $F : \mathbb{P} \to \mathbb{Q}$, are functions $F : \widehat{\mathbb{P}} \to \widehat{\mathbb{Q}}$ which need only preserve *nonempty* joins.

A map $F : \mathbb{P} \to \mathbb{Q}$ in $\mathbf{Aff_2}$ is really a nonempty-join preserving function from $\widehat{\mathbb{P}}$ to $\widehat{\mathbb{Q}}$, so takes a (denotation of a) nondeterministic process with computation paths in $\mathbb{P}$ as input and yields a (denotation of a) non-derministic process with computation paths in $\mathbb{Q}$ as output. Because the map need only preserve nonempty joins it is at liberty to ignore the input process in giving non-trivial output. Because the map preserves all nonempty joins, if a computation path of the resulting output process requires computation of the input process, then it only requires a single computation path of the input process. The input process does not need to be copied to explore the range of computation paths it might follow.

A map in $\mathbf{Aff_2}$ is determined by its action on computation paths extended to include the empty path $\perp$. There is an embedding $j_\mathbb{P} :$ $\mathbb{P}_\perp \to \widehat{\mathbb{P}}$ which takes $\perp$ to $\emptyset$ and any $\lfloor p \rfloor$ to $y_\mathbb{P}(p)$. The map $j_\mathbb{P}$ satisfies the universal property that for any monotonic map $F : \mathbb{P}_\perp \to \mathcal{E}$, where $\mathcal{E}$ is a partial order with all nonempty joins, there is a unique nonempty-join preserving map $F^\dagger : \widehat{\mathbb{P}} \to \mathcal{E}$ such that $F = F^\dagger \circ j_\mathbb{P}$:

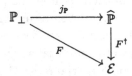

The proof of the universal property rests on the fact that every $X \in \widehat{\mathbb{P}}$ is the nonempty join of the set consisting of all $j_\mathbb{P}(p')$, where $p' \in \mathbb{P}_\perp$, that X dominates—this set is nonempty because it contains $j_\mathbb{P}(\perp)$ $(= \emptyset)$.

By instantiating $\mathcal{E}$ to $\widehat{\mathbb{Q}}$ where $\mathbb{Q}$ is a partial order, we see that maps

$\mathbb{P} \to \mathbb{Q}$ in $\mathbf{Aff_2}$ are in 1-1 correspondence with maps $\mathbb{P}_\perp \to \mathbb{Q}$ in $\mathbf{Lin_2}$, and so to elements in $(\widehat{\mathbb{P}_\perp)^{op}} \times \mathbb{Q}$.

There is a unique (and obvious) way to extend the lifting operation $(-)_\perp$ to a functor from $\mathbf{Aff_2}$ to $\mathbf{Lin_2}$ so that the correspondence

$$\mathbf{Aff_2}(\mathbb{P}, \mathbb{Q}) \cong \mathbf{Lin_2}(\mathbb{P}_\perp, \mathbb{Q})$$

is natural in $\mathbb{P} \in \mathbf{Aff_2}$ and $\mathbb{Q} \in \mathbf{Lin_2}$. This exhibits the functor $(-)_\perp$ as a left adjoint to the inclusion functor $\mathbf{Lin_2} \hookrightarrow \mathbf{Aff_2}$. Composing the two adjoints, we obtain a comonad $(-)_\perp$ on $\mathbf{Lin_2}$ whose coKliesli category is isomorphic to $\mathbf{Aff_2}$. Clearly, a map in $\mathbf{Aff_2}$ also belongs to the subcategory $\mathbf{Lin_2}$ iff it is strict in the sense of preserving $\emptyset$.

With the help of the comonad $(-)_\perp$ we have turned the model of linear logic $\mathbf{Lin_2}$ into a model $\mathbf{Aff_2}$ of *affine* linear logic (a model of intuitionistic linear logic in which the structural rule of weakening is satisfied through the unit of the tensor also being a terminal object—[17]). Its operations are defined in terms of the corresponding operations in $\mathbf{Lin_2}$. For example, its tensor $\otimes$ is defined so that

$$(\mathbb{P} \otimes \mathbb{Q})_\perp = \mathbb{P}_\perp \times \mathbb{Q}_\perp \,,$$

a product, and so tensor in $\mathbf{Lin_2}$, of partial orders $\mathbb{P}_\perp$ and $\mathbb{Q}_\perp$. Its function space is given by $(\mathbb{P}_\perp)^{op} \times \mathbb{Q}$. The category $\mathbf{Aff_2}$ has the same products as $\mathbf{Lin_2}$. It does not have coproducts, though we will later see a form of prefixed sum which is useful in giving semantics to process languages.

4.4 Processes as presheaves

In order to take account of the branching structure of nondeterministic processes we move from a representation of a process as characteristic function from computation-path shapes to $\emptyset$ or $\mathbf{1}$ in the partial order $\mathbf{2}$, and explore the variation where we measure the presence of a computation-path shape in a process by a set in the category $\mathbf{Set}$ of sets.

It is useful for later to broaden our understanding of shapes of computation paths to be objects in a small category $\mathbb{P}$. In our applications, the category $\mathbb{P}$ is thought of as a path category, consisting of shapes of paths, where a map $e : p \to p'$ expresses how the path p extends to the path p'. Let $\mathbb{P}$ be a small category. The category of *presheaves over* $\mathbb{P}$, written $\widehat{\mathbb{P}}$, is the category $[\mathbb{P}^{op}, \mathbf{Set}]$ with objects the functors from $\mathbb{P}^{op}$

(the opposite category) to the category of sets, and maps the natural transformations between them.

A presheaf $X : \mathbb{P}^{op} \to \mathbf{Set}$ specifies for a typical path p the set $X(p)$ of computation paths of shape p. The presheaf X acts on a map $e : p \to p'$ in $\mathbb{P}$ to give a function $X(e)$ saying how p'-paths in X restrict to p-paths in X—several paths may restrict to the same path. In this way a presheaf can model the nondeterministic branching of a process.

A presheaf category has all limits and colimits given pointwise, at a particular object, by the corresponding limits or colimits of sets. In particular, a presheaf category has all sums (coproducts) of presheaves; the sum $\Sigma_{i \in I} X_i$ of presheaves X_i, $i \in I$, over $\mathbb{P}$ has a contribution $\Sigma_{i \in I} X_i(p)$, the disjoint union of sets, at p an object of $\mathbb{P}$. The empty sum of presheaves is the presheaf $\emptyset$ with empty contribution at each p in $\mathbb{P}$. In process terms, a sum of presheaves represents a nondeterministic sum of processes.

4.4.1 A linear category of presheaf models

A category of presheaves, $\widehat{\mathbb{P}}$, is accompanied by the *Yoneda embedding*, a functor $y_{\mathbb{P}} : \mathbb{P} \to \widehat{\mathbb{P}}$, which fully and faithfully embeds $\mathbb{P}$ in the category of presheaves. For every object p of $\mathbb{P}$, the Yoneda embedding yields $y_{\mathbb{P}}(p) = \mathbb{P}(-, p)$. Presheaves isomorphic to images of objects of $\mathbb{P}$ under the Yoneda embedding are called *representables*.

Via the Yoneda embedding we can regard $\mathbb{P}$ essentially as a full subcategory of $\widehat{\mathbb{P}}$. Moreover $\widehat{\mathbb{P}}$ is characterized (up to equivalence of categories) as the free colimit completion of $\mathbb{P}$. In other words, the Yoneda embedding $y_{\mathbb{P}}$ satisfies the universal property that for any functor $F : \mathbb{P} \to \mathcal{E}$, where $\mathcal{E}$ is a category with all colimits, there is a colimit preserving functor $G : \widehat{\mathbb{P}} \to \mathcal{E}$, determined to within isomorphism, such that $F \cong G \circ y_{\mathbb{P}}$:

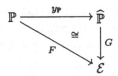

The proof rests on the fact that any presheaf is a colimit of representables—see *e.g.*, [22] P.43. We will describe G above as $F \cdot -$.

In particular, we can take $\mathcal{E}$ to be a presheaf category $\widehat{\mathbb{Q}}$. As the universal property suggests, colimit-preserving functors between presheaf categories are useful. Define the category $\mathbf{Lin_{Set}}$ to consist of small categories $\mathbb{P}$, $\mathbb{Q}$, $\cdots$, with maps $G : \mathbb{P} \longrightarrow \mathbb{Q}$ the colimit-preserving functors from $\widehat{\mathbb{P}}$ to $\widehat{\mathbb{Q}}$.

By the universal property, colimit-preserving functors $G : \widehat{\mathbb{P}} \longrightarrow \widehat{\mathbb{Q}}$ correspond to within isomorphism to functors $F : \mathbb{P} \longrightarrow \widehat{\mathbb{Q}}$, and such functors are in 1-1 correspondence with profunctors $\bar{F} : \mathbb{P} \longrightarrow \mathbb{Q}$. Recall that the category of profunctors from $\mathbb{P}$ to $\mathbb{Q}$, written $\mathbf{Prof}(\mathbb{P}, \mathbb{Q})$, is the functor category $[\mathbb{P} \times \mathbb{Q}^{op}, \mathbf{Set}]$, which clearly equals the category of presheaves $\widehat{\mathbb{P}^{op} \times \mathbb{Q}}$, and is isomorphic to the functor category $[\mathbb{P}, \widehat{\mathbb{Q}}]$. We thus have the chain of equivalences:

$$\mathbf{Lin_{Set}}(\mathbb{P}, \mathbb{Q}) \simeq [\mathbb{P}, \widehat{\mathbb{Q}}] \cong \mathbf{Prof}(\mathbb{P}, \mathbb{Q}) = \widehat{\mathbb{P}^{op} \times \mathbb{Q}} \, .$$

Exactly analogously with the domain model $\mathbf{Lin_2}$ in Section 4.3.1, one can build a model of linear logic out of $\mathbf{Lin_{Set}}$, though there are now subtleties, as what were previously functors must now be pseudo functors (preserving composition only up to coherent isomorphism). In particular, the involution of linear logic takes a map $F : \mathbb{P} \longrightarrow \mathbb{Q}$ to a map $F^{\perp} : \mathbb{Q}^{op} \longrightarrow \mathbb{P}^{op}$ in $\mathbf{Lin_{Set}}$ via

$$\mathbf{Lin_{Set}}(\mathbb{P}, \mathbb{Q}) \simeq \widehat{\mathbb{P}^{op} \times \mathbb{Q}} \cong \widehat{(\mathbb{Q}^{op})^{op} \times \mathbb{P}^{op}} \simeq \mathbf{Lin_{Set}}(\mathbb{Q}^{op}, \mathbb{P}^{op}) \, .$$

On objects, the tensor product of $\mathbb{P}$ and $\mathbb{Q}$ is given by the product of categories $\mathbb{P} \times \mathbb{Q}$ and the function space from $\mathbb{P}$ to $\mathbb{Q}$ by $\mathbb{P}^{op} \times \mathbb{Q}$. On objects $\mathbb{P}$ and $\mathbb{Q}$, products and coproducts are both given by $\mathbb{P} + \mathbb{Q}$, the sum of categories $\mathbb{P}$ and $\mathbb{Q}$. As for the exponential ! of linear logic, there are many possible choices—see Section 4.8.2.

Just as in the domain case, maps in $\mathbf{Lin_{Set}}$ are most often too restrictive. Maps in $\mathbf{Lin_{Set}}$ preserve colimits and, in particular, sums and the empty colimit $\emptyset$, a property which is only the case for rather special operations on processes.

4.4.2 An affine-linear category of presheaf models

Many operations associated with process languages do not preserve sums, so arbitrary colimits. Prefixing operations only preserve connected colimits (colimits of nonempty connected diagrams). Prefixing operations derive from the functor $\lfloor - \rfloor : \widehat{\mathbb{P}} \longrightarrow \widehat{\mathbb{P}_{\perp}}$. The lifted category $\mathbb{P}_{\perp}$ comprises a copy of $\mathbb{P}$—the copy in $\mathbb{P}_{\perp}$ of the original object p is written $\lfloor p \rfloor$—to which a new initial object $\perp$ has been adjoined. (This definition extends to categories the earlier definition of lifting on partial orders, Section 4.3.2.) The functor $\lfloor - \rfloor : \widehat{\mathbb{P}} \longrightarrow \widehat{\mathbb{P}_{\perp}}$ adjoins a "root" to a presheaf X in $\widehat{\mathbb{P}}$ in the sense that $\lfloor X \rfloor(\lfloor p \rfloor)$ is $X(p)$, for any p in $\mathbb{P}$, while

$\lfloor X \rfloor(\bot)$ is the singleton set $\{*\}$, the new root being $*$; the restriction maps are extended so that restriction to $\bot$ sends elements to $*$. A map from X to Y in $\widehat{\mathbb{P}}$ is sent to its obvious extension from $\lfloor X \rfloor$ to $\lfloor Y \rfloor$ in $\widehat{\mathbb{P}_{\bot}}$. Presheaves that to within isomorphism can be obtained as images under $\lfloor - \rfloor$ are called *rooted* [19].

Proposition 4.1 *Any presheaf Y in $\widehat{\mathbb{P}_{\bot}}$ has a decomposition as a sum of rooted presheaves*

$$Y \cong \Sigma_{i \in Y(\bot)} \lfloor Y_i \rfloor \; ,$$

where, for $i \in Y(\bot)$, the presheaf Y_i in $\widehat{\mathbb{P}}$ is the restriction of Y to the elements which are sent under Y to the element i over $\bot$, i.e., for $p \in \mathbb{P}$,

$$Y_i(p) = \{x \in Y(\lfloor p \rfloor) \mid (Y e_p)(x) = i\}$$

where $e_p : \bot \longrightarrow \lfloor p \rfloor$ is the unique map from $\bot$ to p in $\mathbb{P}$.

Intuitively, thinking of presheaves as processes, the presheaves Y_i, where $i \in Y(\bot)$, in the decomposition of Y a presheaf over $\mathbb{P}_{\bot}$, are those processes that Y can become after performing the initial action $\bot$.

The *strict Yoneda embedding* $j_{\mathbb{P}_{\bot}} : \mathbb{P}_{\bot} \to \widehat{\mathbb{P}}$, sends $\bot$ to $\emptyset$ and elsewhere acts like $y_{\mathbb{P}}$. The presheaf category $\widehat{\mathbb{P}}$, with the strict Yoneda embedding $j_{\mathbb{P}}$ is a free connected-colimit completion of $\mathbb{P}_{\bot}$. Together they satisfy the universal property that for any functor $F : \mathbb{P} \to \mathcal{E}$, where $\mathcal{E}$ is a category with all connected colimits, there is a connected-colimit preserving functor $F^{\dagger} : \widehat{\mathbb{P}} \to \mathcal{E}$, determined to within isomorphism, such that $F \cong F^{\dagger} \circ j_{\mathbb{P}}$:

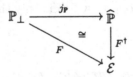

The central observation on which the proof relies is that any presheaf is a connected colimit of representables, $j_{\mathbb{P}}(\lfloor p \rfloor)$ with p in $\mathbb{P}$, together with $j_{\mathbb{P}}(\bot) = \emptyset$, the empty presheaf.

The universal property suggests the importance of connected-colimit preserving functors. Define $\mathbf{Aff_{Set}}$ to be the category consisting of objects small categories $\mathbb{P}, \mathbb{Q}, \cdots$, with maps $G : \mathbb{P} \to \mathbb{Q}$ the connected-colimit preserving functors $G : \widehat{\mathbb{P}} \to \widehat{\mathbb{Q}}$ between the associated presheaf categories, and composition the usual composition of functors. A functor which preserves colimits certainly preserves connected colimits, so

$\mathbf{Lin_{Set}}$ is a subcategory of $\mathbf{Aff_{Set}}$. The two categories, $\mathbf{Lin_{Set}}$ and $\mathbf{Aff_{Set}}$, share the same objects. We can easily characterise those maps in $\mathbf{Aff_{Set}}$ which are in $\mathbf{Lin_{Set}}$:

Proposition 4.2 *Suppose* $F : \widehat{\mathbb{P}} \longrightarrow \widehat{\mathbb{Q}}$ *is a functor which preserves connected colimits. The following properties are equivalent:*

(i) F *preserves all colimits,*

(ii) F *preserves all coproducts (sums),*

(iii) F *is strict, i.e.,* F *sends the empty presheaf to the empty presheaf.*

Because $\widehat{\mathbb{P}}$ is the free connected-colimit completion of $\mathbb{P}_\perp$, we obtain the equivalence

$$\mathbf{Aff_{Set}}(\mathbb{P}, \mathbb{Q}) \simeq [\mathbb{P}_\perp, \widehat{\mathbb{Q}}] \, ,$$

and consequently the equivalence

$$\mathbf{Aff_{Set}}(\mathbb{P}, \mathbb{Q}) \simeq \mathbf{Lin_{Set}}(\mathbb{P}_\perp, \mathbb{Q}) \, .$$

The equivalence is part of an adjunction between $\mathbf{Aff_{Set}}$ and $\mathbf{Lin_{Set}}$ regarded as 2-categories, in which the 2-cells are natural transformations. We can easily extend lifting to a 2-functor $(-)_\perp : \mathbf{Aff_{Set}} \longrightarrow \mathbf{Lin_{Set}}$; for $F : \mathbb{P} \longrightarrow \mathbb{Q}$ in $\mathbf{Aff_{Set}}$, the functor $F_\perp : \mathbb{P}_\perp \longrightarrow \mathbb{Q}_\perp$ in $\mathbf{Lin_{Set}}$ takes $Y \in \widehat{\mathbb{P}_\perp}$ with decomposition $\Sigma_{i \in Y(\perp)} \lfloor Y_i \rfloor$ to $F_\perp(Y) = \Sigma_{i \in Y(\perp)} \lfloor F(Y_i) \rfloor$. Lifting restricts to a 2-comonad on $\mathbf{Lin_{Set}}$ with $\mathbf{Aff_{Set}}$ as its coKleisli category.

The comonad $(-)_\perp$ has turned the model of linear logic $\mathbf{Lin_{Set}}$ into a model $\mathbf{Aff_{Set}}$ of affine linear logic (where the tensor unit is terminal).

4.4.3 Bisimulation

Presheaves are being thought of as nondeterministic processes on which equivalences such as bisimulation are important in abstracting away from inessential differences of behaviour. Bisimulation between presheaves is derived from notion of open map between presheaves [18, 19].

A morphism $h : X \to Y$, between presheaves X and Y, is *open* iff for all morphisms $e : p \to q$ in $\mathbb{P}_\perp$, any commuting square

$$
\begin{array}{ccc}
j_{\mathbb{P}}(p) & \xrightarrow{\ x\ } & X \\
{\scriptstyle j_{\mathbb{P}}(e)} \downarrow & & \downarrow {\scriptstyle h} \\
j_{\mathbb{P}}(q) & \xrightarrow[\ y\ ]{} & Y
\end{array}
$$

can be split into two commuting triangles

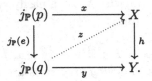

That the square commutes means that the path $h \circ x$ in Y can be extended via e to a path y in Y. That the two triangles commute means that the path x can be extended via e to a path z in X which matches y.

Open maps are a generalisation of functional bisimulations, or zig-zag morphisms, known from transition systems [19]. Presheaves in $\widehat{\mathbb{P}}$ are *bisimilar* iff there is a span of open maps between them.[†]

The preservation of connected colimits by a functor between presheaf categories is sufficient to ensure that it preserves open maps and bisimulation.

Theorem 4.3 *[11, 7] Let $G : \widehat{\mathbb{P}} \to \widehat{\mathbb{Q}}$ be any connected-colimit preserving functor between presheaf categories. Then G preserves open maps and open-map bisimulation.*

4.5 Constructions

We now describe the constructions which form the basis of a denotational semantics for a language for affine-linear processes. The types of the language will be interpreted as objects (in fact, path orders) and the terms, describing processes with free variables, as maps of an affine category. The constructions can be read as being in both the category of domains **Aff$_2$** and the category of presheaf models **Aff$_{Set}$**, sometimes referring to their linear subcategories. By using the neutral language of categories we can describe the operations in either set-up. Below **Aff** can refer to either **Aff$_2$** or **Aff$_{Set}$**, and correspondingly **Lin** to either **Lin$_2$** or **Lin$_{Set}$**. **Aff$_{Set}$** and **Lin$_{Set}$** are 2-categories in which maps are related by natural transformations. There, and so in the general discussion with **Aff**, we can only characterise constructions to within isomorphism of maps. Of course, isomorphism of maps coincides with

---

[†] We have chosen here to develop the definition of open map from the strict Yoneda embedding rather than the Yoneda embedding. Maps between presheaves are open with respect to strict Yoneda iff they are surjective and open with respect to Yoneda.

equality in the categorises of domains. Although the operations will always be with respect to path orders they could all be extended easily to small categories.

4.5.1 Sums and fixed points

Each object $\mathbb{P}$ is associated with (nondeterministic) sum operations, a map $\Sigma : \&_{i \in I} \mathbb{P} \to \mathbb{P}$ in **Aff** taking a tuple $\{X_i \mid i \in I\}$, to the sum (coproduct) $\Sigma_{i \in I} X_i$ in $\widehat{\mathbb{P}}$. The empty sum yields $\emptyset \in \mathbb{P}$. Finite sums, of size k, are typically written as $X_1 + \cdots + X_k$.

For objects $\mathbb{P}$ and $\mathbb{Q}$, the category **Aff**$(\mathbb{P}, \mathbb{Q})$ of maps with natural transformations, being equivalent to $(\widehat{\mathbb{P}_\perp})^{op} \times \mathbb{Q}$, has all colimits and in particular all ω-colimits. Any operation $G : \mathbf{Aff}(\mathbb{P}, \mathbb{Q}) \longrightarrow \mathbf{Aff}(\mathbb{P}, \mathbb{Q})$ which preserves connected colimits will have a fixed point $fix\ G : \mathbb{P} \to \mathbb{Q}$, a map in **Aff**. We will build up the denotation of fixed points out of composition in **Aff**. The composition $G \circ F$ of maps F in $\mathbf{Aff}(\mathbb{P}, \mathbb{Q})$ and G in $\mathbf{Aff}(\mathbb{Q}, \mathbb{R})$, being got as the application $G(F(-))$, preserves connected colimits in the argument F, and colimits in G.[†]

4.5.2 Tensor

The tensor product $\mathbb{P} \otimes \mathbb{Q}$ of path orders $\mathbb{P}$, $\mathbb{Q}$ is given by the set $(\mathbb{P}_\perp \times \mathbb{Q}_\perp) \setminus \{(\perp, \perp)\}$, ordered coordinatewise, in other words, as the product of $\mathbb{P}_\perp$ and $\mathbb{Q}_\perp$ as partial orders but with the bottom element $(\perp, \perp)$ removed.

Let $F : \mathbb{P} \to \mathbb{P}'$ and $G : \mathbb{Q} \to \mathbb{Q}'$. We define $F \otimes G : \mathbb{P} \otimes \mathbb{Q} \to \mathbb{P}' \otimes \mathbb{Q}'$ as the extension (*cf.* Sections 4.3.2 and 4.4.2) $H^\dagger$ of a functor

$$H : (\mathbb{P} \otimes \mathbb{Q})_\perp \to \widehat{\mathbb{P}' \otimes \mathbb{Q}'} \ .$$

Notice that $(\mathbb{P} \otimes \mathbb{Q})_\perp$ is isomorphic to the product as partial orders of $\mathbb{P}_\perp \times \mathbb{Q}_\perp$ in which the bottom element is then $(\perp, \perp)$. With this realisation of $(\mathbb{P} \otimes \mathbb{Q})_\perp$ we can define $H : \mathbb{P}_\perp \times \mathbb{Q}_\perp \to \widehat{\mathbb{P}' \otimes \mathbb{Q}'}$ by taking

$$(H(p, q))(p', q') = \lfloor F(p) \rfloor (p') \times \lfloor G(q) \rfloor (q')$$

for $p \in \mathbb{P}_\perp$, $q \in \mathbb{Q}_\perp$ and $(p', q') \in \mathbb{P}' \otimes \mathbb{Q}'$—on the right we use the product (in **Set**, or **2** where it amounts to the meet).

[†] The story specialises to the category of domains **Aff₂**. In a domain, colimits reduce to joins and connected colimits to nonempty joins. In particular, ω-colimits amount to joins of ω-chains. An operation between domains preserves (connected) colimits iff it preserves (nonempty) joins.

The unit for tensor is the empty path order $\mathbb{O}$.

Objects $X \in \widehat{\mathbb{P}}$ correspond to maps $\tilde{X} : \mathbb{O} \to \mathbb{P}$ sending $\emptyset$ to X. Given $X \in \widehat{\mathbb{P}}$ and $Y \in \widehat{\mathbb{Q}}$ we define $X \otimes Y \in \widehat{\mathbb{P} \otimes \mathbb{Q}}$ to be the element pointed to by $\tilde{X} \otimes \tilde{Y} : \mathbb{O} \to \mathbb{P} \otimes \mathbb{Q}$.

4.5.3 Function space

The function space of path orders $\mathbb{P} \multimap \mathbb{Q}$ is given by the product of partial orders $(\mathbb{P}_\perp)^{op} \times \mathbb{Q}$. Thus the elements of $\mathbb{P} \multimap \mathbb{Q}$ are pairs, which we write suggestively as $(p \mapsto q)$, with $p \in \mathbb{P}_\perp, q \in \mathbb{Q}$, ordered by

$$(p' \mapsto q') \le (p \mapsto q) \iff p \le p' \ \& \ q' \le q$$

—note the switch in order on the left.

We have the following chain of isomorphisms between partial orders:

$$\mathbb{P} \otimes \mathbb{Q} \multimap \mathbb{R} = (\mathbb{P} \otimes \mathbb{Q})_\perp{}^{op} \times \mathbb{R} \cong \mathbb{P}_\perp{}^{op} \times \mathbb{Q}_\perp{}^{op} \times \mathbb{R} \cong \mathbb{P} \multimap (\mathbb{Q} \multimap \mathbb{R}) .$$

This gives an isomorphism between $\widehat{\mathbb{P} \otimes \mathbb{Q} \multimap \mathbb{R}}$ and $\mathbb{P} \multimap \widehat{(\mathbb{Q} \multimap \mathbb{R})}$. Thus there is a 1-1 correspondence *curry* from maps $\mathbb{P} \otimes \mathbb{Q} \to \mathbb{R}$ to maps $\mathbb{P} \to (\mathbb{Q} \multimap \mathbb{R})$ in **Aff**; its inverse is called *uncurry*. We obtain *linear application, app* : $(\mathbb{P} \multimap \mathbb{Q}) \otimes \mathbb{P} \to \mathbb{Q}$, as *uncurry*$(1_{\mathbb{P} \multimap \mathbb{Q}})$.

We shall write $u\,t$ for the application of u of type $\mathbb{P} \multimap \mathbb{Q}$ to t of type $\mathbb{P}$. The ability to curry justifies the formation of terms $\lambda x.u$ of type $\mathbb{P} \multimap \mathbb{Q}$ by lambda abstraction where u of type $\mathbb{Q}$ is a term with free variable x of type $\mathbb{P}$. Because of linearity constraints on the occurrence of variables, we will have that an application $(\lambda x.u)t$ will be isomorphic to a substitution $u[t/x]$—see Lemma 2.

4.5.4 Products

The product of path orders $\mathbb{P} \& \mathbb{Q}$ is given by the disjoint union of $\mathbb{P}$ and $\mathbb{Q}$. An object of $\widehat{\mathbb{P} \& \mathbb{Q}}$ can be identified with a pair (X, Y), with $X \in \widehat{\mathbb{P}}$ and $Y \in \widehat{\mathbb{Q}}$, which provides the projections $\pi_1 : \mathbb{P} \& \mathbb{Q} \to \mathbb{P}$ and $\pi_2 : \mathbb{P} \& \mathbb{Q} \to \mathbb{Q}$. More general, not just binary, products $\&_{i \in I} \mathbb{P}_i$ with projections π_j, for $j \in I$, are defined similarly. From the universal property of products, a collection of maps $F_i : \mathbb{P} \to \mathbb{P}_i$, for $i \in I$, can be tupled together to form a unique map $\langle F_i \rangle_{i \in I} : \mathbb{P} \to \&_{i \in I} \mathbb{P}_i$ with the property that $\pi_j \circ \langle F_i \rangle_{i \in I} = F_j$ for all $j \in I$. The empty product is given by $\mathbb{O}$ and as the terminal object is associated with unique maps $\mathbb{P} \to \mathbb{O}$, constantly $\emptyset$, for any path order $\mathbb{P}$. Finite products, of size k, are most often written as $\mathbb{P}_1 \& \cdots \& \mathbb{P}_k$.

Because there are empty objects we can define maps in **Lin** from products to tensors of path orders. For instance, in the binary case, $\sigma : \mathbb{P}\&\mathbb{Q} \to \mathbb{P} \otimes \mathbb{Q}$ in **Lin** is specified by

$$(X, Y) \mapsto (X \otimes \emptyset) + (\emptyset \otimes Y) .$$

The composition of such a map with the diagonal map of the product, *viz.*

$$\delta_{\mathbb{P}} : \mathbb{P} \xrightarrow{\ diag\ } \mathbb{P}\&\mathbb{P} \xrightarrow{\ \sigma\ } \mathbb{P} \otimes \mathbb{P}$$

gives a weak form of diagonal map taking X to $(X \otimes \emptyset) + (\emptyset \otimes X)$. General weak diagonal maps

$$\delta_{\mathbb{P}_k} : \mathbb{P} \longrightarrow \mathbb{P} \otimes \cdots \otimes \mathbb{P}$$

in **Lin**, from $\mathbb{P}$ to k copies of $\mathbb{P}$ tensored together, are defined analogously. They will play a role later in the semantics of a general affine linear language; weak diagonal maps allow the same argument to be used in several different, though incompatible, ways.

4.5.5 Prefixed sums

The category **Aff** does not have coproducts. However, we can build a useful sum in **Aff** with the help of the coproduct of **Lin** and lifting. Let $\mathbb{P}_\alpha$, for $\alpha \in A$, be a family of path orders. As their prefixed sum, $\Sigma_{\alpha \in A} \alpha \mathbb{P}_\alpha$, we take the disjoint union of the path orders $\Sigma_{\alpha \in A} \mathbb{P}_{\alpha \perp}$, over the underlying set $\bigcup_{\alpha \in A} \{\alpha\} \times (\mathbb{P}_\alpha)_\perp$; the latter path order forms a coproduct in **Lin** with the obvious injections $in_\beta : \mathbb{P}_{\beta \perp} \longrightarrow \Sigma_{\alpha \in A} \alpha \mathbb{P}_\alpha$, for $\beta \in A$. The *injections* $\beta : \mathbb{P}_\beta \to \Sigma_{\alpha \in A} \alpha \mathbb{P}_\alpha$ in **Aff**, for $\beta \in A$, are defined to be the composition $\beta = in_\beta(\lfloor - \rfloor)$. This construction is not a coproduct in **Aff**. However, it does satisfy a weaker property analogous to the universal property of a coproduct. Suppose $F_\alpha : \mathbb{P}_\alpha \to \mathbb{Q}$ are maps in **Aff** for all $\alpha \in A$. Then, there is a mediating map

$$F : \Sigma_{\alpha \in A} \alpha \mathbb{P}_\alpha \longrightarrow \mathbb{Q}$$

in **Lin** determined to within isomorphism such that

$$F \circ \alpha \cong F_\alpha$$

for all $\alpha \in A$.

Suppose that the family of maps $F_\alpha : \mathbb{P}_\alpha \to \mathbb{Q}$, with $\alpha \in A$, has the property that each F_α is constantly $\emptyset$ whenever $\alpha \in A$ is different from

β and that F_β is $H : \mathbb{P}_\beta \to \mathbb{Q}$. Write $H_{\mathbb{Q}\beta} : \Sigma_{\alpha \in A} \alpha \mathbb{P}_\alpha \to \mathbb{Q}$ for a choice of mediating map in **Lin**. Then

$$H_{\mathbb{Q}\beta}(\beta Y) \cong H(Y) \,, \; H_{\mathbb{Q}\beta}(\alpha Z) = \emptyset \text{ if } \alpha \neq \beta \,, H_{\mathbb{Q}\beta}(\Sigma_{i \in I} X_i) \cong \Sigma_{i \in I} H_{\mathbb{Q}\beta}(X_i) \,,$$

where $Y \in \widehat{\mathbb{P}_\beta}$, $Z \in \widehat{\mathbb{P}_\alpha}$ and $X_i \in \widehat{\Sigma_{\alpha \in A} \alpha \mathbb{P}_\alpha}$ for all $i \in I$. In particular, for empty sums, $H_{\mathbb{Q}\beta}(\emptyset) = \emptyset$.

For a general family $F_\alpha : \mathbb{P}_\alpha \to \mathbb{Q}$, with $\alpha \in A$, we can describe the action of the mediating morphism, to within isomorphism, on $X \in \widehat{\Sigma_{\alpha \in A} \alpha \mathbb{P}_\alpha}$ as $F(X) = \Sigma_{\alpha \in A} (F_\alpha)_{\mathbb{Q}\alpha}(X)$.

If a term u of type $\mathbb{Q}$ with free variable x of type $\mathbb{P}$ denotes $H : \mathbb{P}_\beta \longrightarrow \mathbb{Q}$ in **Aff** and t is of type $\Sigma_{\alpha \in A} \alpha \mathbb{P}_\alpha$, then we shall write

$$[t > \alpha x \Rightarrow u]$$

for $H_{\mathbb{Q}\beta}(t)$. This term $[t > \alpha x \Rightarrow u]$ tests or matches t denoting an element of a prefixed sum against the pattern αx and passes the results of successful matches for x on to u; the possibly multiple results of successful matches are then summed together.

Because prefixed sum is not a coproduct we do not have that tensor distributes over prefixed sum. However there is a map in **Aff**,

$$dist : \mathbb{Q} \otimes \Sigma_{\alpha \in A} \alpha \mathbb{P}_\alpha \longrightarrow \Sigma_{\alpha \in A} \alpha (\mathbb{Q} \otimes \mathbb{P}_\alpha) \,,$$

expressing a form of distributivity. The map $dist$ is given as the extension $H^\dagger$ of the functor

$$H : \mathbb{Q}_\perp \times (\Sigma_{\alpha \in A} \alpha \mathbb{P}_\alpha)_\perp \to \Sigma_{\alpha \in A} \alpha (\mathbb{Q} \otimes \mathbb{P}_\alpha)$$

where

$$H(q, (\alpha, p)) = y_{\Sigma_{\alpha \in A} \alpha (\mathbb{Q} \otimes \mathbb{P}_\alpha)}(\alpha, (q, p)) \text{ and } H(q, \perp) = \emptyset \,.$$

Unary prefixed sums in **Aff**, when the indexing set is a singleton, are an important special case as they amount to lifting.

4.5.6 Recursive type definitions

Suppose that we wish to model a process language rather like CCS but where processes are passed instead of discrete values, subject to the linearity constraint that when a process is received it can be run at most once. Assume the synchronised communication occurs along channels forming the set A. The path orders can be expected to be the "least" to satisfy the following equations:

$$\mathbb{P} = \tau \mathbb{P} + \Sigma_{a \in A} a!\mathbb{C} + \Sigma_{a \in A} a?\mathbb{F} \,, \qquad \mathbb{C} = \mathbb{P} \otimes \mathbb{P} \,, \qquad \mathbb{F} = (\mathbb{P} \multimap \mathbb{P}) \,.$$

The three components of process paths $\mathbb{P}$ represent paths beginning with a silent (τ) action, an output on a channel ($a!$), resuming as a concretion path (in $\mathbb{C}$), and an input from a channel ($a?$), resuming as an abstraction path (in $\mathbb{F}$). It is our choice of path for abstractions which narrows us to an *affine-linear* process-passing language, one where the input process can be run at most once to yield a single (computation) path.

We can solve such recursive equations for path orders by several techniques, ranging from sophisticated methods providing inductive and coinductive characterisations [8], to simple methods essentially based on inductive definitions. Paralleling techniques on information systems [20], path orders under the order

$$\mathbb{P} \trianglelefteq \mathbb{Q} \iff \mathbb{P} \subseteq \mathbb{Q} \ \& \ (\forall p, p' \in \mathbb{P}. \ p \leq_{\mathbb{P}} p' \iff p \leq_{\mathbb{Q}} p')$$

form a (large) complete partial order with respect to which all the constructions on path orders we have just seen can be made Scott continuous. Solutions to equations like those above are then obtained as (simultaneous) least fixed points.

4.6 An affine-linear language for processes

Assume that path orders are presented using the constructions with the following syntax:

$$\mathbb{T} ::= \mathbb{O} \mid \mathbb{T}_1 \otimes \mathbb{T}_2 \mid \mathbb{T}_1 \multimap \mathbb{T}_2 \mid \Sigma_{\alpha \in A} \alpha \mathbb{T}_\alpha \mid \mathbb{T}_1 \& \mathbb{T}_2$$
$$\mid P \mid \mu_j P_1, \cdots, P_k.(\mathbb{T}_1, \cdots, \mathbb{T}_k)$$

All the construction names have been met earlier with the exception of the notation for recursively defined path orders. Above P is drawn from a set of variables used in the recursive definition of path orders; $\mu_j P_1, \cdots, P_k.(\mathbb{T}_1, \cdots, \mathbb{T}_k)$ stands for the j-component (so $1 \leq j \leq k$) of the least solution to the defining equations

$$P_1 = \mathbb{T}_1, \ \cdots, \ P_k = \mathbb{T}_k \ ,$$

in which the expressions $\mathbb{T}_1, \cdots, \mathbb{T}_k$ may contain $P_1, \cdots, P_k$. We shall write $\mu P_1, \cdots, P_k.(\mathbb{T}_1, \cdots, \mathbb{T}_k)$ as an abbreviation for

$$(\mu_1 P_1, \cdots, P_k.(\mathbb{T}_1, \cdots, \mathbb{T}_k), \cdots, \mu_k P_1, \cdots, P_k.(\mathbb{T}_1, \cdots, \mathbb{T}_k)) \ .$$

In future we will often use vector notation and, for example, write $\mu \overrightarrow{P}.\overrightarrow{\mathbb{T}}$ for the expression above, and confuse a closed expression for a path order with the path order itself.

The operations of Sections 4.3 and 4.5 form the basis of a syntax of terms which will be subject to typing and linearity constraints:

$$t, u, v, \cdots ::= \quad x, y, z, \cdots \qquad \qquad \text{(Variables)}$$

$t, u, v, \cdots ::=$ $\quad x, y, z, \cdots$	(Variables)
$\emptyset \mid \Sigma_{i \in I} t_i \mid$	(Sums)
$rec\ x.t \mid$	(Recursive definitions)
$\lambda x.t \mid u\ v \mid$	(Abstraction, application)
$\alpha t \mid [t > \alpha x \Rightarrow u] \mid$	(Injections and match)
$(t, u) \mid [t > (x, -) \Rightarrow u] \mid$	
$\quad\quad [t > (-, x) \Rightarrow u] \mid$	(Pairing and match)
$t \otimes u \mid [t > x \otimes y \Rightarrow u]$	(Tensor and match)

The language, first introduced in [30], is similar to that in [1], being based on a form of pattern matching. Accordingly, variables in the pattern, like x in the pattern of $[t > \alpha x \Rightarrow u]$, are binding occurrences and bind later occurrences of the variable in the body, u in this case. We shall take for granted an understanding of free and bound variables, and substitution on raw terms. In examples we will allow ourselves to use $+$ both in writing sums of terms and prefixed sums of path orders.

Let $\mathbb{P}_1, \cdots, \mathbb{P}_k$ be closed expressions for path orders and assume that the variables $x_1, \cdots, x_k$ are distinct. A syntactic judgement

$$x_1 : \mathbb{P}_1, \cdots, x_k : \mathbb{P}_k \vdash t : \mathbb{Q}$$

stands for a map

$$[\![x_1 : \mathbb{P}_1, \cdots, x_k : \mathbb{P}_k \vdash t : \mathbb{Q}]\!] : \mathbb{P}_1 \otimes \cdots \otimes \mathbb{P}_k \to \mathbb{Q}$$

in **Aff**. We shall typically write Γ, or Δ, for an environment list $x_1 : \mathbb{P}_1, \cdots, x_k : \mathbb{P}_k$. We shall most often abbreviate the denotation map to

$$\mathbb{P}_1 \otimes \cdots \otimes \mathbb{P}_k \xrightarrow{\ t\ } \mathbb{Q}, \text{ or even } \Gamma \xrightarrow{\ t\ } \mathbb{Q}.$$

Here k may be 0 so the environment list in the syntactic judgement is empty and the corresponding tensor product the empty path order $\mathbb{O}$.

An affine-linear language will restrict copying and so substitutions of a common term into distinct variables. The counterpart in the models is the absence of a suitable diagonal map from objects $\mathbb{P}$ to $\mathbb{P} \otimes \mathbb{P}$. For example the function $X \mapsto X \otimes X$ from $\widehat{\mathbb{P}}$ to $\widehat{\mathbb{P} \otimes \mathbb{P}}$ is not in general a map in **Aff**.[†] Consider a term $t(x, y)$, with its free variables x and y

---

† To see this, for example in **Aff₂**, assume that $\mathbb{P}$ is the discrete order on the set $\{a, b\}$. Then the nonempty sum $x = y_{\mathbb{P}}(a) + y_{\mathbb{P}}(b)$ is not sent to

$$(y_{\mathbb{P}}(a) \otimes y_{\mathbb{P}}(a)) + (y_{\mathbb{P}}(b) \otimes y_{\mathbb{P}}(b)) = y_{\mathbb{P} \otimes \mathbb{P}}(a, a) + y_{\mathbb{P} \otimes \mathbb{P}}(b, b)$$

shown explicitly, for which

$$x : \mathbb{P}, y : \mathbb{P} \vdash t(x, y) : \mathbb{Q} \,,$$

corresponding to a map $\mathbb{P} \otimes \mathbb{P} \xrightarrow{t(x,y)} \mathbb{Q}$ in **Aff**. This does not generally entail that

$$x : \mathbb{P} \vdash t(x, x) : \mathbb{Q}$$

—there may not be a corresponding map in **Aff**, for example if $t(x, y) = x \otimes y$. There is however a condition on how the variables x and y occur in t which ensures that the judgement $x : \mathbb{P} \vdash t(x, x) : \mathbb{Q}$ holds and that it denotes the map in **Aff** obtained as the composition

$$\mathbb{P} \xrightarrow{\delta} \mathbb{P} \otimes \mathbb{P} \xrightarrow{t(x,y)} \mathbb{Q}$$

—using the weak diagonal map seen earlier in Section 4.5.4. (For example, in the term $x + y$, where x and y have the same type $\mathbb{P}$, only computation at one of the arguments x and y is possible, and it is legitimate to diagonalise to $x + x$ to obtain an affine-linear map.) Syntactically, this is assured if the variables x and y are *not crossed* in t according to the following definition:

Definition 4.4 Let t be a raw term. Say a set of variables V is *crossed* in t iff there are subterms of t of the form

a tensor $s \otimes u$, an application $(s \; u)$, or a match $[v > u \Rightarrow s]$

for which t has free occurrences of variables from V appearing in both s and u.

For example, variables x and y are crossed in $x \otimes y$, but variables x and y are not crossed in $(x + y) \otimes z$. Note that a set of variables V is crossed in a term t if V contains variables x, y, not necessarily distinct, so that $\{x, y\}$ is crossed in t. We are mainly interested in when sets of variables are *not crossed* in a term. A set of variables $\{x_1, \cdots, x_k\}$ not being crossed in a term t ensures that computation paths at arguments $x_1, \cdots, x_k$ are in conflict—at most one can contribute to the computation path of t. Sets of variables of the same type which are not crossed in a term will behave like single variables with regard to substitutions.

The term-formation rules are listed below alongside their interpretations as constructors on morphisms, taking the morphisms denoted by

as would be needed to preserve non-empty sums, but instead to

$$x \otimes x = y_{\mathbb{P} \otimes \mathbb{P}}(a, a) + y_{\mathbb{P} \otimes \mathbb{P}}(b, b) + y_{\mathbb{P} \otimes \mathbb{P}}(a, b) + y_{\mathbb{P} \otimes \mathbb{P}}(b, a)$$

with extra "cross terms."

the premises to that denoted by the conclusion (along the lines of [2]). We assume that the variables in any enviroment list which appears are distinct.

Structural rules:

$$\frac{}{x : \mathbb{P} \vdash x : \mathbb{P}} \quad , \text{ interpreted as } \quad \frac{}{\mathbb{P} \xrightarrow{1_{\mathbb{P}}} \mathbb{P}} \cdot$$

$$\frac{\Delta \vdash t : \mathbb{P}}{\Gamma, \Delta \vdash t : \mathbb{P}} \quad , \text{ interpreted as } \quad \frac{\Delta \xrightarrow{t} \mathbb{P}}{\Gamma \otimes \Delta \xrightarrow{\emptyset \otimes 1_{\Delta}} \mathbb{O} \otimes \Delta \cong \Delta \xrightarrow{t} \mathbb{P}} \cdot$$

$$\frac{\Gamma, x : \mathbb{P}, y : \mathbb{Q}, \Delta \vdash t : \mathbb{R}}{\Gamma, y : \mathbb{Q}, x : \mathbb{P}, \Delta \vdash t : \mathbb{R}}, \text{ interpreted via } s : \mathbb{Q} \otimes \mathbb{P} \cong \mathbb{P} \otimes \mathbb{Q} \text{ as}$$

$$\frac{\Gamma \otimes \mathbb{P} \otimes \mathbb{Q} \otimes \Delta \xrightarrow{t} \mathbb{R}}{\Gamma \otimes \mathbb{Q} \otimes \mathbb{P} \otimes \Delta \xrightarrow{1_{\Gamma} \otimes s \otimes 1_{\Delta}} \Gamma \otimes \mathbb{P} \otimes \mathbb{Q} \otimes \Delta \xrightarrow{t} \mathbb{R}} \cdot$$

Recursive path orders:

$$\frac{\Gamma \vdash t : \mathbb{T}_j[\mu \vec{P}.\vec{\mathbb{T}}/\vec{P}]}{\Gamma \vdash t : \mu_j \vec{P}.\vec{\mathbb{T}}} \quad , \quad \frac{\Gamma \vdash t : \mu_j \vec{P}.\vec{\mathbb{T}}}{\Gamma \vdash t : \mathbb{T}_j[\mu \vec{P}.\vec{\mathbb{T}}/\vec{P}]} \cdot$$

where the premise and conclusion of each rule are interpreted as the same map because $\mu_j \vec{P}.\vec{\mathbb{T}}$ and $\mathbb{T}_j[\mu \vec{P}.\vec{\mathbb{T}}/\vec{P}]$ denote equal path orders.

Sums of terms:

$$\frac{}{\Gamma \vdash \emptyset : \mathbb{P}} \quad , \text{ interpreted as } \quad \frac{}{\Gamma \xrightarrow{\emptyset} \mathbb{P}} , \text{ the constantly } \emptyset \text{ map.}$$

$$\frac{\Gamma \vdash t_i : \mathbb{P} \text{ for all } i \in I}{\Gamma \vdash \Sigma_{i \in I} t_i : \mathbb{P}} \quad , \text{ interpreted as } \quad \frac{\Gamma \xrightarrow{t_i} \mathbb{P} \text{ for all } i \in I}{\Gamma \xrightarrow{\langle t_i \rangle_{i \in I}} \&_{i \in I} \mathbb{P} \xrightarrow{\Sigma} \mathbb{P}} \cdot$$

Recursive definitions:

$$\frac{\Gamma, x : \mathbb{P} \vdash t : \mathbb{P} \quad \{y, x\} \text{ not crossed in } t \text{ for all } y \text{ in } \Gamma}{\Gamma \vdash rec\ x.t : \mathbb{P}} ,$$

$$\text{interpreted as } \quad \frac{\Gamma \otimes \mathbb{P} \xrightarrow{t} \mathbb{P}}{\Gamma \xrightarrow{fix\ F} \mathbb{P}} \cdot$$

—see Section 4.5.1, where for $g : \Gamma \longrightarrow \mathbb{P}$ the map $F(g) : \Gamma \longrightarrow \mathbb{P}$ is the composition

$$\Gamma \xrightarrow{\delta} \Gamma \otimes \Gamma \xrightarrow{1_{\Gamma} \otimes g} \Gamma \otimes \mathbb{P} \xrightarrow{t} \mathbb{P} \cdot$$

Abstraction:

$$\frac{\Gamma, x : \mathbb{P} \vdash t : \mathbb{Q}}{\Gamma \vdash \lambda x.t : \mathbb{P} \multimap \mathbb{Q}} \quad , \text{ interpreted as } \quad \frac{\Gamma \otimes \mathbb{P} \xrightarrow{t} \mathbb{Q}}{\Gamma \xrightarrow{curry\, t} (\mathbb{P} \multimap \mathbb{Q})} .$$

Application:

$$\frac{\Gamma \vdash u : \mathbb{P} \multimap \mathbb{Q} \quad \Delta \vdash v : \mathbb{P}}{\Gamma, \Delta \vdash u\, v : \mathbb{Q}} ,$$

$$\text{interpreted as } \quad \frac{\Gamma \xrightarrow{u} (\mathbb{P} \multimap \mathbb{Q}) \quad \Delta \xrightarrow{v} \mathbb{P}}{\Gamma \otimes \Delta \xrightarrow{u \otimes v} (\mathbb{P} \multimap \mathbb{Q}) \otimes \mathbb{P} \xrightarrow{app} \mathbb{Q}} .$$

Injections and match for prefixed sums:

$$\frac{\Gamma \vdash t : \mathbb{P}_\beta, \text{ where } \beta \in A}{\Gamma \vdash \beta t : \Sigma_{\alpha \in A} \alpha \mathbb{P}_\alpha} \quad , \text{ interpreted as } \quad \frac{\Gamma \xrightarrow{t} \mathbb{P}_\beta, \text{ where } \beta \in A}{\Gamma \xrightarrow{t} \mathbb{P}_\beta \xrightarrow{\beta} \Sigma_{\alpha \in A} \alpha \mathbb{P}_\alpha} .$$

$$\frac{\Gamma, x : \mathbb{P}_\beta \vdash u : \mathbb{Q}, \text{ where } \beta \in A. \quad \Delta \vdash t : \Sigma_{\alpha \in A} \alpha \mathbb{P}_\alpha}{\Gamma, \Delta \vdash [t > \beta x \Rightarrow u] : \mathbb{Q}} \quad , \text{ interpreted as }$$

$$\frac{\Gamma \otimes \mathbb{P}_\beta \xrightarrow{u} \mathbb{Q} \quad \Delta \xrightarrow{t} \Sigma_{\alpha \in A} \alpha \mathbb{P}_\alpha}{\Gamma \otimes \Delta \xrightarrow{1_\Gamma \otimes t} \Gamma \otimes \Sigma_{\alpha \in A} \alpha \mathbb{P}_\alpha \xrightarrow{dist} \Sigma_{\alpha \in A} \alpha (\Gamma \otimes \mathbb{P}_\alpha) \xrightarrow{u_{\textcircled{a}\beta}} \mathbb{Q}} .$$

Pairing and matches for products:

$$\frac{\Gamma \vdash t : \mathbb{P} \quad \Gamma \vdash u : \mathbb{Q}}{\Gamma \vdash (t, u) : \mathbb{P} \& \mathbb{Q}} \quad , \text{ interpreted as } \quad \frac{\Gamma \xrightarrow{t} \mathbb{P} \quad \Gamma \xrightarrow{u} \mathbb{Q}}{\Gamma \xrightarrow{\langle t, u \rangle} \mathbb{P} \& \mathbb{Q}} .$$

$$\frac{\Gamma, x : \mathbb{P} \vdash u : \mathbb{R} \quad \Delta \vdash t : \mathbb{P} \& \mathbb{Q}}{\Gamma, \Delta \vdash [t > (x, -) \Rightarrow u] : \mathbb{R}}, \text{ interpreted as }$$

$$\frac{\Gamma \otimes \mathbb{P} \xrightarrow{u} \mathbb{R} \quad \Delta \xrightarrow{t} \mathbb{P} \& \mathbb{Q}}{\Gamma \otimes \Delta \xrightarrow{1_\Gamma \otimes (\pi_1 \circ t)} \Gamma \otimes \mathbb{P} \xrightarrow{u} \mathbb{R}} .$$

$$\frac{\Gamma, x : \mathbb{Q} \vdash u : \mathbb{R} \quad \Delta \vdash t : \mathbb{P} \& \mathbb{Q}}{\Gamma, \Delta \vdash [t > (-, x) \Rightarrow u] : \mathbb{R}}, \text{ interpreted as }$$

$$\frac{\Gamma \otimes \mathbb{Q} \xrightarrow{u} \mathbb{R} \quad \Delta \xrightarrow{t} \mathbb{P} \& \mathbb{Q}}{\Gamma \otimes \Delta \xrightarrow{1_\Gamma \otimes (\pi_2 \circ t)} \Gamma \otimes \mathbb{Q} \xrightarrow{u} \mathbb{R}} .$$

Tensor operation and match for tensor:

$$\frac{\Gamma \vdash t : \mathbb{P} \quad \Delta \vdash u : \mathbb{Q}}{\Gamma, \Delta \vdash t \otimes u : \mathbb{P} \otimes \mathbb{Q}} \quad , \text{ interpreted as } \quad \frac{\Gamma \xrightarrow{t} \mathbb{P} \quad \Delta \xrightarrow{u} \mathbb{Q}}{\Gamma \otimes \Delta \xrightarrow{t \otimes u} \mathbb{P} \otimes \mathbb{Q}} .$$

$$\frac{\Gamma, x : \mathbb{P}, y : \mathbb{Q} \vdash u : \mathbb{R} \quad \Delta \vdash t : \mathbb{P} \otimes \mathbb{Q}}{\Gamma, \Delta \vdash [t > x \otimes y \Rightarrow u] : \mathbb{R}}, \text{ interpreted as }$$

$$\frac{\Gamma \otimes P \otimes Q \xrightarrow{u} R \quad \Delta \xrightarrow{t} P \otimes Q}{\Gamma \otimes \Delta \xrightarrow{1_\Gamma \otimes t} \Gamma \otimes P \otimes Q \xrightarrow{u} R} \ .$$

By a straightforward induction on the derivation of the typing judgement we obtain:

Proposition 4.5 *Suppose* $\Gamma, x : P \vdash t : Q$. *The set* $\{x\}$ *is not crossed in* t.

Exploiting the naturality of the various operations used in the semantic definitions, we can prove a general substitution lemma. It involves the weak diagonal maps $\delta_k : P \longrightarrow P \otimes \cdots \otimes P$ of Section 4.5.4.

Lemma 1 *(Substitution Lemma) Suppose*

$$\Gamma, x_1 : P, \cdots, x_k : P \vdash t : Q$$

and that the set of variables $\{x_1, \cdots, x_k\}$ *is not crossed in* t. *Suppose* $\Delta \vdash u : P$ *where the variables of* Γ *and* Δ *are disjoint. Then,*

$$\Gamma, \Delta \vdash t[u/x_1, \cdots, u/x_k] : Q$$

and

$$[\![\Gamma, \Delta \vdash t[u/x_1, \cdots, u/x_k] : Q]\!] \cong$$
$$[\![\Gamma, x_1 : P, \cdots, x_k : P \vdash t : Q]\!] \quad \circ \quad (1_\Gamma \otimes (\delta_k \circ [\![\Delta \vdash u : P]\!])) \ .$$

In particular, as singleton sets of variables are not crossed in well-formed terms, we can specialise the Substitution Lemma to the following:

Corollary 4.6 *If* $\Gamma, x : P \vdash t : Q$ *and* $\Delta \vdash u : P$, *where the variables of* Γ *and* Δ *are disjoint, then* $\Gamma, \Delta \vdash t[u/x] : Q$ *and*

$$[\![\Gamma, \Delta \vdash t[u/x] : Q]\!] \cong [\![\Gamma, x : P \vdash t : Q]\!] \circ (1_\Gamma \otimes [\![\Delta \vdash u : P]\!]) \ .$$

As consequences of Corollary 4.6, linear application amounts to substitution, and recursions unfold in the expected way:

Lemma 2 *Suppose* $\Gamma \vdash (\lambda x.t) \, u : Q$. *Then,* $\Gamma \vdash t[u/x] : Q$ *and*

$$[\![\Gamma \vdash (\lambda x.t) \, u : Q]\!] \cong [\![\Gamma \vdash t[u/x] : Q]\!] \ .$$

Lemma 3 *Suppose* $\Gamma \vdash rec \ x.t : P$. *Then* $\Gamma \vdash t[rec \ x.t/x] : P$ *and*

$$[\![\Gamma \vdash rec \ x.t : P]\!] \cong [\![\Gamma \vdash t[rec \ x.t/x] : P]\!] \ .$$

The next lemma follows directly from the universal properties of pre-fixed sum (the last property because the mediating map is in **Lin**, so preserves sums):

Lemma 4 *Properties of prefix match:*

$$[\![\Gamma \vdash [\beta t > \beta x \Rightarrow u] : Q]\!] \cong [\![\Gamma \vdash u[t/x] : Q]\!] ,$$
$$[\![\Gamma \vdash [\alpha t > \beta x \Rightarrow u] : Q]\!] = \emptyset \quad \text{if } \alpha \neq \beta ,$$
$$[\![\Gamma \vdash [\Sigma_{i \in I} t_i > \beta x \Rightarrow u] : Q]\!] \cong \Sigma_{i \in I} [\![\Gamma \vdash [t_i > \beta x \Rightarrow u] : Q]\!] .$$

General patterns

We can write terms more compactly by generalising the patterns in matches. General patterns are built up according to

$$p ::= x \mid \emptyset \mid \alpha p \mid p \otimes q \mid (p, -) \mid (-, p) .$$

A match on a pattern $[u > p \Rightarrow t]$ binds the free variables of the pattern p to the resumptions after following the path specified by the pattern in u; because the term t may contain these variables freely the resumptions may influence the computation of t. Such a match is understood inductively as an abbreviation for a term in the metalanguage:

$$[u > x \Rightarrow t] \equiv (\lambda x.t)\, u , \qquad [u > \emptyset \Rightarrow t] \equiv t ,$$
$$[u > \alpha p \Rightarrow t] \equiv [u > \alpha x \Rightarrow [x > p \Rightarrow t]] \quad \text{for a fresh variable } x,$$
$$[u > (p, -) \Rightarrow t] \equiv [u > (x, -) \Rightarrow [x > p \Rightarrow t]] \quad \text{for a fresh variable } x,$$
$$[u > (-, p) \Rightarrow t] \equiv [u > (-, x) \Rightarrow [x > p \Rightarrow t]] \quad \text{for a fresh variable } x,$$
$$[u > p \otimes q \Rightarrow t] \equiv [u > x \otimes y \Rightarrow [x > p \Rightarrow [y > q \Rightarrow t]]] \text{ for fresh } x, y.$$

Let $\lambda x \otimes y.t$ stand for $\lambda w.[w > x \otimes y \Rightarrow t]$, where w is a fresh variable, and write $[u_1 > p_1, \cdots, u_k > p_k \Rightarrow t]$ to abbreviate $[u_1 > p_1 \Rightarrow [\cdots [u_k > p_k \Rightarrow t] \cdots]$.

4.7 Examples

The affine-linear language is remarkably expressive, as the following examples show. Through having denotations in **Aff$_{\text{Set}}$**, all operations expressible in the language will automatically preserve open-map bisimulation.

4.7.1 CCS

As in CCS, assume a set of labels A, a complementation operation producing $\bar{a}$ from a label a, with $\bar{\bar{a}} = a$, and a distinct label τ. In the metalanguage we can specify the path order $\mathbb{P}$ as the solution to[†]

$$\mathbb{P} = \tau\mathbb{P} + \Sigma_{a \in A} a\mathbb{P} + \Sigma_{a \in A} \bar{a}\mathbb{P} .$$

So $\mathbb{P}$ is given as $\mu P.\tau P + \Sigma_{a \in A} aP + \Sigma_{a \in A} \bar{a}P$. There are injections from $\mathbb{P}$ into its expression as a prefixed sum given as τt, at and $\bar{a}t$ for $a \in A$ and a term t of type $\mathbb{P}$. The CCS parallel composition can be defined as the following term of type $\mathbb{P} \otimes \mathbb{P} \multimap \mathbb{P}$ in the metalanguage:

$$Par = rec\ P.\ \lambda x \otimes y.\ \Sigma_{\alpha \in A \cup \{\tau\}}[x > \alpha x' \Rightarrow \alpha(P(x' \otimes y))]+$$
$$\Sigma_{\alpha \in A \cup \{\tau\}}[y > \alpha y' \Rightarrow \alpha(P(x \otimes y'))]+$$
$$\Sigma_{a \in A}[x > ax',\ y > \bar{a}y' \Rightarrow \tau(P(x' \otimes y'))] .$$

The other CCS operations are easy to encode, though recursive definitions in CCS have to be restricted to fit within the affine language. Interpreted in $\mathbf{Aff_2}$ two CCS terms will have the same denotation iff they have same traces (or execution sequences). By virtue of having been written down in the metalanguage the operation of parallel composition will preserve open-map bisimulation when interpreted in $\mathbf{Aff_{Set}}$; for this specific $\mathbb{P}$, open-map bisimulation coincides with strong bisimulation. In $\mathbf{Aff_{Set}}$ we can recover the expansion law by the properties of prefix match—Lemma 4. In detail, write $X|Y$ for $Par\ X \otimes Y$, where X and Y are terms of type $\mathbb{P}$. Suppose

$$X = \Sigma_{\alpha \in A \cup \{\tau\}}\Sigma_{i \in I(\alpha)}\alpha X_i , \qquad Y = \Sigma_{\alpha \in A \cup \{\tau\}}\Sigma_{j \in J(\alpha)}\alpha Y_j .$$

Using Lemma 2, and then that the matches distribute over nondeterministic sums,

$$X|Y \cong \Sigma_{\alpha \in A \cup \{\tau\}}[X > \alpha x' \Rightarrow \alpha(x'|Y)] + \Sigma_{\alpha \in A \cup \{\tau\}}[Y > \alpha y' \Rightarrow \alpha(X|y')]$$
$$+ \Sigma_{a \in A}[X > ax',\ Y > \bar{a}y' \Rightarrow \tau(x'|y')]$$
$$\cong \Sigma_{\alpha \in A \cup \{\tau\}}\Sigma_{i \in I(\alpha)}\alpha(X_i|Y) + \Sigma_{\alpha \in A \cup \{\tau\}}\Sigma_{j \in J(\alpha)}\alpha(X|Y_j)$$
$$+ \Sigma_{a \in A}\Sigma_{i \in I(a),j \in J(\bar{a})}\tau(X_i|Y_j) .$$

In similar ways it is easy to express CSP in the affine-linear language along the lines of [4], and any parallel composition given by a synchronisation algebra [31].

[†] In examples, for readablility, we will generally write recursive definitions of types and processes as equations.

4.7.2 A linear higher-order process language

Recall the path orders for processes, concretions and abstractions for a higher-order language in Section 4.5.6. We are chiefly interested in the parallel composition of processes, $Par_{\mathbb{P},\mathbb{P}}$ of type $\mathbb{P} \otimes \mathbb{P} \multimap \mathbb{P}$. But parallel composition is really a family of mutually dependent operations also including components such as $Par_{\mathbb{F},\mathbb{C}}$ of type $\mathbb{F} \otimes \mathbb{C} \multimap \mathbb{P}$ to say how abstractions compose in parallel with concretions *etc.* All these components can be tupled together in a product using $\&$, and parallel composition defined as a simultaneous recursive definition whose component at $\mathbb{P} \otimes \mathbb{P} \multimap \mathbb{P}$ satisfies

$$P|Q = \Sigma_\alpha[P > \alpha x \Rightarrow \alpha(x|Q)] +$$
$$\Sigma_\alpha[Q > \alpha y \Rightarrow \alpha(P|y)] +$$
$$\Sigma_a[P > a?f, \; Q > a!(s \otimes r) \Rightarrow \tau((f\,s)|r)] +$$
$$\Sigma_a[P > a!(s \otimes r), \; Q > a?f \Rightarrow \tau(r|(f\,s))] \,,$$

where we have chosen suggestive names for the injections and, for instance, $P|Q$ abbreviates $Par_{\mathbb{P},\mathbb{P}}(P \otimes Q)$. In the summations $a \in A$ and α ranges over $a!, a?, \tau$ for $a \in A$.

4.7.3 Mobile ambients with public names

We can translate the Ambient Calculus with public names [6] into the affine-linear language, following similar lines to the linear process-passing language above. Assume a fixed set of ambient names $n, m, \cdots \in N$. The syntax of ambients is extended beyond just processes (P) to include concretions (C) and abstractions (F), following [5]:

$$P ::= \emptyset \mid P|P \mid rep\,P \mid n[P] \mid in\,n\,P \mid out\,n\,P \mid open\,n!\,P \mid$$
$$\tau\,P \mid mvin\,n!\,C \mid mvout\,n!\,C \mid open\,n?\,P \mid mvin\,n?\,F \mid x$$
$$C ::= P \otimes P$$
$$F ::= \lambda x.P$$

The notation for actions departs a little from that of [5]. Here some actions are marked with ! and others with ?—active (or inceptive) actions are marked by ! and passive (or receptive) actions by ?. We say actions α and β are *complementary* iff one has the form $open\,n!$ or $mvin\,n!$ while the other is $open\,n?$ or $mvin\,n?$ respectively. Complementary actions can synchronise together to form a τ-action. We adopt a slightly different notation for concretions ($P \otimes R$ instead of $\langle P \rangle R$) and abstractions ($\lambda x.P$

instead of $(x)P$ to make their translation into the affine-linear language clear.

The usual conventions are adopted for variables. Terms are assumed to be *linear*, in that a variable appears on at most one side of any parallel compositions within the term, and subterms of the form $repP$ have no free variables. A replication $repP$ is intended to behave as $P \mid repP$ so readily possesses a recursive definition in the affine-linear language.

Suitable path orders for ambients are given recursively by:

$$\mathbb{P} = \tau\mathbb{P} + \Sigma_{n \in N} in \ n \ \mathbb{P} + \Sigma_{n \in N} out \ n \ \mathbb{P} + \Sigma_{n \in N} open \ n!\mathbb{P}+$$

$$\Sigma_{n \in N} mvin \ n!\mathbb{C} + \Sigma_{n \in N} mvout \ n!\mathbb{C} + \Sigma_{n \in N} open \ n?\mathbb{P} + \Sigma_{n \in N} mvin \ n?\mathbb{F}$$

$$\mathbb{C} = \mathbb{P} \otimes \mathbb{P}$$

$$\mathbb{F} = \mathbb{P} \multimap \mathbb{P}$$

The eight components of the prefixed sum in the equation for $\mathbb{P}$ correspond to eight forms of ambient actions: τ, $in \ n$, $out \ n$, $open \ n!$, $mvin \ n!$, $mvout \ n!$, $open \ n?$ and $mvin \ n?$. We obtain the prefixing operations as injections into the appropriate component of $\mathbb{P}$ as a prefixed sum.

Parallel composition is really a family of operations, one of which is a binary operation between processes but where in addition there are parallel compositions of abstractions with concretions, and even abstractions with processes and concretions with processes. The family of operations

$$(-|-) : \mathbb{F} \otimes \mathbb{C} \multimap \mathbb{P}, \quad (-|-) : \mathbb{C} \otimes \mathbb{F} \multimap \mathbb{P},$$
$$(-|-) : \mathbb{F} \otimes \mathbb{P} \multimap \mathbb{F}, \quad (-|-) : \mathbb{P} \otimes \mathbb{F} \multimap \mathbb{F},$$
$$(-|-) : \mathbb{C} \otimes \mathbb{P} \multimap \mathbb{C}, \quad (-|-) : \mathbb{P} \otimes \mathbb{C} \multimap \mathbb{C}$$

are defined in a simultaneous recursive definition as follows:
Processes in parallel with processes:

$$P|Q = \Sigma_\alpha[P > \alpha x \Rightarrow \alpha(x|Q)]+$$
$$\Sigma_\alpha[Q > \alpha y \Rightarrow \alpha(P|y)]+$$
$$\Sigma_n[P > open \ n!x, \ Q > open \ n?y \Rightarrow \tau(x|y)]+$$
$$\Sigma_n[P > open \ n?x, \ Q > open \ n!y \Rightarrow \tau(x|y)]+$$
$$\Sigma_n[P > mvin \ n?f, \ Q > mvin \ n!(s \otimes r) \Rightarrow \tau((f \ s)|r)]+$$
$$\Sigma_n[P > mvin \ n!(s \otimes r), \ Q > mvin \ n?f \Rightarrow \tau(r|(f \ s))] \ .$$

Abstractions in parallel with concretions:

$$F|C = [C > s \otimes r \Rightarrow (F \ s)|r] \ .$$

Abstractions in parallel with processes:

$$F|P = \lambda x.((F\ x)|P)\ .$$

Concretions in parallel with processes:

$$C|P = [C > s \otimes r \Rightarrow s \otimes (r|P)]\ .$$

The remaining cases are given symmetrically.

Presheaves X, Y over $\mathbb{P}$ will have decompositions into rooted components:

$$X \cong \Sigma_\alpha \Sigma_{i \in X(\alpha)} \alpha X_i\ , \qquad Y \cong \Sigma_\alpha \Sigma_{j \in X(\alpha)} \alpha Y_j$$

—here α ranges over ambient actions. By the properties of prefix-match (Lemma 4), their parallel composition satisfies the expansion law

$$X|Y \cong \Sigma_\alpha \Sigma_{i \in X(\alpha)} \alpha(X_i|Y) + \Sigma_\alpha \Sigma_{j \in Y(\alpha)} \alpha(X|Y_j) +$$

$$\Sigma_n \Sigma_{i \in X(open\ n!),j \in Y(open\ n?)} \tau(X_i|Y_j) + \Sigma_n \Sigma_{i \in X(open\ n?),j \in Y(open\ n!)} \tau(X_i|Y_j) +$$

$$\Sigma_n \Sigma_{i \in X(mvin\ n!),j \in Y(mvin\ n?)} \tau(X_i|Y_j) + \Sigma_n \Sigma_{i \in X(mvin\ n?),j \in Y(mvin\ n!)} \tau(X_i|Y_j)$$

Ambient creation can be defined recursively in the affine-linear language:

$$m[P] = [P > \tau x \Rightarrow \tau m[x]] +$$

$$\Sigma_n [P > in\ n\ x \Rightarrow mvin\ n!(m[x] \otimes \emptyset)] +$$

$$\Sigma_n [P > out\ n\ x \Rightarrow mvout\ n!(m[x] \otimes \emptyset)] +$$

$$[P > mvout\ m!(s \otimes r) \Rightarrow \tau(s|m[r])] +$$

$$open\ m?P + mvin\ m?\ \lambda y.m[P|y]\ .$$

The denotations of ambients are determined by their capabilities: an ambient $m[P]$ can perform the internal (τ) actions of P, enter a parallel ambient ($mvin\ n!$) if called upon to do so by an *in n*-action of P, exit an ambient n ($mvout\ n!$) if P so requests through an *out n*-action, be exited if P so requests through an *mvout m!*-action, be opened ($open\ m?$),or be entered by an ambient ($mvin\ m?$); initial actions of other forms are restricted away. Ambient creation is at least as complicated as parallel composition. This should not be surprising given that ambient creation corresponds intuitively to putting a process behind (so in parallel with) a wall or membrane which if unopened mediates in the communications the process can do, converting some actions to others and restricting some others away. The tree-containment structure of ambients is captured in the chain of *open m?*'s that they can perform.

By the properties of prefix-match, there is an expansion theorem for ambient creation. For X with decomposition

$$X = \Sigma_\alpha \Sigma_{i \in X(\alpha)} \alpha X_i \ ,$$

where α ranges over atomic actions of ambients,

$$m[X] \cong \Sigma_{i \in X(\tau)} \tau \ m[X_i]$$
$$\Sigma_n \Sigma_{j \in X(in \ n)} mvin \ n!(m[X_j] \otimes \emptyset) +$$
$$\Sigma_n \Sigma_{k \in X(out \ n)} mvout \ n!(m[X_k] \otimes \emptyset)$$
$$\Sigma_{s \in X(mvout \ m!)} [X_s > s \otimes r \Rightarrow \tau(s|m[r])] +$$
$$open \ m?X + mvin \ m?(\lambda y.m[X|y]) \ .$$

4.7.4 Nondeterministic dataflow

The affine linear language allows us to define processes of the kind encountered in treatments of nondeterministic dataflow.

Define $\mathbb{P}$ recursively so that

$$\mathbb{P} = a\mathbb{P} + b\mathbb{P} \ .$$

$\mathbb{P}$ consists of finite streams (or sequences) of a's and b's.

The recursively defined process $A : \mathbb{P} \multimap \mathbb{P}$ selects and outputs a's from a stream of a's and while ignoring all b's:

$$A = \lambda x. \ [x > ax' \Rightarrow a(A \ x')] + [x > bx' \Rightarrow z_1 \otimes (A \ x')]$$

The recursively defined process $F : \mathbb{P} \otimes \mathbb{P}$ produces two parallel streams of a's and b's as output such that it outputs the same number of a's and b's to both streams:

$$F = [F > z_1 \otimes z_2 \Rightarrow (az_1) \otimes (az_2) + (bz_1) \otimes (bz_2)]$$

The recursively defined process $S : \mathbb{P} \multimap (\mathbb{P} \otimes \mathbb{P})$ separates a stream of a's and b's into two streams, the first consisting solely of a's and the second solely of b's:

$$S = \lambda x. \quad [x > ax', (S \ x') > z_1 \otimes z_2 \Rightarrow (az_1) \otimes z_2] +$$
$$[x > bx', (S \ x') > z_1 \otimes z_2 \Rightarrow z_1 \otimes (bz_2)]$$

A subcategory of $\mathbf{Aff_{Set}}$ supports a "trace operation" to represent processes with feedback loops (see [13]). The trace operation is, however, not definable in the present affine-linear language. It can be shown by induction on the typing derivation of a term that:

Proposition 4.7 *Suppose* $\Gamma \vdash t : \mathbb{Q}$ *where* $\Gamma \equiv x_1 : \mathbb{P}_1, \cdots, x_k : \mathbb{P}_k$. *Let* p *be path in* $(\mathbb{P}_1 \otimes \cdots \otimes \mathbb{P}_k)_\perp$ *and* q *be a path in* $\mathbb{Q}$. *The presheaf denotation of a term* $\Gamma \vdash t : \mathbb{Q}$, *applied to* $j_{\mathbb{P}_1 \otimes \cdots \otimes \mathbb{P}_k}(p)$ *as presheaf, has a nonempty contribution at* q *iff the trace-set denotation of* $\Gamma \vdash t : \mathbb{Q}$, *applied to* $j_{\mathbb{P}_1 \otimes \cdots \otimes \mathbb{P}_k}(p)$ *as a trace-set, contains* q.

Thus, supposing that the trace operation of [13] were definable in the presheaf semantics, we would obtain a compositional relational semantics of nondeterministic dataflow with feedback, shown impossible by the Brock-Ackerman anomaly [3].

4.8 Nonlinearity

Of course code can be copied, and this may lead to maps which are not linear. According to the discipline of linear logic, nonlinear maps from $\mathbb{P}$ to $\mathbb{Q}$ are introduced as linear maps from $!\mathbb{P}$ to $\mathbb{Q}$—the exponential $!$ applied to $\mathbb{P}$ allows arguments from $\mathbb{P}$ to be copied or discarded freely.

In the domain model of linear logic $!\mathbb{P}$ can be taken to be the finite-join completion of $\mathbb{P}$. Then, the nonlinear maps, maps $!\mathbb{P} \longrightarrow \mathbb{Q}$ in $\mathbf{Lin_2}$, correspond to Scott continuous functions $\widehat{\mathbb{P}} \longrightarrow \widehat{\mathbb{Q}}$. A close analogue for presheaf models is to interpret $!\mathbb{P}$ as the finite-colimit completion of $\mathbb{P}$. Note that now $!\mathbb{P}$ is a category, and no longer just a partial order. With this understanding of $!\mathbb{P}$, it can be shown that $\widehat{\mathbb{P}}$ with the inclusion functor $!\mathbb{P} \longrightarrow \widehat{\mathbb{P}}$ is the free filtered colimit completion of $!\mathbb{P}$—see [21]. It follows that maps $!\mathbb{P} \longrightarrow \mathbb{Q}$ in $\mathbf{Lin_{Set}}$ correspond, to within isomorphism, to continuous (*i.e.*, filtered colimit preserving) functors $\widehat{\mathbb{P}} \longrightarrow \widehat{\mathbb{Q}}$. But, unfortunately, continuous functors from $\widehat{\mathbb{P}}$ to $\widehat{\mathbb{Q}}$ need not send open maps to open maps. This raises the question of whether other choices of exponential fit in better with bisimulation.

Bear in mind the intuition that objects of $\mathbb{P}$ correspond to the shapes of computation path a process, represented as a presheaf in $\widehat{\mathbb{P}}$, might perform. An object of $!\mathbb{P}$ should represent a computation path of an assembly of processes each with computation-path shapes in $\mathbb{P}$—the assembly of processes can then be the collection of copies of a process, possibly at different states. If we take $!\mathbb{P}$ to be the finite colimit completion of $\mathbb{P}$, an object of $!\mathbb{P}$ as a finite colimit would express how paths coincide initially and then branch. One way to understand this object as a computation path of an assembly of processes, is that the assembly of processes is not fixed once and for all. Rather the assembly grows as further copies are invoked, and that these copies can be made of a processes *after* they have run for a while. The copies can then themselves be run and the resulting processes copied. In this way, by keeping track of the origins of copies, we can account for the identifications of sub-paths.

This intuition suggests exploring other less liberal ways of copying, without, for example, being able to copy after some initial run. We will discover candidates for exponentials !ℙ based on computation-path shapes of simple assemblies of processes, ones built out of indexed families. We start with an example.

4.8.1 An example

First observe the hopeful sign that maps which are not linear may still preserve bisimulation. For example, a functor yielding a presheaf $H(X, Y)$, for presheaves X and Y over ℙ, which is "bilinear" or "affine bilinear," in the sense that it is linear (*i.e.*, colimit preserving) or affine linear (*i.e.*, connected-colimit preserving) in each argument separately, when diagonalised to the functor giving $H(X, X)$ for X in $\widehat{ℙ}$, will still preserve open maps and bisimulation. A well-known example of a bilinear functor is the product operation on presheaves [18]; with one argument fixed, the product is left adjoint to the exponentiation in the presheaf category, and so product preserves colimits, and thus open maps, in each argument. On similar lines, it can be shown that the tensor operations in **Lin**_{Set} and **Aff**_{Set} are bilinear and affine bilinear, respectively. With this encouragement we look for alternative interpretations of the exponential !, where the nonlinear maps !ℙ ⟶ ℚ in **Lin**_{Set} preserve open maps.

Because sum preserves open maps, by the remarks above, the functor *copy* taking a presheaf X over ℙ to the presheaf

$$copy(X) = 1 + X + X^2 + X^3 + \cdots + X^k + \cdots$$

over

$$!ℙ = 1 + ℙ + ℙ^2 + ℙ^3 + \cdots + ℙ^k + \cdots$$

will preserve open maps. Here the superscripts abbreviate repeated applications of tensor in **Lin**_{Set}. So $ℙ^k$ is the product of k copies of the partial order ℙ, in which the objects are k-tuples of objects of ℙ—in particular, 1 is the partial order consisting solely of the empty tuple called 1 above. The presheaf X^k comprises k copies of X tensored together, so that $X^k\langle p_1, \cdots, p_k \rangle = X(p_1) \times \cdots \times X(p_k)$.

By supplying "coefficients" we can obtain various nonlinear maps. An appropriate form of polynomial is given by a functor

$$F : !\mathbb{P} \longrightarrow \widehat{\mathbb{Q}} ,$$

which splits up into a family of functors

$$F_k : \mathbb{P}^k \longrightarrow \widehat{\mathbb{Q}} , \text{ for } k \in \omega .$$

We can extend F to a functor $F[-] = F \cdot copy(-) : \widehat{\mathbb{P}} \longrightarrow \widehat{\mathbb{Q}}$. For $X \in \widehat{\mathbb{P}}$,

$$F[X] = F_0 + F_1 \cdot X + F_2 \cdot X^2 + F_3 \cdot X^3 + \cdots + F_k \cdot X^k + \cdots$$

Because $F \cdot -$ is colimit-preserving it preserves open maps. So does *copy*. Hence $F[-]$ preserves open maps.

Note, that the original polynomial F is not determined to within isomorphism by the functor $F[-]$ it induces. (We can only hope for such uniqueness if we restrict to polynomials which are symmetric, *i.e.*, such that $F_k \cong F_k \circ \pi$ for all permutations π of the k arguments.)

We write **Poly**$(\mathbb{P}, \mathbb{Q})$ for the functor category $[!\mathbb{P}, \widehat{\mathbb{Q}}]$ of polynomials from $\mathbb{P}$ to $\mathbb{Q}$. In order to compose polynomials, $F \in$ **Poly**$(\mathbb{P}, \mathbb{Q})$ and $G \in$ **Poly**$(\mathbb{Q}, \mathbb{R})$, we first define $F^! \in$ **Poly**$(\mathbb{P}, !\mathbb{Q})$ by taking

$$F^!\langle p_1, \cdots, p_n \rangle \langle q_1, \cdots, q_k \rangle = \Sigma_{\mu\langle s_1, \cdots, s_k \rangle = \langle p_1, \cdots, p_n \rangle} F s_1 q_1 \times \cdots \times F s_k q_k ,$$

when $\langle p_1, \cdots, p_n \rangle \in !\mathbb{P}$ and $\langle q_1, \cdots, q_k \rangle \in !\mathbb{Q}$. The operation $\mu :$ $!!\mathbb{P} \longrightarrow \mathbb{P}$ flattens, by concatenation, a tuple $\langle s_1, \cdots, s_k \rangle$ of tuples $s_r = \langle s_{r1}, \cdots, s_{rm_r} \rangle$, for $1 \leq r \leq k$, down to a tuple

$$\mu\langle s_1, \cdots, s_k \rangle = \langle s_{11}, \cdots, s_{1m_1}, s_{21}, \cdots s_{k1}, \cdots, s_{km_k} \rangle .$$

So, the sum is indexed by all ways to partition $\langle p_1, \cdots, p_n \rangle$ into tuples $\langle s_1, \cdots, s_k \rangle$. Now, we can define the composition of polynomials to be

$$G \circ F = G \cdot (F^! -) \in \textbf{Poly}(\mathbb{P}, \mathbb{R}) .$$

At $\langle p_1, \cdots, p_n \rangle$ in $!\mathbb{P}$,

$$G \circ F\langle p_1, \cdots, p_n \rangle = \Sigma_{\mu\langle s_1, \cdots, s_k \rangle = \langle p_1, \cdots, p_n \rangle} G_k \cdot F s_1 \times \cdots \times F s_k ,$$

where $F s_1 \times \cdots F s_k$, built using the tensor of **Lin**, is such that

$$F s_1 \times \cdots F s_k \langle q_1, \cdots, q_k \rangle = F s_1 q_1 \times \cdots F s_k q_k$$

for $\langle q_1, \cdots, q_k \rangle$ in $!\mathbb{Q}$. The composition of polynomials is only defined to within isomorphism; they form a bicategory **Poly**, rather than a category.

Note that $!\mathbb{O} = \mathbb{1}$. In the special case where $F : !\mathbb{O} \longrightarrow \widehat{\mathbb{Q}}$, so that F

merely points to a presheaf X in $\widehat{\mathbb{Q}}$, the composition $G{\circ}F$ of a polynomial $G :\mathord{!}\mathbb{Q} \longrightarrow \widehat{\mathbb{R}}$ with the polynomial F is isomorphic to $G[X]$. So certainly compositions of this form preserve open maps and bisimulation.

More generally, polynomials in $\mathbf{Poly}(\mathbb{P}, \mathbb{Q})$ and $\mathbf{Poly}(\mathbb{Q}, \mathbb{R})$ correspond to presheaves in $(\widehat{\mathord{!}\mathbb{P})^{op}} \times \mathbb{Q}$ and $(\widehat{\mathord{!}\mathbb{Q})^{op}} \times \mathbb{R}$, respectively. So under this correspondence polynomials are related by open maps and bisimulation. It can be shown that the composition of polynomials in general preserves open maps, so bisimulation, between polynomials.

However, the present interpretation of ! fails as a candidate for the exponential of linear logic. This is because $\mathbf{Poly}$ is not cartesian-closed in any reasonable sense. It easy to see that there is an isomorphism of categories

$$\mathbf{Poly}(\mathbb{R}, \mathbb{P}\&\mathbb{Q}) \cong \mathbf{Poly}(\mathbb{R}, \mathbb{P}) \times \mathbf{Poly}(\mathbb{R}, \mathbb{Q}) ,$$

natural in $\mathbb{R}$ in $\mathbf{Lin_{Set}}$, showing the sense in which $\mathbb{P}\&\mathbb{Q}$, given by juxtaposition, remains a product in the bicategory of polynomials. There is also clearly an isomorphism of functor categories

$$[\mathord{!}\mathbb{P}\times\mathord{!}\mathbb{Q}, \widehat{\mathbb{R}}] \cong [\mathord{!}\mathbb{P}, ((\widehat{\mathord{!}\mathbb{Q})^{op}} \times \mathbb{R})] .$$

But, in general, $\mathord{!}(\mathbb{P}\&\mathbb{Q})$ and $\mathord{!}\mathbb{P}\times\mathord{!}\mathbb{Q}$ are not isomorphic, so that $(\mathord{!}\mathbb{Q})^{op}\times\mathbb{R}$ is not a function space for the polynomials with respect to $-\&-$. The difficulty boils down to a lack of symmetry in the current definition of $\mathord{!}\mathbb{P}$, where tuples like $\langle p_1, \cdots, p_k \rangle$ and its permutations $\langle p_{\pi(1)}, \cdots, p_{\pi(k)} \rangle$ are not necessarily related by any maps. Nor for that matter, are there any maps from a tuple $\langle p_1, \cdots, p_k \rangle$ to a larger tuple $\langle p_1, \cdots, p_k, \cdots, p_m \rangle$, even though intuitively the larger tuple would be a path of a larger assembly of processes, so arguably an extension of the smaller tuple in which further copies have been invoked.

To allow different kinds of polynomial, polynomials which can take account of the symmetry there exists between different copies and also permit further copies to be invoked as needed, we broaden the picture.

4.8.2 General polynomials

The example suggests that we take assemblies of processes to be families where we can reindex copies, precisely how being prescribed in $\mathbb{U}$, a subcategory of sets in which the maps are the possible reindexings. A $\mathbb{U}$-family of a category $\mathcal{A}$ comprises $\langle A_i \rangle_{i \in I}$ where $i \in I$, with I an object of $\mathbb{U}$, index objects A_i in $\mathcal{A}$. A map of families $(f, e) : \langle A_i \rangle_{i \in I} \to \langle A'_j \rangle_{j \in J}$ consists of a reindexing function $f : I \to J$ in $\mathbb{U}$ and $e = \langle e_i \rangle_{i \in I}$, a family of maps $e_i : A_i \to A'_{f(i)}$ in $\mathcal{A}$. With the obvious composition we obtain $\mathcal{F}_U(\mathcal{A})$, the category of U-families.

Imitating the example, we define the category of polynomials

$$\mathbf{Poly}_U(\mathbb{P}, \mathbb{Q})$$

from $\mathbb{P}$ to $\mathbb{Q}$, to be the functor category $[\mathcal{F}_U(\mathbb{P}), \mathbb{Q}]$. Under sufficient conditions, that U is small, has a singleton and dependent sums (a functor $\Sigma : \mathcal{F}_U(\mathbb{U}) \to \mathbb{U}$ collapsing any family of sets in $\mathbb{U}$ to a set in $\mathbb{U}$), we can compose polynomials in the manner of the coKleisli construction. For this we need to turn $\mathcal{F}_U$ into a functor on polynomials for which we need a "distributive law" converting a family of presheaves into a presheaf over families of paths. It can be shown that provided all the maps in U (the possible reindexings) are injective, composition of polynomials preserves open maps and bisimulation. Provided U contains the empty set, we can specialise composition, as in the example, to obtain a functor $F[-] : \widehat{\mathbb{P}} \longrightarrow \widehat{\mathbb{Q}}$ from a polynomial F in $\mathbf{Poly}_U(\mathbb{P}, \mathbb{Q})$.

The example is now seen as the special case in which U consists of subsets, possibly empty, of positive natural numbers $\{1, \cdots, n\}$ with identities as the only maps. In the special case in which U is the full subcategory of **Set** consisting of the empty set and a singleton, polynomials amount to functors $\mathbb{P}_\perp \longrightarrow \widehat{\mathbb{Q}}$ so to maps in **Aff**$_{\mathbf{Set}}$. If we take U to be $\mathbb{I}$ (finite sets with injections) or $\mathbb{B}$ (finite sets with bijections), we can repair an inadequacy in the example; then, $\mathcal{F}_U(\mathbb{P}\&\mathbb{Q})$ and $\mathcal{F}_U(\mathbb{P}) \times \mathcal{F}_U(\mathbb{Q})$ are isomorphic, so that we obtain a function space for the polynomials with respect to the product $-\&-$. Both $\mathcal{F}_\mathbb{I}$ and $\mathcal{F}_\mathbb{B}$ are good candidates for the exponential !—they also behave well with respect to bisimulation.

There is a fly in the ointment however. The complete mathematical story, in which one would see the polynomials as maps in a coKleisli construction, uses bicategories and at least pseudo (co)monads on biequivalent 2-categories. At the time of writing (December 2001) the theory of pseudo monads, even the definitions, is not sufficiently developed.

4.9 Related work

This article presents two domain theories for concurrent computation. One uses domains of a traditional kind, though in a non-traditional way through being based on computation paths (though the path-based domain theory here was anticipated in Matthew Hennessy's work on domain models of concurrency [14]). In the other domain theory, domains are understood as presheaf categories, accompanied by the bisimulation equivalence got from open maps.

Just as there are alternatives to the domain theory of Dana Scott, in particular, the stable domain theory of Gérard Berry, so are there alternative denotational semantics of the affine-linear language. Mikkel Nygaard and I have shown how to give an event-structure semantics to the affine-linear language; both types and terms denote event structures. (Event structures are a key "independence model" for computation, one in which the concurrency of events is represented by their causal independence.) The domain theory can be seen as analogous to the stable domain theory of Berry. The presheaf semantics here and that based on event structures differ at function spaces. In the fragment of the affine-linear language without function spaces, the event-structure semantics gives an informative representation of the definable presheaves; elements of a definable presheaf correspond to finite configurations of an event structure, with restriction in the presheaf matched by restriction to a subconfiguration in the event structure. Tensor corresponds to a simple parallel composition of event structures got by disjoint juxtaposition. Unfortunately the event structures definable in the affine-linear language can be shown too impoverished to coincide with those of the event-structure semantics of CCS, given for example in [31].

Mikkel Nygaard and I are developing an operational semantics for the affine linear language [26]. This work has also led us to an expressive nonlinear language with a simple operational semantics; its denotational semantics is based on a choice of exponential from Section 4.8. One aim is to give an operational account of open-map bisimulation on higher-order processes.

The semantics here does not cover name generation as in Milner's π-Calculus. Although one can give a presheaf semantics to the pi-Calculus [9], we do not presently know how to extend this to also include higher-order processes.

This article has demonstrated that linearity as formalised in linear logic can play a central role in developing a domain theory suitable for concurrent computation. At the same time, the mathematical neutrality of the domain theories here begins to show how concurrency need not remain the rather separate study it has become. There are unresolved issues in extending the work of this article towards a fully-fledged domain theory, one able to cope more completely with the range of models for concurrency. Among them is the question of how to extend this work to include name generation, its relation with operational semantics, and the place of independence models such as event structures.

Acknowledgements

A good deal of the background for this work was developed with Gian Luca Cattani for his PhD [7]. Discussions with Martin Hyland and John Power have played a crucial role in the ongoing work on nonlinearity. I am grateful for discussions with my PhD student Mikkel Nygaard.

Bibliography

[1] S. Abramsky. Computational interpretations of linear logic. *Theoretical Computer Science*, 111, 1-2, 3–57, 1993.

[2] T. Braüner. An Axiomatic Approach to Adequacy. BRICS Dissertation Series DS-96-4, 1996.

[3] J. Brock & W. Ackerman. Scenarios: A model of non-determinate computation. In *Proc. of Formalization of Programming Concepts*, LNCS 107, 1981.

[4] S.D. Brookes. On the relationship of CCS and CSP. In *Proc. of ICALP'83*, LNCS 154, 1983.

[5] L. Cardelli & A. Gordon. A commitment relation for the ambient calculus. Note ambient-commitment.pdf at http://research.microsoft.com/ adg/Publications/, 2000.

[6] L. Cardelli & A. Gordon. Anytime, Anywhere. Modal logics for mobile ambients. In *Proc. of POPL'00*, 2000.

[7] G. L. Cattani. PhD thesis, CS Dept., University of Aarhus, BRICS-DS-99-1, 1999.

[8] G. L. Cattani, M. Fiore & G. Winskel. A Theory of Recursive Domains with Applications to Concurrency. In *Proc. of LICS '98*.

[9] G. L. Cattani, I. Stark, & G. Winskel. Presheaf Models for the π-Calculus. In *Proc. of CTCS '97*, LNCS 1290, 1997.

[10] G. L. Cattani & G. Winskel. Presheaf Models for Concurrency. In *Proc. of CSL' 96*, LNCS 1258, 1997.

[11] G. L. Cattani & G. Winskel. Profunctors, open maps and bisimulation. Manuscript, 2000.

[12] G. L. Cattani, A. J. Power & G. Winskel. A categorical axiomatics for bisimulation. In *Proc. of CONCUR'98*, LNCS 1466, 1998.

[13] T. Hildebrandt, P. Panangaden & G. Winskel. Relational semantics of nondeterministic dataflow. In *Proc. of CONCUR'98*, LNCS 1466, 1998.

[14] M. Hennessy. A Fully Abstract Denotational Model for Higher-Order Processes. *Information and Computation*, 112:55–95, 1994.

[15] M. Hennessy & G.D. Plotkin. Full abstraction for a simple parallel programming language. In *Proc. of MFCS'79*, LNCS 74, 1979.

[16] C.A.R. Hoare. A model for communicating sequential processes. *Tech. Report PRG-22, University of Oxford Computing Lab.*, 1981.

[17] B. Jacobs. Semantics of weakening and contraction. *Annals of Pure and Applied Logic*, 69:73–106, 1994.

[18] A. Joyal & I. Moerdijk. A completeness theorem for open maps. *Annals of Pure and Applied Logic*, 70:51–86, 1994.

[19] A. Joyal, M. Nielsen & G. Winskel. Bisimulation from open maps. *Information and Computation*, 127:164–185, 1996.

[20] G.Winskel & K.Larsen. Using information systems to solve recursive domain equations effectively. LNCS 173, 1984.

[21] G.M. Kelly. *Basic concepts of enriched category theory*. London Math. Soc. Lecture Note Series 64, CUP, 1982.

[22] S. Mac Lane & I. Moerdijk. *Sheaves in Geometry and Logic*. Springer-Verlag, 1992..

[23] R. Milner. *A Calculus of Communicating Systems*. LNCS 92, 1980.

[24] R. Milner. *Communication and concurrency*. Prentice Hall, 1989.

[25] M. Nielsen, G.D. Plotkin & G. Winskel. Petri nets, Event structures and Domains, part 1. Theoretical Computer Science, vol. 13, 1981.

[26] M. Nygaard & G. Winskel. Linearity in distributed computation. In *Proc. of LICS '02*.

[27] G. Winskel. A representation of completely distributive algebraic lattices. Report of the Computer Science Dept., Carnegie-Mellon University, 1983.

[28] G. Winskel. An introduction to event structures. In *Proc. of REX summerschool in temporal logic, 'May 88*, LNCS 354, 1988.

[29] G. Winskel. A presheaf semantics of value-passing processes. In *Proceedings of CONCUR'96*, LNCS 1119, 1996.

[30] G. Winskel. A linear metalanguage for concurrency. In *Proceedings of AMAST'98*, LNCS 1548, 1999.

[31] G. Winskel & M. Nielsen. Models for concurrency. In *Handbook of Logic in Computer Science, Vol.4*, OUP, 1995.

Part two
Refereed Articles

5

An Axiomatic Approach to Structural Rules for Locative Linear Logic

Jean-Marc Andreoli

Xerox Research Centre Europe, Grenoble, France and
Institut de Mathématiques de Luminy, France

Abstract

This paper proposes a generic, axiomatic framework to express and study structural rules in resource conscious logics derived from Linear Logic. The proposed axioms aim at capturing minimal concepts, operations and relations in order to build an inference system which extends that of Linear Logic by the introduction of structure and structural rules, but still preserves in a very natural way the essential properties of any logical inference system: Cut elimination and Focussing. We consider here finite but unbounded structures, generated from elementary structures called "places". The set of places is "isotropic", in that no single place has a distinguished role in the structures, and each structure can be "transported" into an isomorphic structure by applying a permutation of places to the places from which it is built. The essential role of these places in the definition of Logic has been shown in Ludics [6], and leads to a *locative* reading of the traditional logical concepts (formulas, sequents, proofs), which is adopted here. All the logical connectives (multiplicatives, additives, exponentials) are expressed here in a locative manner.

5.1 Introduction

Linear Logic [5] essentially differs from Classical Logic by the removal of the structural axioms of Contraction and Weakening. However, it retains other structural axioms, such as Exchange, which determine the essential properties of the whole system. Thus, from a sequent calculus point of view, Linear Logic sequents are built from a single constructor (usually represented as the "comma") which is structurally considered associative-commutative, and hence, so are all the binary connectives. The structure of commutative monoid underlying phase semantics also results from this choice.

Various systems explore other syntactic structures for the sequents and other structural axioms. For example, Non-Commutative logic [1, 10], a conservative extension of both Linear Logic and Cyclic Logic [11], introduces order varieties as the structure for sequents (characterised by the axioms of "See-Saw" and "co-See-Saw"). The expression of the inference rules of the sequent system in this case relies in fact on two kinds of structures, order varieties and orders, where orders are order varieties viewed from a single point (that of the principal formula in an inference). Order varieties are "isotropic", in that any point in a sequent can, at any time, be chosen as occurrence of the principal formula, thus preserving the perfect symmetry of the system (this is essentially the "focussing" result of [10], not to be confused with the Focussing property [2, 3] which concerns commutative as well as non commutative Linear Logic and which has deeper implications). This duplicity in structure between order and order varieties corresponds in fact to a powerful intuition: when choosing a principal formula to perform an inference, one takes a specific viewpoint which alters the perception of the context. This intuition is the basis of all the development presented here.

We present a class of logics, called Coloured Linear Logics, also based on two kinds of structures, called "varieties" and "presentations" (a generalisation of, respectively, the order varieties and orders of Non Commutative Logic), together with abstract operators relating them. However, instead of attempting to choose *a priori* what varieties and presentations are, as in Non Commutative Logic, these structures are left undefined. Thus concrete instances of Coloured Linear Logics can be created from the pattern by different choices of structures and operators. It is nevertheless possible to define a generic inference system, using abstract varieties and presentations, and incorporating abstract structural rules. Its inferences simply capture the basic intuition that presentations are varieties "viewed from a specific viewpoint", that of the principal formula being decomposed in an inference.

In order for this system to satisfy the essential properties of Cut elimination and Focussing, characterising a real logic as more than an arbitrary set of rules, it is necessary to constrain the variety-presentation frameworks of Coloured Linear Logics. These constraints are expressed as a set of axioms which variety-presentation frameworks must satisfy.

For purpose of clarity, these axioms are introduced and discussed right away in Section 5.2, without any explicit reference to their role, however essential, in the sequent calculus. They give rise to a generic notion of substitution, with an associativity result which is the cornerstone of the whole system. The true motivations for the choice of these axioms appear in Section 5.3, where the sequent calculus is presented: the axioms ensure the fundamental properties (Cut Elimination, Focussing) which we expect from the calculus.

In a Coloured Linear Logic, as in Linear Logic, the sequent calculus deals with formula occurrences rather than formulas themselves: the structure of the sequents (presentations or varieties) is naturally defined over occurrences, and two occurrences of the same formula must not be confused. Occurrences are often kept implicit in the formulation of sequent calculi, but it is useful to make them explicit, especially to be able to keep track of the dependencies between the occurrences of formulas in the conclusion sequent of an inference and the occurrences in its premisses. One way of expliciting these dependencies is to use a "locative" version of the calculus, where each formula occurrence appearing in a proof has a unique address, and the addresses keep track of the formula/sub-formula dependencies. From that point of view, the "loci" of Ludics [6] provide an appropriate representation for addresses. The sequent calculus of Coloured Linear Logics is defined below in an entirely locative manner, using "loci" only instead of the usual syntax of formulas.

Note that this paper does not directly provide a solution to the general problem of structure in Linear Logic. It only offers a convenient framework in which to study this problem. The constraints expressed by the axioms of that framework are minimal, and leave many choices undecided as to the structures themselves and the structural rules (essentially any context-free linear rewrite rule on varieties is a good candidate for a structural rule). This degree of freedom is conceptually unsatisfactory: Logic aims at Necessity, and there should be no room for arbitrary choices. This issue can only be resolved by taking other perspectives to the problem, for example proof nets, games or denotational semantics, which could provide hints as to whether the whole approach makes sense (i.e. how relevant is the fundamental intuition supporting the distinction introduced in Coloured Linear Logics between presentations and varieties), and what the ultimate structure in a resource conscious logic should be.

5.2 Variety-Presentation Frameworks

5.2.1 Axioms

We make use of two basic abstract structures: presentations (ω) and varieties (α). Varieties are used below as the structural basis for sequents, and presentations for sequents presented from the point of view of one of their formula occurrences. We also make use of the following abstract operators on varieties and presentations, with the following signature (where $\wp(E)$ denotes the set of subsets of E):

Occurrence-set	μ : VARIETY $\cup$ PRESENTATION
	$\mapsto$ $\|\mu\|$: $\wp$(PRESENTATION)

A *place* is a presentation which is its own unique occurrence:

$$\text{PLACE} =_{\text{def}} \{ \, x \in \text{PRESENTATION} \mid |x| = \{x\} \, \}$$

Note here that places are a derived concept, not a primary one.

Composition	ω_1 : PRESENTATION , ω_2 : PRESENTATION
	$\mapsto$ $\omega_1 * \omega_2$: VARIETY
Decomposition	α : VARIETY , x : PLACE $\mapsto$ $(\alpha)_x$: PRESENTATION
Relaxation	α_1 : VARIETY $\preceq$ α_2 : VARIETY
Void	$\bigcirc$: PRESENTATION

We define the following axioms:

- **[Occurrence-set]**:
  The occurrence set of any presentation or variety μ is finite and contains only places. The set of places is countably infinite.

$$|\mu| \in \wp_f(\text{PLACE}) \qquad \text{PLACE} \equiv \mathbb{N}$$

- **[Composition]**:
  For any presentations ω_1, ω_2,

$$|\omega_1 * \omega_2| = |\omega_1| \odot |\omega_2|$$
$$\omega_1 * \omega_2 = \omega_2 * \omega_1$$

where $\odot$ denotes the symmetric difference operator on sets, ie. $A \odot B = (A \cup B) \setminus (A \cap B)$, which is associative-commutative with neutral $\emptyset$. Note that we could have restricted this axiom to the case where $|\omega_1| \cap |\omega_2| = \emptyset$ (hence $|\omega_1 * \omega_2| = |\omega_1| \cup |\omega_2|$), since that is the only case encountered later. Only, the more general formulation given here simplifies the proofs of the theorems.

- **[Decomposition]**:

  For any variety α, place x and presentation ω:

  $$x \in |\alpha| \;\Rightarrow\; x * (\alpha)_x = \alpha$$
  $$x \notin |\omega| \;\wedge\; \omega * x = \alpha \;\Rightarrow\; \omega = (\alpha)_x$$

  This implies, by **[Composition]**, that if $x \in |\alpha|$ then $|(\alpha)_x| = |\alpha| \setminus \{x\}$. Hence, for a given x, the mapping $\alpha \mapsto (\alpha)_x$ (for any variety α having occurrence x) and $\omega \mapsto \omega * x$ (for any presentation ω not having occurrence x) are inverse of each other.

- **[Commutation]**:

  For any variety α, presentations ω_1, ω_2 and places x_1, x_2,

  $$x_1 \neq x_2 \;\wedge\; |\omega_1| \cap |\omega_2| = \emptyset \;\wedge\; x_1 \in |\alpha| \setminus |\omega_2| \;\wedge\; x_2 \in |\alpha| \setminus |\omega_1| \;\Rightarrow$$
  $$((\alpha)_{x_1} * \omega_1)_{x_2} * \omega_2 = ((\alpha)_{x_2} * \omega_2)_{x_1} * \omega_1$$

  From the previous axioms, it is easy to show that, under the stated condition, the two sides of the equality have the same occurrence set. This axiom asserts that they are equal.

- **[Relaxation]**:

  For any varieties α_1, α_2, presentation ω and place x,

  $$\alpha_1 \preceq \alpha_2 \;\Rightarrow\; |\alpha_1| = |\alpha_2|$$
  $$x \in |\alpha_1| \;\wedge\; \alpha_1 \preceq \alpha_2 \;\Rightarrow\; (\alpha_1)_x * \omega \preceq (\alpha_2)_x * \omega$$

  Hence, relaxation applies only to varieties with the same occurrence set and is compatible with decomposition/composition.

- **[Void]**:

  For any presentation ω:

  $$|\omega| = \emptyset \;\Leftrightarrow\; \omega = \bigcirc$$

  Hence presentation $\bigcirc$ has no occurrence and it is the only one. Given a presentation ω, the variety $\omega * \bigcirc$ is written $\overline{\omega}$, and, by **[Composition]**, we have $|\overline{\omega}| = |\omega|$.

- **[Pair]**:

  There exists at least one presentation with two occurrences (or, equivalently, at least one variety with three occurrences). Assuming that the previous axioms hold, it can be shown that this axiom is equivalent to the following condition: for any variety α, there exists at least one presentation ω such that $\alpha = \overline{\omega}$.

Definition 5.1 *A* variety-presentation framework *is a pair of sets* VARI-ETY,PRESENTATION *equipped with occurrence-set, composition, decom-*

*position operators, relaxation relation and void element, satisfying all
the axioms above.*

The last two axioms ([**Void**] and [**Pair**]) are only used in the treatment
of the exponentials (Section 5.3.5). The multiplicative-additive case only
relies on the other axioms.

5.2.2 Examples

In the following examples of variety-presentation frameworks, we assume
given an arbitrary countably infinite set $\mathcal{P}$ the elements of which are
called points. In each example, we define two mappings: one which
maps each presentation or variety μ into a finite subset $\downarrow\mu$ of $\mathcal{P}$ called
the support set of μ, and one which maps each point p of $\mathcal{P}$ into a
presentation $\uparrow p$ called the promotion of p, such that the support set of
the promotion of p is exactly the singleton $\{p\}$.

$$\downarrow : \text{VARIETY} \cup \text{PRESENTATION} \longmapsto \wp_f(\mathcal{P})$$
$$\uparrow : \mathcal{P} \longmapsto \text{PRESENTATION}$$
$$\forall p \in \mathcal{P} \quad \downarrow\uparrow p = \{p\}$$

The occurrence set $|\mu|$ of a variety or presentation μ is then defined as
the set of promotions of the points of the support set of μ.

$$|\mu| = \{ \uparrow p \mid p \in \downarrow\mu \}$$

Hence, the places are exactly the promotions of the points of $\mathcal{P}$. Indeed,

- If μ is a place, then $\{\mu\} = |\mu| = \{\uparrow p \mid p \in \downarrow\mu\}$ hence $\mu = \uparrow p$ for some
  point p.
- If p is a point, then $|\uparrow p| = \{\uparrow q \mid q \in \downarrow\uparrow p\} = \{\uparrow q \mid q \in \{p\}\} = \{\uparrow p\}$
  hence $\uparrow p$ is a place.

The following examples are introduced by specifying the presentations
and varieties, together with the support and promotion mappings.

5.2.2.1 Sets

The most basic example of a variety-presentation framework is given by
Linear Logic, and captures its underlying structure of multiset.

- Presentations and varieties are simply finite subsets of $\mathcal{P}$ with no
  structure at all. The support set of a variety or presentation is itself
  while the promotion of a point p is the singleton $\{p\}$.

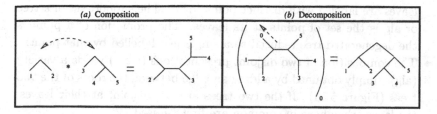

Fig. 5.1. The basic operators of presentations and varieties on trees and algs

- Composition is simply symmetric difference. Decomposition is asymmetric difference. Relaxation is equality.

$$\omega_1 * \omega_2 = \omega_1 \odot \omega_2 \qquad (\alpha)_x = \alpha \setminus x \qquad \alpha_1 \preceq \alpha_2 \Leftrightarrow \alpha_1 = \alpha_2$$

5.2.2.2 Orders and order varieties

Non Commutative Logic also provides an example of variety-presentation framework, which, in fact, was the starting point of this paper.

- A presentation (resp. variety) is a pair $\langle D, G \rangle$ where D is a finite subset of $\mathcal{P}$ and $G \subset D^2$ (resp. $G \subset D^3$) is the graph of an order (resp. order variety) over D. The set D is the support set of the presentation or variety. The promotion of a point p is the pair $(\{p\}, \emptyset)$, ie. the trivial order over the singleton $\{p\}$.
- The composition $\omega_1 * \omega_2$ of two presentations with disjoint support sets D_1, D_2 is defined as in [10] (composition of orders). If D_1, D_2 are not disjoint, then $\omega_1 * \omega_2$ is defined as $\omega_1' * \omega_2'$ where ω_i' is the presentation ω_i restricted to $D_i \setminus (D_1 \cap D_2)$. By construction, ω_1', ω_2' are disjoint. In other words, when composing two orders, they are first restricted to the points where they do not overlap.
- The decomposition $(\alpha)_x$ of a variety α along a place $x \in |\alpha|$ is defined as $(\alpha)_p$ in [10], where p is the point of the support set of α whose promotion is x.
- Relaxation between varieties $\alpha_1 = (D_1, G_1)$ and $\alpha_2 = (D_2, G_2)$ is simply defined as $D_1 = D_2$ and $G_1 \subseteq G_2$.

5.2.2.3 Trees and algs

Many examples of variety-presentation frameworks are based on graph-like structures, possibly decorated with "colours", eg. trees and algs.

- Presentations and varieties are binary trees (resp. ternary algs, ie. rootless trees) whose internal nodes are labeled with colours and whose leaves are labeled with distinct points of $\mathcal{P}$. The support set of a tree or alg is the set of points at its leaves. The promotion of a point is the degenerated tree reduced to a single leaf labelled by that point.

- The composition of two disjoint presentations (trees) yields a variety (alg), simply obtained by adding an edge between the roots of the two trees (Figure 5.1*a*). If the two trees are not disjoint at their leaves, the leaves they have in common are first "erased".

- The decomposition of a variety (alg) along one of its occurrences (leaf) yields a presentation (tree), obtained by selecting the chosen occurrence as the root (Figure 5.1*b*).

- The Relaxation relation between varieties can be that induced by any set of linear graph rewriting rules (linearity ensures that the alg structure is preserved).

In particular, consider the variety-presentation framework of trees and algs with two colours $\parallel$ and $<$, and with the following Relaxation rules

Associativity $\parallel$	$(x \parallel y) * (z \parallel t)$	$\preceq$	$(y \parallel z) * (t \parallel x)$
Associativity $<$	$(x < y) * (z < t)$	$\preceq$	$(y < z) * (t < x)$
Commutativity $\parallel$	$(x \parallel y) * z$	$\preceq$	$(x \parallel z) * y$
Entropy	$(x \parallel y) * z$	$\preceq$	$(x < y) * z$

Among these rules, Associativity and Commutativity are reversible, while Entropy is not. It is easy to show, using the analysis of [10], that this variety-presentation framework yields exactly the same logic as the framework of order varieties and orders (recalled in the previous example), or that of series-parallel order varieties and orders, namely Non Commutative Logic.

5.2.2.4 *Cyclic Colourings*

Let Δ be an arbitrary set of colours, equipped with an order relation $\preceq$ and an involutive operation $c \mapsto c^*$ which is compatible with $\preceq$ (ie. $c_1 \preceq c_2 \Rightarrow c_1^* \preceq c_2^*$). If D is a finite set, let $D^{!3}$ denote the set of triples of *pairwise distinct* elements of D. A cyclic colouring over D is a mapping ρ from $D^{!3}$ into Δ, such that for all $p_1, p_2, p_3 \in D^{!3}$,

$$\rho(p_2, p_3, p_1) = \rho(p_1, p_2, p_3) \ \wedge \ \rho(p_1, p_3, p_2) = \rho(p_1, p_2, p_3)^*$$

Let O be a distinguished object outside $\mathcal{P}$ (eg. $\mathcal{P}$ itself).

- A variety (resp. presentation) is a pair (D, ρ) where $D \subset \mathcal{P} \cup \{O\}$ such that $O \notin D$ (resp. $O \in D$) and ρ is a cyclic colouring over D. The support set of the variety (or presentation) is $D \setminus \{O\}$. For any point p, the promotion of p is the presentation $(\{O,p\},\emptyset)$ based on the unique cyclic colouring over $\{O,p\}$ (the empty one, since $\{O,p\}^{!3} = \emptyset$).

- The composition of two presentations $\omega_1 = (D_1,\rho_1)$ and $\omega_2 = (D_2,\rho_2)$ where $D_1 \cap D_2 = \{O\}$ is the variety (D,ρ), where $D = D_1 \cup D_2 \setminus \{O\}$ and ρ is defined for any $p_1,p_2,p_3 \in D^{!3}$ by

$$\rho(p_1,p_2,p_3) = \rho_j(q_1,q_2,q_3)$$
$$\text{where } \forall i = 1,2,3 \;\; q_i = (\text{if } p_i \in D_j \text{ then } p_i \text{ else } O)$$

and $j \in \{1,2\}$ is such that at least two elements out of $\{p_1,p_2,p_3\}$ belong to D_j. Under the assumption that $D_1 \cap D_2 = \{O\}$, the index j is uniquely defined, q_1,q_2,q_3 are pairwise distinct, and it is easy to show that ρ is a cyclic colouring over D. Thus, for example,

$$p_1,p_2 \in D_1 \wedge p_3 \in D_2 \Rightarrow \rho(p_1,p_2,p_3) = \rho_1(p_1,p_2,O)$$
$$p_1,p_2,p_3 \in D_1 \Rightarrow \rho(p_1,p_2,p_3) = \rho_1(p_1,p_2,p_3)$$

If $D_1 \cap D_2 \neq \{O\}$, then $\omega_1 * \omega_2$ is defined as $\omega_1' * \omega_2'$ where ω_i' is the presentation ω_i restricted to $D_i' = D_i \setminus (D_1 \cap D_2 \setminus \{O\})$. By construction, $D_1' \cap D_2' = \{O\}$.

- The decomposition of a variety $\alpha = (D,\rho)$ along one of its occurrences $x = (\{O,p\},\emptyset)$ with $p \in D \setminus \{O\}$ is the presentation (D',ρ') where $D' = \{O\} \cup D \setminus \{p\}$ and ρ' is defined for any $p_1,p_2,p_3 \in D'^{!3}$ by

$$\rho'(p_1,p_2,p_3) = \rho(q_1,q_2,q_3)$$
$$\text{where } \forall i = 1,2,3 \;\; q_i = (\text{if } p_i = O \text{ then } p \text{ else } p_i)$$

- The Relaxation relation is induced by the comparator $\preceq$ on colours, ie. for varieties $\alpha = (D,\rho)$ and $\alpha' = (D',\rho')$

$$\alpha \preceq \alpha' \;\; \Leftrightarrow \;\; D = D' \wedge \forall p_1,p_2,p_3 \in D^{!3} \;\; \rho(p_1,p_2,p_3) \preceq \rho'(p_1,p_2,p_3)$$

5.2.2.5 Cyclic Relations

A special case of the previous example (Cyclic colourings) is given by the set of colours $\Delta = \{\bullet, \blacktriangleleft, \blacktriangleright, \blacklozenge\}$ with

$$\bullet \preceq \begin{matrix} \blacktriangleright \\ \blacktriangleleft \end{matrix} \preceq \blacklozenge \quad\Bigg| \quad \begin{matrix} \bullet^* = \bullet \\ \blacklozenge^* = \blacklozenge \end{matrix} \quad\Bigg| \quad \begin{matrix} \blacktriangleright^* = \blacktriangleleft \\ \blacktriangleleft^* = \blacktriangleright \end{matrix}$$

In that case, it is easy to show that there is an isomorphism Λ between cyclic colourings and ternary cyclic relations.

Proposition 5.2 *Let Λ be the mapping from the set of cyclic colourings over D into the set of ternary relations over D defined, for any cyclic* *colouring ρ, and any $p_1, p_2, p_3 \in D^{!3}$, by*

$$\Lambda(\rho)(p_1, p_2, p_3) \quad \Leftrightarrow \quad \blacktriangleleft \preceq \rho(p_1, p_2, p_3)$$

Λ *is injective and its image is exactly the set of ternary cyclic relations over D. For any such relation $\mathcal{R}$, and any $p_1, p_2, p_3 \in D^{!3}$, we have*

$\Lambda^{-1}(\mathcal{R})(p_1, p_2, p_3) =$
$\max_{\preceq}((\text{if } \mathcal{R}(p_1, p_2, p_3) \text{ then } \blacktriangleleft \text{ else } \bullet), (\text{if } \mathcal{R}(p_1, p_3, p_2) \text{ then } \blacktriangleright \text{ else } \bullet))$

Thus, we obtain a logic $\mathcal{L}$ of arbitrary ternary cyclic relations (not necessarily cyclic orders as in Cyclic Logic). $\mathcal{L}$ is an extension of both Linear Logic, Cyclic Logic and Non Commutative Logic (ie. provable formulas in these logics are also provable in $\mathcal{L}$). The extension is conservative for Linear Logic and Cyclic Logic (ie. provable formulas in $\mathcal{L}$ which belong to these logics are also provable in them), but it is not conservative for Non Commutative Logic. For example, it is possible to show that the following formula

$$a \,\mathord{\invamp}_\bullet\, (b \,\mathord{\invamp}_\blacktriangleleft\, c) \,\mathord{\invamp}_\bullet\, ((a^\perp \otimes_\blacktriangleleft c^\perp) \,\mathord{\invamp}_\blacktriangleleft\, (\perp \otimes_\blacktriangleleft b^\perp))$$

is provable in $\mathcal{L}$. It also belongs to Non Commutative Logic (identify $\blacktriangleleft$ with $<$ and $\bullet$ with $\|$) but is not provable there. Indeed, the proof in $\mathcal{L}$ requires a Relaxation that generates a ternary cyclic relation which is not an order variety, nor can it be further relaxed into an order variety.

5.2.2.6 Structads

Another axiomatisation of structural rules in Linear Logic is proposed by [8], based on the notion of "structads" which are species [7] equipped with a composition operator very close to the composition operator of variety-presentation frameworks presented here. In fact, the two axiomatisations, discovered independently, are very similar, at least when restricted to the structads on a single polarity (called C-structads in [8]). Thus, it is quite straightforward to derive, from each of the examples of variety-presentation frameworks above, a corresponding C-structad, and, vice versa, from each of the numerous examples of C-structads in [8], a corresponding variety-presentation framework.

There are some differences between the two approaches though. Here, we follow a strict policy for the management of places, consistent with our locative perspective. No identification of structures, modulo "canonical isomorphisms", is needed (technically, this comes from the fact that we do not use the direct sum but the symmetric difference for the occurrence set of the composition of two presentations). We don't even need

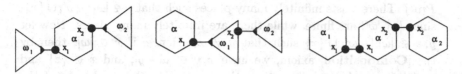

Fig. 5.2. Illustrations of the [**Commutation**] axiom

the "transport" operation of species, since it can be reconstructed from the composition and decomposition operations (through the notion of "relocator", introduced below).

5.2.3 Assignments, Substitutions

Given a variety-presentation framework (hence satisfying the axioms of Section 5.2.1), it is possible to define a generic notion of substitution on varieties and presentations. Note that, as usual with the ubiquitous notion of substitution, expliciting it is a rather bureaucratic task, but reveals interesting features about the deep structure of the objects at hand.

Definition 5.3 *Given a variety* α, *a presentation* ω *and a place* x, *the expression* $\alpha[x \rightarrow \omega]$ *denotes either the variety* $(\alpha)_x * \omega$ *if* x *occurs in* α *or* α *otherwise, and is called the substitution of* x *by* ω *in* α.

Substitutions can also be applied to presentations.

Definition 5.4 *Let* ω, ω' *be presentations and* x, y *be places. The substitution of* x *by* ω' *in* ω *via* y *is the presentation*

$$\omega[x \rightarrow_y \omega'] = ((\omega * y)[x \rightarrow \omega'])_y$$

Although y occurs in the expression defining $\omega[x \rightarrow_y \omega']$, there exists an infinite subset of places y for which the result does not depend on y. Indeed:

Proposition 5.5 *Let* ω, ω' *be presentations, and* x *be a place. Then* $\omega[x \rightarrow_y \omega']$ *is independent of the place* y *such that* $y \notin |\omega| \cup |\omega'| \cup \{x\}$. *Furthermore, such a place exists (hence we can write* $\omega[x \rightarrow \omega']$ *without subscript).*

Proof There exists infinitely many places such that $y \notin |\omega| \cup |\omega'| \cup \{x\}$, since $|\omega|, |\omega'|$ are finite, while there are infinitely many places. Now let $y, z \notin |\omega| \cup |\omega'| \cup \{x\}$ such that $y \neq z$. Hence, if $x \in |\omega|$, then by the [**Composition**] axiom, we have $x, y \in |\omega * y|$, and $x \notin \{z\}$ and $y \notin |\omega'|$ and $\{z\} \cap |\omega'| = \emptyset$, so that we are in the conditions of the [**Commutation**] axiom:

$$
\begin{aligned}
\omega[x \to {}_y\omega'] * z &= ((\omega * y)[x \to \omega'])_y * z & \text{by Definition} \\
&= ((\omega * y)_x * \omega')_y * z & \text{by Definition} \\
&= ((\omega * y)_y * z)_x * \omega' & \text{by the [\textbf{Commutation}] axiom} \\
&= (\omega * z)_x * \omega' & \text{by the [\textbf{Decomposition}] axiom} \\
&= \omega[x \to {}_z\omega'] * z & \text{by Definition.}
\end{aligned}
$$

Hence $\omega[x \to {}_y\omega'] = \omega[x \to {}_z\omega']$ by the [**Decomposition**] axiom. The case where $x \notin |\omega|$ is treated similarly. $\qquad\square$

We can now give several reformulations of the [**Commutation**] axiom, implicitly containing the [**Composition**] and [**Decomposition**] axioms.

Lemma 5 *Let $\alpha, \alpha_1, \alpha_2$ be varieties, x_1, x_2 places and ω_1, ω_2 presentations. If $\omega_1 = x_1$ or $\omega_2 = x_2$ or ($x_1 \neq x_2$ and $|\omega_1| \cap |\omega_2| = \emptyset$ and $x_1 \notin |\omega_2|$ and $x_2 \notin |\omega_1|$) then*

$$\alpha[x_1 \to \omega_1][x_2 \to \omega_2] = \alpha[x_2 \to \omega_2][x_1 \to \omega_1] \tag{5.1}$$

If $\omega_2 = x_2$ or ($x_2 \in |\omega_1| \setminus |\alpha|$ and $|\omega_2| \cap |\alpha| = \emptyset$) then

$$\alpha[x_1 \to \omega_1][x_2 \to \omega_2] = \alpha[x_1 \to \omega_1[x_2 \to \omega_2]] \tag{5.2}$$

If $x_1 \in |\alpha| \cap |\alpha_1|$ and $x_2 \in |\alpha| \cap |\alpha_2|$ and $|\alpha_1| \cap |\alpha_2| = \emptyset$ then

$$\alpha_1[x_1 \to (\alpha_2[x_2 \to (\alpha)_{x_2}])_{x_1}] = \alpha_2[x_2 \to (\alpha_1[x_1 \to (\alpha)_{x_1}])_{x_2}] \tag{5.3}$$

Proof Simple case-by-case analysis. Figure 5.2 illustrates the main case in each of the three statements. Note that (5.1) holds even if $x_1, x_2 \notin |\alpha|$. $\qquad\square$

In the sequel, we are interested in the substitution of multiple places by presentations. For this, we introduce the notion of presentation assignment, to specify what has to be substituted by what, and of enumeration of places, to specify in which order the substitutions should take place.

Definition 5.6 *An enumeration of the places is a bijection from* $\mathbb{N}^{\circ}$ *into the set of the places. A presentation assignment (or assignment for short) is a mapping* φ *from places to presentations. Let* D *be a set of places.*

- φ *is said to be* admissible *on* D *if* $\forall x, y \in D \ x \neq y \Rightarrow$
  $\varphi(x) = x \ \lor \ \varphi(y) = y \ \lor \ (x \notin |\varphi(y)| \ \land \ y \notin |\varphi(x)|).$
- φ *is said to be* linear *on* D *if*
  $\forall x, y \in D \ x \neq y \Rightarrow |\varphi(x)| \cap |\varphi(y)| = \emptyset.$
- *A* relocator *of* D *is a permutation* σ *of the places which is admissible on* D.
- $\varphi\lfloor_D$ *is the assignment defined for all place* x *by*
  $\varphi\lfloor_D (x) = \text{if } x \in D \text{ then } x \text{ else } \varphi(x).$

We can now define a generalised notion of substitution.

Definition 5.7 *Let* φ *be an assignment,* X *an enumeration of the places and* α *a variety. Let* $(i_k)_{k=1...p}$ *be the (unique) increasing (finite) sequence of* $\mathbb{N}^{\circ}$ *such that* $|\alpha| = \{X_{i_k}\}_{k=1...p}$. *The raw* φ-*substitution of* α *ordered by* X *is the variety*

$$\varphi \blacktriangleright_X \alpha = \alpha[X_{i_1} \to \varphi(X_{i_1})] \ldots [X_{i_p} \to \varphi(X_{i_p})]$$

Lemma 5(5.1) states that, under certain conditions, the order in which two substitutions are performed is irrelevant. This can be extended to any number of substitutions as follows:

Proposition 5.8 *Let* φ *be an assignment and* α *a variety. If* φ *is linear and admissible on* $|\alpha|$ *then* $\varphi \blacktriangleright_X \alpha$ *is independent of the enumeration of the places* X. *Furthermore, such an enumeration exists (hence we can write* $\varphi \blacktriangleright \alpha$ *without any subscript).*

Proof Essentially, we have to show that for any permutation μ over $\{1, \ldots, p\}$, we have

$$\alpha[X_{i_1} \to \varphi(X_{i_1})] \cdots [X_{i_p} \to \varphi(X_{i_p})] \ = \ \alpha[X_{i_{\mu(1)}} \to \varphi(X_{i_{\mu(1)}})] \cdots$$
$$\cdots [X_{i_{\mu(p)}} \to \varphi(X_{i_{\mu(p)}})] \quad (5.4)$$

Since any permutation of $\{1, \ldots, p\}$ is the composition of transpositions $n/n - 1$, it is therefore sufficient to prove (5.4) when μ is such a transposition, and we can furthermore assume that $n = p$ (the other cases are obvious consequences of this one). Hence we have to prove

$$\beta[X_{i_p} \to \varphi(X_{i_p})][X_{i_{p-1}} \to \varphi(X_{i_{p-1}})] \;=\; \beta[X_{i_{p-1}} \to \varphi(X_{i_{p-1}})]$$
$$[X_{i_p} \to \varphi(X_{i_p})]$$

where $\beta = \alpha[X_{i_1} \to \varphi(X_{i_1})]\dots[X_{i_{p-2}} \to \varphi(X_{i_{p-2}})]$, which is a straightforward application of Lemma 5(5.1) and the linearity and admissibility of φ on $|\alpha|$. $\qquad\square$

We need the following lemmas:

Lemma 6 *Let σ be a permutation of the places and D a set of places.*

$$\sigma \text{ is a relocator of } D \;\Leftrightarrow\; \forall x \in \sigma(D) \cap D \quad \sigma x = x$$

Proof Simple manipulation of the definition. $\qquad\square$

Lemma 7 *Let φ be an assignment, r a place and α a variety. If $r \in |\alpha|$ and φ is linear and admissible on $|\alpha|$, then so is $\varphi|_r$ and*

$$\varphi \blacktriangleright \alpha = (\varphi|_r \blacktriangleright \alpha)[r \to \varphi(r)]$$

Proof Straightforward application of Proposition 5.8, where the enumeration of the places is chosen in such a way that r is the last element of $|\alpha|$. The only difficulty is to show that $\varphi|_r$ is linear and admissible on $D = |\alpha|$. Admissibility is a direct consequence of the admissibility of φ on D. Linearity is a bit less straightforward. If $\varphi(r) = r$ then $\varphi|_r = \varphi$, which is linear on D. Now assume $\varphi(r) \neq r$ and let $x \neq y \in D$.

- If $x \neq r$ and $y \neq r$, then $\varphi|_r (x) = \varphi(x)$ and $\varphi|_r (y) = \varphi(y)$, and by linearity of φ on D we have $|\varphi|_r (x)| \cap |\varphi|_r (y)| = \emptyset$.
- If $x = r$ (hence $y \neq r$), then $\varphi|_r (x) = r$ and $\varphi|_r (y) = \varphi(y)$. If $\varphi(y) = y$ then, obviously, $r \notin |\varphi(y)|$. If $\varphi(y) \neq y$, since $\varphi(r) \neq r$, then by admissibility of φ on D, we have $r \notin |\varphi(y)|$. Hence, in both cases $|\varphi|_r (x)| \cap |\varphi|_r (y)| = \emptyset$.

$$\square$$

Lemma 8 *Let φ be an assignment and α a variety. If φ is linear and admissible on $|\alpha|$, then*

$$|\varphi \blacktriangleright \alpha| = \bigcup_{x \in |\alpha|} |\varphi(x)|$$

Proof This is shown by induction on the cardinality of the set $D = \{x \in |\alpha| / \varphi(x) \neq x\}$. If $D = \emptyset$, then $\varphi = \text{Id}$ on $|\alpha|$ and

$$|\varphi \blacktriangleright \alpha| = |\alpha| = \bigcup_{x \in |\alpha|} \{x\} = \bigcup_{x \in |\alpha|} |\varphi(x)|$$

Now, if D is not empty, choose $r \in D$. By the induction hypothesis, we have

$$|\varphi\!\downarrow_r \blacktriangleright \alpha| = \bigcup_{x \in |\alpha|} |\varphi\!\downarrow_r (x)| = \underbrace{\bigcup_{x \in |\alpha| \setminus \{r\}} |\varphi(x)| \cup \{r\}}_{A}$$

Since $\varphi\!\downarrow_r$ is linear on D, this union is direct, hence $r \notin A$, and by Lemma 7, we have

$$|\varphi \blacktriangleright \alpha| = |(\varphi\!\downarrow_r \blacktriangleright \alpha)[r \to \varphi(r)]| = (A \uplus \{r\}) \setminus \{r\} \odot |\varphi(r)| = A \odot |\varphi(r)|$$

By linearity of φ, we have $A \cap |\varphi(r)| = \emptyset$, hence the result. $\qquad\square$

Lemma 9 *Let φ be an assignment, α a variety and σ a relocator of $|\alpha|$.*

- *φ is linear on $\sigma(|\alpha|)$ if and only if $\varphi\sigma$ is linear on $|\alpha|$.*
- *If φ and $\varphi\sigma$ are linear and admissible on, respectively, $\sigma(|\alpha|)$ and $|\alpha|$, then*

$$\varphi \blacktriangleright (\sigma \blacktriangleright \alpha) = (\varphi\sigma) \blacktriangleright \alpha$$

Proof This is shown by induction on the cardinality of the set $D = \{x \in |\alpha| / \varphi(\sigma x) \neq \sigma x\}$. If $D = \emptyset$ then $\varphi\sigma = \sigma$ on $|\alpha|$ and $\varphi = \text{Id}$ on $\sigma(|\alpha|)$, and

$$\varphi \blacktriangleright (\sigma \blacktriangleright \alpha) = \sigma \blacktriangleright \alpha = (\varphi\sigma) \blacktriangleright \alpha$$

Now, if D is not empty, choose $r \in D$. Since φ is linear and admissible on $\sigma(|\alpha|)$, by Lemma 7, so is $\varphi\!\downarrow_{\sigma r}$, and hence $\varphi\!\downarrow_{\sigma r} \sigma$ is linear on $|\alpha|$. Let's now show that $\varphi\!\downarrow_{\sigma r} \sigma$ is also admissible on $|\alpha|$. Let $x \neq y \in |\alpha|$ such that $\varphi\!\downarrow_{\sigma r} (\sigma x) \neq x$ and $\varphi\!\downarrow_{\sigma r} (\sigma y) \neq y$.

- If $x \neq r$ and $y \neq r$, then $\varphi\!\downarrow_{\sigma r} (\sigma x) = \varphi(\sigma x)$ and $\varphi\!\downarrow_{\sigma r} (\sigma y) = \varphi(\sigma y)$. By admissibility of $\varphi\sigma$ on $|\alpha|$, we have $x \notin |\varphi\!\downarrow_{\sigma r} (\sigma y)|$ and $y \notin |\varphi\!\downarrow_{\sigma r} (\sigma x)|$.
- If $x = r$ (hence $y \neq r$), then $\varphi\!\downarrow_{\sigma r} (\sigma x) = \sigma r$ and $\varphi\!\downarrow_{\sigma r} (\sigma y) = \varphi(\sigma y)$.
  - Suppose $y = \sigma r$, hence $y \in \sigma(|\alpha|) \cap |\alpha|$, hence, by Lemma 6, $\sigma y = y$, hence $y = r$, contradiction. Thus, $y \notin |\varphi\!\downarrow_{\sigma r} (\sigma x)|$.

- If $\varphi(\sigma r) \neq r$, since $\varphi(\sigma y) \neq y$, by admissibility of $\varphi\sigma$ on $|\alpha|$, we have $x = r \notin |\varphi|_{\sigma r} (\sigma y)|$.
- If $\varphi(\sigma r) = r$, hence $r \in |\varphi(\sigma r)|$ and, by linearity of $\varphi\sigma$ on $|\alpha|$ we have $x = r \notin |\varphi|_{\sigma r} (\sigma y)|$.

Thus, $\varphi|_{\sigma r}$ and $\varphi|_{\sigma r} \sigma$ are linear and admissible on, respectively, $\sigma(|\alpha|)$ and $|\alpha|$. Furthermore, it is easy to show that for all $x \in |\alpha|$, we have

$$(\varphi|_{\sigma r} \sigma)|_r (x) = (\varphi\sigma)|_r (x) \qquad (5.5)$$

Hence,

$$
\begin{aligned}
\varphi \blacktriangleright (\sigma \blacktriangleright \alpha) &= (\varphi|_{\sigma r}\blacktriangleright (\sigma \blacktriangleright \alpha))[\sigma r \to \varphi(\sigma r)] \quad \text{by Lemma 7} \\
&= ((\varphi|_{\sigma r} \sigma) \blacktriangleright \alpha)[\sigma r \to \varphi(\sigma r)] \quad \text{by induction} \\
&= ((\varphi|_{\sigma r} \sigma)|_r\blacktriangleright \alpha)[r \to \sigma r][\sigma r \to \varphi(\sigma r)] \\
&\qquad \text{by Lemma 7} \\
&= ((\varphi\sigma)|_r\blacktriangleright \alpha)[r \to \sigma r][\sigma r \to \varphi(\sigma r)] \quad \text{by (5.5)} \\
&= ((\varphi\sigma)|_r\blacktriangleright \alpha)[r \to \varphi(\sigma r)] = (\varphi\sigma) \blacktriangleright \alpha \\
&\qquad \text{by Lemma 5(5.2) and Lemma 7}
\end{aligned}
$$

$\square$

Definition 5.9 *Let D be a set of places. Let φ be an assignment, X an enumeration of the places, σ a relocator of D and α a variety with $|\alpha| = D$. The φ-substitution of α via σ ordered by X is the variety*

$$\varphi \triangleright_{X;\sigma} \alpha = \varphi\sigma^{-1} \blacktriangleright_X (\sigma \blacktriangleright \alpha)$$

Since, by definition, a relocator of $|\alpha|$ is admissible (and obviously linear) on $|\alpha|$, by Proposition 5.8, the notation $\sigma \blacktriangleright \alpha$ need not be subscripted by X. Furthermore, we have $|\sigma \blacktriangleright \alpha| = \sigma(|\alpha|)$, and hence,

$$\varphi \triangleright_{X;\sigma} \alpha = \alpha[X_{i_1} \to X_{j_1}] \ldots [X_{i_p} \to X_{j_p}][X_{j_1} \to \varphi(X_{i_1})] \ldots [X_{j_p} \to \varphi(X_{i_p})]$$

where $\sigma(|\alpha|) = \{X_{j_1}, \ldots, X_{j_p}\}$ with $(j_k)_{k=1\ldots p}$ strictly increasing and $X_{i_k} = \sigma^{-1}(X_{j_k})$ for $k = 1 \ldots p$ (the substitutions of σ can be performed in any order). In other words, the occurrences of α are first relocated and then the *raw* substitution is performed. The relocator can be chosen in such a way that $\varphi\sigma^{-1}$ be admissible on $\sigma(|\alpha|)$, in which case, assuming φ is linear, $\varphi \triangleright_{X;\sigma} \alpha$ becomes entirely independent of X. Interestingly, it can also be shown independent of the "mediator" σ:

Proposition 5.10 *Let φ be an assignment and α a variety such that φ is linear on $|\alpha|$. The variety $\varphi \triangleright_{X;\sigma} \alpha$ is independent of the enumeration of the places X and of the relocator σ such that $\varphi\sigma^{-1}$ is admissible on $\sigma(|\alpha|)$. Furthermore, such a relocator exists (hence we can write $\varphi \triangleright \alpha$ without any subscript).*

$$|\varphi \triangleright \alpha| = \bigcup_{x \in |\alpha|} |\varphi(x)|$$

Moreover, when φ is admissible on $|\alpha|$, then $\varphi \triangleright \alpha = \varphi \blacktriangleright \alpha$.

Proof Let $D = |\alpha|$ (which is finite).

- We first show that for any finite set of places D', there exists a relocator of D outside D', i.e. such that $\sigma(D) \cap D' = \emptyset$. Indeed, since D, D' are finite while there are infinitely many places, we can choose a subset D'' of places with the same cardinality as $D \cap D'$ and disjoint from $D \cup D'$. Let μ be a bijection from $D \cap D'$ into D''. Define σ as the mapping from places to places such that for all place x, if x is in $D \cap D'$ then $\sigma x = \mu x$, if x is in D'' then $\sigma x = \mu^{-1}x$, and if x is neither in $D \cap D'$ nor in D'' then $\sigma x = x$. By construction, σ is a permutation of the places.

$$\sigma(D) = \sigma(D \setminus D') \cup \sigma(D \cap D') = (D \setminus D') \cup D'' \quad \text{hence}$$
$$\begin{cases} \sigma(D) \cap D = D \setminus D' \\ \sigma(D) \cap D' = \emptyset \end{cases}$$

  By construction, σ is the identity on $D \setminus D'$. Hence, by Lemma 6, σ is a relocator of D outside D'.

- Now, choose a relocator σ of D outside $\bigcup_{x \in D} |\varphi(x)|$. Hence $\sigma(D) \cap \bigcup_{x \in D} |\varphi(x)| = \emptyset$. Hence, $\forall y \in \sigma(D)$ $\sigma(D) \cap |\varphi\sigma^{-1}(y)| = \emptyset$. Therefore, σ is a relocator of D such that $\varphi\sigma^{-1}$ is admissible on $\sigma(D)$.

- Finally, consider two relocators σ, σ' of D such that $\varphi\sigma^{-1}$ is admissible on $\sigma(D)$ (and idem for σ'). Since φ is linear on D, so is $\varphi\sigma^{-1}$ on $\sigma(D)$ (and idem for σ').

  - If $\sigma(D) \cap \sigma'(D) = \emptyset$ then, by Lemma 6, $\sigma\sigma'^{-1}$ is a relocator (hence admissible) on $\sigma'(D)$. Moreover, $\sigma\sigma'^{-1}\sigma' = \sigma$ is admissible on D (since it is a relocator). Hence, by Lemma 9, we have

$$\sigma\sigma'^{-1} \blacktriangleright (\sigma' \blacktriangleright \alpha) = \sigma \blacktriangleright \alpha \tag{5.6}$$

By hypothesis, $\varphi\sigma^{-1}$ and $\varphi\sigma^{-1}\sigma\sigma'^{-1} = \varphi\sigma'^{-1}$ are admissible on, respectively, $\sigma(D)$ and $\sigma'(D)$. Hence, by Lemma 9, we have

$$\varphi\sigma^{-1} \blacktriangleright (\sigma\sigma'^{-1} \blacktriangleright (\sigma' \blacktriangleright \alpha)) = \varphi\sigma'^{-1} \blacktriangleright (\sigma' \blacktriangleright \alpha) \qquad (5.7)$$

Combining (5.6) and (5.7) we get

$$\varphi\sigma^{-1} \blacktriangleright (\sigma \blacktriangleright \alpha) = \varphi\sigma'^{-1} \blacktriangleright (\sigma' \blacktriangleright \alpha)$$

– If $\sigma(D) \cap \sigma(D') \neq \emptyset$, we can choose a relocator σ'' of D outside $\sigma(D) \cup \sigma'(D) \cup \bigcup_{x \in D} |\varphi(x)|$. Hence, as above, $\varphi\sigma''^{-1}$ is admissible on $\sigma''(D)$ and $\sigma(D) \cap \sigma''(D) = \sigma'(D) \cap \sigma''(D) = \emptyset$. Hence, as in the previous case, we have

$$\varphi\sigma^{-1} \blacktriangleright (\sigma \blacktriangleright \alpha) = \varphi\sigma''^{-1} \blacktriangleright (\sigma'' \blacktriangleright \alpha) = \varphi\sigma'^{-1} \blacktriangleright (\sigma' \blacktriangleright \alpha)$$

$$\square$$

Lemma 10 *Let φ be an assignment, α a variety, r, s places. If φ is linear on $|\alpha|$ and $r \in |\alpha|$ and $s \notin |\alpha| \cup \bigcup_{x \in |\alpha| \setminus \{r\}} |\varphi(x)|$, then*

$$\varphi \triangleright \alpha = (\varphi|_s \triangleright (\alpha[r \to s]))[s \to \varphi(r)]$$

Proof Left to the reader. $\square$

Up to now, we have applied multi-substitutions to varieties. It is easy to extend this to presentations as follows.

Definition 5.11 *Let φ be an assignment and ω a presentation. Let y be a place. The φ-substitution of ω via y is the presentation*

$$\varphi \triangleright_y \omega = (\varphi|_y \triangleright (\omega * y))_y$$

Note that this definition assumes that $\varphi|_y$ is linear on $|\omega * y|$. This is always true for some values of y and, for these values, the result is independent of y:

Proposition 5.12 *Let φ be an assignment and ω a variety such that φ is linear on $|\omega|$. The presentation $\varphi \triangleright_y \omega$ is independent of the place y such that $y \notin |\omega| \cup \bigcup_{x \in |\omega|} |\varphi(x)|$. Furthermore, such a place exists (hence we can write $\varphi \triangleright \omega$ without any subscript).*

$$|\varphi \triangleright \omega| = \bigcup_{x \in |\omega|} |\varphi(x)|$$

Proof There exist infinitely many places such that $y \notin |\omega| \cup \bigcup_{x \in |\omega|} |\varphi(x)|$, since $|\omega|$ as well as each $|\varphi(x)|$ are finite, while there are infinitely many places. Now, let $y, z \notin |\omega| \cup \bigcup_{x \in |\omega|} |\varphi(x)|$. Hence

$$
\begin{aligned}
(\varphi \rhd_z \omega) * y &= (\varphi\lfloor_z \rhd(\omega * z))[z \to y] &&\text{by Definition 5.11} \\
&= (\varphi\lfloor_{y,z} \rhd((\omega * z)[z \to y])) \\
&\quad [y \to \varphi\lfloor_{y,z}(z)][z \to y] &&\text{by Lemma 10} \\
&= \varphi\lfloor_y \rhd(\omega * y) = (\varphi \rhd_y \omega) * y &&\text{by Definition 5.11}
\end{aligned}
$$

$\square$

To complete the loop, we now apply multi-substitutions to assignments themselves.

Definition 5.13 *Let φ, ψ be assignments. The assignment $\varphi \rhd \psi$ is defined for all place x by $(\varphi \rhd \psi)(x) = \varphi \rhd (\psi(x))$ if φ is linear on $|\psi(x)|$ and some arbitrary value otherwise.*

Lemma 11 *Let φ, ψ be assignments and D a set of places. If ψ is linear on D and φ on $\bigcup_{x \in D} |\psi(x)|$ then $\varphi \rhd \psi$ is linear on D.*

Proof Left to the reader. $\square$

We can now state the fundamental theorem of this section:

Theorem 5.14 *Let φ, ψ be assignments and α a variety such that ψ is linear on $|\alpha|$ and φ is linear on $\bigcup_{x \in |\alpha|} |\psi(x)|$.*

$$
\varphi \rhd (\psi \rhd \alpha) = (\varphi \rhd \psi) \rhd \alpha
$$

More generally, $\rhd$ is an associative operation on linear assignments (on all the places).

Proof Proceed in two steps:

(i) Assume φ is admissible on $D = |\psi \rhd \alpha| = \bigcup_{x \in |\alpha|} |\psi(x)|$ and proceed by induction. If φ is not identity on D, choose $r \in D$ such that $\varphi(r) \neq r$. Since $r \in D$, there exists a place $x \in |\alpha|$ such that $r \in |\psi(x)|$. Since ψ is linear on $|\alpha|$, for all $y \in |\alpha| \setminus \{x\}$ we have $r \in |\psi(x)|$ hence $r \notin |\psi(y)|$ hence $\varphi\lfloor_r = \varphi$ on $|\psi(y)|$ hence

$$
(\varphi\lfloor_r \rhd \psi)(y) = (\varphi \rhd \psi)(y) \tag{5.8}
$$

Now, we have $\varphi \blacktriangleright (\psi \triangleright \alpha) =$

$$
\begin{aligned}
& (\varphi\lfloor_r \blacktriangleright (\psi \triangleright \alpha))[r \to \varphi(r)] && \text{by Lemma 7} \\
=\ & ((\varphi\lfloor_r \triangleright \psi) \triangleright \alpha)[r \to \varphi(r)] && \text{by induction hyp.} \\
=\ & ((\varphi\lfloor_r \triangleright \psi)\lfloor_z \triangleright (\alpha[x \to z])) \\
& [z \to (\varphi\lfloor_r \triangleright \psi)(x)][r \to \varphi(r)] && \text{by Lemma 10 for some } z \\
=\ & ((\varphi\lfloor_r \triangleright \psi)\lfloor_z \triangleright (\alpha[x \to z])) \\
& [z \to (\varphi\lfloor_r \blacktriangleright \psi(x))[r \to \varphi(r)]] && \text{by Lemma 5(5.2)} \\
=\ & ((\varphi\lfloor_r \triangleright \psi)\lfloor_z \triangleright (\alpha[x \to z])) \\
& [z \to (\varphi \blacktriangleright \psi(x))] && \text{by Lemma 7} \\
=\ & ((\varphi \triangleright \psi)\lfloor_z \triangleright (\alpha[x \to z])) \\
& [z \to (\varphi \triangleright \psi)(x)] && \text{by (5.8)} \\
=\ & (\varphi \triangleright \psi) \triangleright \alpha && \text{by Lemma 10}
\end{aligned}
$$

(ii) This result can easily be generalised to the case where φ is not assumed admissible on D, using a relocator σ and the result in the admissible case:

$$
\varphi \triangleright (\psi \triangleright \alpha) = \varphi\sigma^{-1} \blacktriangleright (\sigma \blacktriangleright (\psi \triangleright \alpha)) = (\varphi\sigma^{-1} \triangleright (\sigma \triangleright \psi)) \triangleright \alpha
$$
$$
\forall x \in |\alpha| \ \ (\varphi\sigma^{-1} \triangleright (\sigma \triangleright \psi))(x) = \varphi\sigma^{-1} \blacktriangleright (\sigma \blacktriangleright (\psi(x))) =
$$
$$
\varphi \triangleright (\psi(x)) = (\varphi \triangleright \psi)(x)
$$

$\square$

We often use the following direct consequence of the theorem.

Corollary 5.15 *Let φ be an assignment and ω_1, ω_2 presentations such that $|\omega_1| \cap |\omega_2| = \emptyset$ and φ is linear on $|\omega_1| \cup |\omega_2|$. Then*

$$
\varphi \triangleright (\omega_1 * \omega_2) = (\varphi \triangleright \omega_1) * (\varphi \triangleright \omega_2)
$$

Proof Let ψ be a presentation assignment linear on $|\omega_1 \cup \omega_2|$ and let x_1, x_2 be two distinct arbitrary places outside $|\psi \triangleright \omega_1| \cup |\psi \triangleright \omega_2|$. Let μ be an assignment such that $\mu(x_i) = \omega_i$ (for $i = 1, 2$). Thus μ is linear on $|x_1 * x_2|$. By definition and [**Decomposition**], we have:

$$
\psi \triangleright (x_1 * x_2) = (x_1 * x_2)[x_1 \to \psi(x_1)][x_2 \to \psi(x_2)] = \psi(x_1) * \psi(x_2) \quad (5.9)
$$

Hence, $\varphi \triangleright (\omega_1 * \omega_2) = \varphi \triangleright (\mu(x_1) * \mu(x_2)) =$

$$
\begin{aligned}
& \varphi \triangleright (\mu \triangleright (x_1 * x_2)) && \text{by (5.9) where } \psi = \mu \\
=\ & (\varphi \triangleright \mu) \triangleright (x_1 * x_2) && \text{by Theorem 5.14} \\
=\ & (\varphi \triangleright \mu)(x_1) * (\varphi \triangleright \mu)(x_2) \\
=\ & (\varphi \triangleright \omega_1) * (\varphi \triangleright \omega_2) && \text{by (5.9) where } \psi = \varphi \triangleright \mu
\end{aligned}
$$

$\square$

5.3 Coloured Linear Logics

In this section, we assume given a variety-presentation framework, and we build a generic sequent calculus using only the operations of that framework. Essentially, a formula is a syntactic construct specifying a structure (here a presentation) on the sets of its sub-formulas, while a sequent is a syntactic construct specifying a structure (here a variety) on its component formulas. This distinction is particularly thin, and we propose a formulation of the sequent calculus where the two notions of sequents and formulas are blurred. This avoids the introduction of yet another concept like "composition modes" [9] or explicit operations on which structural rules are stated as in the classical non associative Lambek calculus [4].

5.3.1 A Locative Sequent Calculus

In fact, all we need is the capacity to explicitly address each formula occurring in a proof, and, for each address, to determine the top-most connective of the formula at that address. For addresses, we use the *loci*, introduced in Ludics [6] with a similar purpose.

Definition 5.16 (Loci) *We assume given an arbitrary (countably infinite) set whose elements are called* biases. *A locus is a finite sequence of biases.*

In the sequel, we assume given a bijection between the set of loci and that of places (of the variety-presentation framework). Such a bijection always exists, since the set of places as well as that of loci are countably infinite. We make use of the following notations:

- ϵ denotes the place corresponding to the empty locus.
- For any place x, $\dot{x}$ denotes the mapping which maps any place y into the place (written $\dot{x}y$) for which the locus is the concatenation of those for x and y.
- For any place x, $\ddot{x}$ denotes the "inverse" mapping of $\dot{x}$, ie.
  $\ddot{x}(y) = $ if $(y = \dot{x}z)$ then z else y

Definition 5.17 (Independence)

- *The prefix order on loci induces a corresponding arborescent order on places:*
  $$x \leq y \iff \exists z \; y = \dot{x}z$$

- *A set D of places is said to be* independent *if*
  $$\forall x, y \in D \ (x \leq y \Rightarrow x = y)$$

We make use below of the following result which is a direct consequence of Theorem 5.14 (and $\ddot{x}\dot{x} = Id$).

Proposition 5.18 *For any presentation ω, and for any linear assignment φ on $|\omega|$, if $x \notin |\varphi \triangleright \omega|$ and $|\varphi \triangleright \omega| \cup \{x\}$ is independent, then $\varphi\ddot{x}$ is linear and admissible on $|\dot{x} \blacktriangleright \omega|$ and*

$$\varphi \triangleright \omega = \varphi\ddot{x} \blacktriangleright (\dot{x} \blacktriangleright \omega)$$

This provides an effective way to reduce any substitution to two raw substitutions. For any place x satisfying the conditions, $\dot{x}$ acts as the relocator (although it is not strictly speaking a relocator, since it is not a permutation).

Example 5.1

In the examples, we use the simple variety-presentation framework of sets. Furthermore, we assume that the decimal digits $1, \ldots, 9$ are biases, so that strings of digits, such as 256 or 1, represent loci, hence places. Consider the presentation (set) $\omega = \{12, 13, 2\}$ and the presentation assignment $\varphi = [12 : \{13, 2\}, 13 : \{14, 12\}, 2 : \{3\}]$ (with arbitrary values for the other places). The computation of $\omega' = \varphi \triangleright \omega$ requires a relocation, because $12, 13$ are in $|\omega|$ and 13 appears in $\varphi(12)$. Using a relocation, we obtain $\omega' = \{13, 2, 14, 12, 3\}$. We can also use the proposition above, choosing, say, $x = 57$. We thus have $\varphi\ddot{x} = [5712 : \{13, 2\}, 5713 : \{14, 12\}, 572 : \{3\}]$ which applies to $\dot{x} \blacktriangleright \omega = \{5712, 5713, 572\}$ and yields ω' by raw substitution. Note that the choice $x = 1$ is forbidden (not independent from $12 \in \varphi(13)$) since $12, 113$ would be in $|\dot{x} \blacktriangleright \omega|$ and 12 would appear in $\varphi\ddot{x}(113)$.   □

Definition 5.19 (Connectives) *A (generalised)* connective *is a set R of presentations such that the set $|R| = \bigcup_{\omega \in R} |\omega|$ is independent. A* polarity *is an element of the set $\{+, -\}$. A polarised* connective *is a pair written R^s where R is a connective and s a polarity. We write $|R^s| = |R|$. If c is a polarised connective, and s a polarity, then c^s denotes the polarised connective obtained from c by multiplying its polarity by s (with the obvious multiplication on polarities).*

Essentially, a connective R corresponds in Ludics to a set of ramifications, except that we do not require here that $|R|$ contain only biases,

Let ω be a presentation and φ be a presentation assignment. Let x, x' be places and R be a connective.

- Logical rule **Negative** case: if $x \notin |\omega|$

$$\frac{\cdots \qquad \omega * (\dot{x} \triangleright \tau) \qquad \cdots}{\omega * x} \; \mathbf{L}(x, R^-)$$

with one premiss for each $\tau \in R$.

- Logical rule **Positive** case: if $\tau \in R$ and φ is linear on $|\tau|$ and $x \notin |\varphi \triangleright \tau|$

$$\frac{\cdots \qquad \varphi(y) * \dot{x}y \qquad \cdots}{(\varphi \triangleright \tau) * x} \; \mathbf{L}(x, R^+, \tau)$$

with one premiss for each $y \in |\tau|$.

- Identities:

$$\text{if } x \neq x' \qquad \frac{}{x * x'} \; \mathbf{A}(x, x') \qquad \Bigg| \qquad \text{if } |\omega| \cap |\omega'| = \emptyset \qquad \frac{\omega * x \qquad \omega' * x}{\omega * \omega'} \; \mathbf{C}(x)$$

- Structural rule: if $\alpha \preceq \alpha'$

$$\frac{\alpha'}{\alpha} \; \mathbf{s}$$

Fig. 5.3. The generic locative inference system of Coloured Linear Logics (multiplicative-additive case)

but only that it be independent (which is weaker). In Ludics, a single ramification represents a multiplicative connective, while here this is the role of a presentation. As in Ludics, additivity is incorporated by considering sets of ramifications (here sets of presentations). The generic locative inference system of Coloured Linear Logics in the multiplicative-additive case is given in Figure 5.3. Its inferences operate on sequents which are simply varieties.

- The negative rule $[\mathbf{L}(x, R^-)]$ decomposes the connective R^-. It mimics the negative rule of Ludics, ie. a combination of "par" and "with" of Linear Logic. It has one premiss for each element of R (possibly none) and each premiss inherits the same context (that of the conclusion).

- The positive rule $[\mathbf{L}(x, R^+, \tau)]$ decomposes the connective R^+. It mimics the positive rule of Ludics, ie. a combination of "tensor" and "plus" of Linear Logic. Each instance of the rule requires a choice of $\tau \in R$ (recorded in the label). It has one premiss for each place occurring in τ (possibly none). Each premiss inherits a piece of the context of the conclusion, filtered by τ.

- The identities, [A] (Axiom) and [C] (Cut) are the usual ones. Note that the two premisses of the Cut rule share the same cut place.

- The structural rule [S] allows inferences which "strengthen" their varieties (when read bottom-up, or "weaken" them when read top-down), such comparisons in strength being defined by the [**Relaxation**] relation of the variety-presentation framework.

For each place x occurring in the conclusion of an inference, and for each premiss of that inference, the structure of loci allows to trace exactly how x is transported from the conclusion into that premiss: x may be deleted, untouched, or expanded (ie. become $\dot{x}y$). Thus, in a Logical inference, the conclusion always consists of two disjoint sets of places: a singleton containing the *principal* place, which is recorded in the label, and the *context*. In each premiss, the context places are either untouched or deleted, while the principal place is expanded. An important property, which is always implicit in the traditional non-locative formulations of sequent calculi, is that in each premiss, the places deriving from the principal one (expanded) and those coming from the context (when not deleted) are still disjoint. Here, as places are made explicit, we need an explicit assumption to ensure this separation property.

Definition 5.20 (Inference trees) *An inference tree is a tree labeled with varieties, obtained by consistently assembling instances of the inference rules of Figure 5.3, and such that*

- *in each Cut inference, the cut place is different from and independent of each of the places occurring in the conclusion;*
- *the occurrence set of the root sequent (variety) is independent.*

The separation property is then a direct consequence of the following one:

Proposition 5.21 *The occurrence set of any sequent (variety) in an inference tree is independent. Any sub-tree of an inference tree is an inference tree.*

This is shown by induction on the depth of the tree.

5.3.2 Cut Elimination

The Cut rule $[\mathbf{C}(x)]$ can be written equivalently (although less symmetrically) as follows (under the same side-conditions):

$$\frac{\alpha \qquad x * \omega}{\alpha[x \twoheadrightarrow \omega]} \; \mathrm{C}(x)$$

The axioms of variety-presentation frameworks have been chosen in such a way as to make possible the definition of a Cut-elimination procedure, which can be seen as the computation of the expression $\alpha[x \twoheadrightarrow \omega]$ in the conclusion of the Cut, using the axioms of the framework. This procedure is defined below in the usual way (with reduction rules corresponding to the usual cases of commutative conversion, symmetric and asymmetric reductions). The locative calculus is untyped, so Cut reduction may fail, but we introduce later the appropriate notion of types with the essential property that well typed inference trees (real proofs) do not reduce to error. Reduction is defined by rewrite rules of the form

$$\frac{\overset{\vdots \pi_a}{\omega_a * z} \qquad \overset{\vdots \pi_b}{\omega_b * z}}{\omega_a * \omega_b} \; \mathrm{C}(z) \qquad \rightsquigarrow \qquad \pi$$

where the lowest inference of the input inference tree is a Cut, and the output inference tree depends on the input sub-trees π_a, π_b.

Convention: in the inference trees, we make use of "dummy" inferences whose premiss and conclusion are equal. The optional argument in the label of such inferences is only used for reference purpose (eg. to justify the equality).

5.3.2.1 Commutative conversions

Commutative conversions deal with the case where one premiss of the Cut (eg. the left one) is preceded by an inference which does not introduce the cut place. Hence, with a negative Logical inference $[\mathbf{L}(x, R^-)]$ for example, we must have $\omega_a * z = \omega * x$ and $x \neq z$ (hence $z \in |\omega|$ and $x \in |\omega_a|$) for some presentation ω. This justifies the equality (e) in the following reduction rule.

$$\frac{\cfrac{\cdots\ \omega * (\dot{x} \blacktriangleright \tau)\ \cdots}{\cfrac{\omega * x}{\omega_a * z}\ \text{(e)}}\ \mathbf{L}(x, R^-)\qquad\qquad \omega_b * z}{\omega_a * \omega_b}\ \mathbf{C}(z)$$

$$\rightsquigarrow$$

$$\cfrac{\cdots\ \cfrac{\cfrac{\cfrac{\omega * (\dot{x} \blacktriangleright \tau)\qquad \omega_b * z}{\omega * (\dot{x} \blacktriangleright \tau)[z \rightarrow \omega_b]}\ \mathbf{C}(z)}{\omega[z \rightarrow \omega_b] * (\dot{x} \blacktriangleright \tau)}\ \text{(e1)}\ \cdots}{\omega[z \rightarrow \omega_b] * x}\ \mathbf{L}(x, R^-)}{\omega_a * \omega_b}\ \text{(e2)}$$

Identity (e1), it is a direct consequence of Corollary 5.15. As for (e2), we have: $\omega[z \rightarrow \omega_b] * x = (\omega * x)[z \rightarrow \omega_b]$ by Corollary 5.15 and therefore $\omega[z \rightarrow \omega_b] * x = (\omega_a * z)[z \rightarrow \omega_b] = \omega_a * \omega_b$ by identity (e) and [**Decomposition**].

The other commutative conversion cases are treated similarly.

5.3.2.2 Symmetric reduction: L-L case

Symmetric reductions deal with the case where each premiss of the Cut is preceded by an inference which does introduce the cut place. We first consider the L-L case where this introduction is performed by Logical inferences $[\mathbf{L}(z, c_a)]$ and $[\mathbf{L}(z, c_b)]$ on both sides (where c_a, c_b are polarised connectives). If connectives c_a, c_b are not dual of each other, the reduction fails. This is the only such case of failure.

$$\cfrac{\cfrac{\cdots}{\omega_a * z}\ \mathbf{L}(z, c_a)\qquad \cfrac{\cdots}{\omega_b * z}\ \mathbf{L}(z, c_b)}{\omega_a * \omega_b}\ \mathbf{C}(z)\qquad \rightsquigarrow\qquad \text{ERROR if } c_a, c_b \text{ are not dual}$$

We now assume that π_a, π_b end with dual Logical inferences, respectively, $[\mathbf{L}(z, R^+, \tau)]$ and $[\mathbf{L}(z, R^-)]$, for some R and $\tau \in R$. In that case, we can write the following reduction rule.

$$\frac{\cfrac{\cdots\ \varphi(y)*\dot{z}y\ \cdots}{(\varphi\triangleright\tau)*z}\ \mathbf{L}(z,R^+,\tau)\qquad\cfrac{\cdots\ \omega*(\dot{z}\blacktriangleright\tau)\ \cdots}{\omega*z}\ \mathbf{L}(z,R^-)}{\omega*(\varphi\triangleright\tau)}\ \mathbf{C}(z)$$

$$\leadsto$$

$$\frac{\cfrac{\cfrac{\cfrac{\varphi(y_1)*\dot{z}y_1\quad\omega*(\dot{z}\blacktriangleright\tau)}{\varphi(y_2)*\dot{z}y_2\quad\omega*(\dot{z}\blacktriangleright\tau)[\dot{z}y_1\rightarrow\varphi(y_1)]}\ \mathbf{C}(\dot{z}y_1)}{\varphi(y_3)*\dot{z}y_3\quad\omega*(\dot{z}\blacktriangleright\tau)[\dot{z}y_1\rightarrow\varphi(y_1)][\dot{z}y_2\rightarrow\varphi(y_2)]}\ \mathbf{C}(\dot{z}y_2)}{\vdots\vdots\vdots}\ \mathbf{C}(\ldots)}{\cfrac{\varphi(y_n)*\dot{z}y_n\quad\omega*(\dot{z}\blacktriangleright\tau)[\dot{z}y_1\rightarrow\varphi(y_1)]\ldots[\dot{z}y_{n-1}\rightarrow\varphi(y_{n-1})]}{\omega*(\dot{z}\blacktriangleright\tau)[\dot{z}y_1\rightarrow\varphi(y_1)]\ldots[\dot{z}y_n\rightarrow\varphi(y_n)]}\ \mathbf{C}(\dot{z}y_n)}}{\omega*(\varphi\triangleright\tau)}\ (e)$$

where $|\tau|=\{y_1,\ldots,y_n\}$. The identity (e) is justified by Proposition 5.18:

$$\omega*(\dot{z}\blacktriangleright\tau)[\dot{z}y_1\rightarrow\varphi(y_1)]\ldots[\dot{z}y_n\rightarrow\varphi(y_n)]=\omega*(\varphi\dot{z}\blacktriangleright(\dot{z}\blacktriangleright\tau))$$
$$=\omega*(\varphi\triangleright\tau)$$

5.3.2.3 Symmetric reduction: A-A case

In the A-A case of symmetric reductions, the cut place is introduced in both premisses of the Cut by an Identity axiom. In that case we have the following reduction rule.

$$\frac{\cfrac{}{x*z}\ \mathbf{A}(x,z)\qquad\cfrac{}{y*z}\ \mathbf{A}(y,z)}{x*y}\ \mathbf{C}(z)\qquad\leadsto\qquad\cfrac{}{x*y}\ \mathbf{A}(x,y)$$

5.3.2.4 Asymmetric reduction

In the case of an asymmetric reductions, the cut place is introduced in one premiss of the Cut by an Identity axiom and in the other by a Logical rule, eg. $[\mathbf{L}(z,R^+)]$ (the specific choice of element of R is ommitted here as it plays no role). In that case we have the following reduction rule.

$$\frac{\overset{\cdots}{\omega * z}\ \text{L}(z, R^+)\qquad \frac{}{y * z}\ \text{A}(y, z)}{\omega * y}\ \text{C}(z)$$

$$\rightsquigarrow$$

$$\frac{\overset{\cdots}{\omega * z}\ \text{L}(z, R^+)\qquad \frac{\cdots\ \frac{\frac{}{\dot{z}u * \dot{y}u}\ \text{A}(\dot{z}u, \dot{y}u)\ \cdots}{(\dot{z} \blacktriangleright \tau) * y}\ \text{L}(y, R^+, \tau)\ \cdots}{y * z}\ \text{L}(z, R^-)}{\omega * y}\ \text{C}(z)$$

where, in the right sub-tree of the output inference tree, the lower logical inference has one premiss for each $\tau \in R$ and, for each such τ, the upper logical inference has one premiss for each $u \in |\tau|$.

The case where the left premiss of the input Cut is the conclusion of a negative Logical inference is treated similarly. In both cases, there is a slight difference with traditional Cut reduction. Traditionally, there are two possibilities to deal with asymmetric reductions of the sort considered here:

- One solution is to write a reduction rule of the form

$$\frac{\overset{\vdots}{\Gamma, F}\qquad \frac{}{F, F^\perp}\ \text{Id}}{\Gamma, F}\ \text{Cut} \qquad \rightsquigarrow \qquad \frac{\vdots}{\Gamma, F}$$

But this is not correct from a locative point of view. Indeed, if we make the formula occurrences explicit, we get

$$\frac{\overset{\vdots}{\Gamma, F_1}\qquad \frac{}{F_2, F_1^\perp}\ \text{Id}}{\Gamma, F_2}\ \text{Cut} \qquad \rightsquigarrow \qquad \frac{\vdots}{\Gamma, F_1}$$

Although the output proof has the same type as the input proof, it does not have the same conclusion (one occurrence of F has been substituted by another).

- The other solution is to reduce beforehand the identity axioms in the input proof so as to make them scope over atoms only. In this way, the situation where the cut formula is principal formula of a logical rule on one side and participates in an identity axiom on the other side simply cannot happen, as that would require the formula to be both atomic and compound. Reduction of the identity axioms is done by rules of the following form.

$$\frac{}{F \otimes G \,,\; F^{\perp} \,\mathbin{⅋}\, G^{\perp}} \text{ Id} \quad \rightsquigarrow \quad \frac{\dfrac{}{F, F^{\perp}} \text{ Id} \quad \dfrac{}{G, G^{\perp}} \text{ Id}}{\dfrac{F \otimes G \,,\; F^{\perp} \,,\; G^{\perp}}{F \otimes G \,,\; F^{\perp} \,\mathbin{⅋}\, G^{\perp}} \mathbin{⅋}} \otimes$$

In the solution presented here, the reduction of identity axioms is performed on demand, when the Cut reduction needs it: the output inference tree is obtained from the input one by performing one step of reduction of the identity above the right premiss of the input Cut, and it is the label of the Logical inference of the other premiss of the input Cut which determines how to perform that reduction.

5.3.2.5 Termination of Cut Elimination

Theorem 5.22 *Cut reduction on inference trees always terminates. If the input inference tree π has no proper axioms and a conclusion α, then any maximal sequence of reductions of π terminates either in error or in an inference tree with no proper axioms, no Cut inference and the same conclusion α.*

Proof We give here a sketch of the proof, which follows quite naturally the classical one for pure Linear Logic. For a given Cut inference $[\mathbf{C}(x)]$, let D be the set of places y such that a logical inference with principal place y appears above that Cut inference, and $x \leq y$ (ie. the places where the cut place is actually expanded). Define the degree of a Cut inference as the number of elements of D, and its depth as the size of the inference tree above it. In commutative conversions, the Cut inference is replaced by Cut inferences with the same degree, but lower depth. In symmetric reductions, the Cut inference is replaced by Cut inferences with a lower degree. So, in both cases, some counter is strictly decreased. The only problem comes from asymmetric reductions, which increases the depth of the reduced Cut inference without reducing its degree. Consider for example the inference tree π where τ is a presentation and $|\tau| = \{a, b\}$:

$$\frac{\dfrac{\omega * (\dot{x} \triangleright \tau)}{\omega * x} \text{ L}(x, \{\tau\}^{-}) \qquad \dfrac{}{y * x} \text{ A}(x, y)}{\omega * y} \text{ C}(x)$$

By asymmetric reduction applied to $[\mathbf{C}(x)]$, it reduces to π' (which is in fact "bigger" than π):

$$\cfrac{\cfrac{\cfrac{}{\dot{x}a * \dot{y}a}\;{\scriptstyle A(\pm a,\,\dot{y}a)} \quad \cfrac{}{\dot{x}b * \dot{y}b}\;{\scriptstyle A(\pm b,\,\dot{y}b)}}{\cfrac{(\dot{y} \triangleright \tau) * x}{y * x}\;{\scriptstyle L(y,\,\{\tau\}^-)}}\;{\scriptstyle L(x,\,\{\tau\}^+)}}{\cfrac{\omega * (\dot{x} \triangleright \tau)}{\omega * x}\;{\scriptstyle L(x,\,\{\tau\}^-)} \qquad}{\omega * y}\;{\scriptstyle C(x)}$$

The degree of $[\mathbf{C}(x)]$ is unchanged and its depth increased. But then, the only way to proceed with π', as far as $[\mathbf{C}(x)]$ is concerned, is by a commutative conversion followed by a symmetric reduction, yielding

$$\cfrac{\cfrac{}{\dot{x}b * \dot{y}b}\;{\scriptstyle A(\pm b,\,\dot{y}b)} \quad \cfrac{\cfrac{}{\dot{x}a * \dot{y}a}\;{\scriptstyle A(\pm a,\,\dot{y}a)} \qquad \omega * (\dot{x} \triangleright \tau)}{\omega * (\dot{x} \triangleright \tau)[\dot{x}a \rightarrow \dot{y}a]}\;{\scriptstyle C(\pm a)}}{\cfrac{\omega * (\dot{y} \triangleright \tau)}{\omega * y}\;{\scriptstyle L(y,\,\{\tau\}^-)}}\;{\scriptstyle C(\pm b)}$$

Thus, $[\mathbf{C}(x)]$ is ultimately replaced by two Cut inferences of lower degree. $\qquad\square$

Note that we make no claim here about strong normalisation (all maximal reductions terminate with the same value). This is true only with restrictions:

- Consider the following commutative conversion:

$$\cfrac{\cfrac{\omega * x}{\omega_a * z}\;{\scriptstyle L(x,\,\emptyset^-)} \quad \cfrac{\vdots}{\omega_b * z}\;{\scriptstyle :\pi}}{\omega_a * \omega_b}\;{\scriptstyle C(z)} \quad \rightsquigarrow \quad \cfrac{\cfrac{\omega[z \triangleright \omega_b] * x}{(\omega * x)[z \triangleright \omega_b]}\;{\scriptstyle L(x,\,\emptyset^-)}}{\omega_a * \omega_b}$$

Here, the output inference tree cannot be re-written further (it does not contain any Cut inference). If, on the other hand, π rewrites into error, we obtain two maximal reductions, one leading to error and the other one to a success. The same argument also applies to symmetric or asymmetric reductions. This problem is eliminated by restricting to an outer-most reduction strategy, which forces the reduction of π to happen before the reduction above.

- Consider also the following commutative conversions (which differ only by the index $i = 1$ or $i = 2$):

$$\cfrac{\cfrac{}{x_1 * z}\;{\scriptstyle L(x_1,\,\emptyset^-)} \quad \cfrac{}{x_2 * z}\;{\scriptstyle L(x_2,\,\emptyset^-)}}{x_1 * x_2}\;{\scriptstyle C(z)} \quad \rightsquigarrow \quad \cfrac{}{x_1 * x_2}\;{\scriptstyle L(x_i,\,\emptyset^-)}$$

If the left premiss is commuted first, we obtain the output tree with $i = 1$, while if the second premiss is commuted first, we obtain the output tree with $i = 2$. Although the two trees have the same conclusion, they are different. Again, here, a left-most reduction strategy avoids the problem.

5.3.3 Focussing

Definition 5.23 *Let R, R' be (non-polarised) connectives, $\omega \in R$ a presentation and $x \in |\omega|$ a place. The expansion of R at ω, x by R' is the connective*

$$R \triangleright_{\omega,x} R' = R \setminus \{\omega\} \cup \{\omega[x \rightarrow (\dot{x} \triangleright \omega')] \mid \omega' \in R'\}$$

In other words, $R \triangleright_{\omega,x} R'$ is obtained by replacing ω in R by all the presentations obtained by substituting in ω the place x by some presentation of R' prefixed by x. Focussing, like Cut elimination, is a procedure which executes rewriting steps on inference trees.

5.3.3.1 Compositions

Compositions (one negative, one positive) deal with the case where two consecutive Logical inferences of the same polarity are such that the principal place of the upper one is an expansion of the principal place x of the lower one (ie. is of the form $\dot{x}a$). For example, in the negative case, this situation is characterised by the identity (e) below. σ denotes the presentation of the connective R labelling the lower inference, the corresponding premiss of which is the conclusion of the upper inference.

$$
\cfrac{
\cfrac{
\overbrace{\cdots \omega' * (\dot{x}\dot{a} \triangleright \tau') \cdots}^{\text{For each } \tau' \in R'}
}{\omega' * \dot{x}a} \; \text{L}(\dot{x}a, R'^{-})
}{
\cfrac{
\overbrace{\cdots \omega * (\dot{x} \triangleright \tau) \cdots}^{\text{For each } \tau \in R \setminus \{\sigma\}}
\qquad \qquad
\omega * (\dot{x} \triangleright \sigma)
}{\omega * x} \; \text{L}(x, R^{-})
} \; \text{(e)}
$$

$\rightsquigarrow$

$$
\cfrac{
\overbrace{\cdots \omega * (\dot{x} \triangleright \tau) \cdots}^{\text{For each } \tau \in R \setminus \{\sigma\}}
\cdots \;
\cfrac{
\overbrace{\omega' * (\dot{x}\dot{a} \triangleright \tau')}^{\text{For each } \tau' \in R'}
}{\omega * (\dot{x} \triangleright (\sigma[a \rightarrow (\dot{a} \triangleright \tau')]))} \; \text{(e')}
\; \cdots
}{\omega * x} \; \text{L}(x, R''^{-})
$$

where $R'' = R \triangleright_{\sigma,a} R'$. Identity (e') is justified as follows:

$$
\begin{aligned}
\omega' * (\dot{x}\dot{a} \triangleright \tau') &= (\omega' * \dot{x}a)[\dot{x}a \rightarrow (\dot{x}\dot{a} \triangleright \tau')] \\
&= (\omega * (\dot{x} \triangleright \sigma))[\dot{x}a \rightarrow (\dot{x}\dot{a} \triangleright \tau')] & \text{by identity (e)} \\
&= \omega * ((\dot{x} \triangleright \sigma)[\dot{x}a \rightarrow (\dot{x}\dot{a} \triangleright \tau')]) & \text{by Corollary 5.15} \\
&= \omega * (\dot{x} \triangleright (\sigma[a \rightarrow (\dot{a} \triangleright \tau')])) & \text{easy to show}
\end{aligned}
$$

The positive Composition case is treated similarly.

5.3.3.2 Commutations

Commutations deal with the case where two consecutive Logical inferences of the same or opposite polarity are such that the principal place y of the upper one is independent of the principal place x of the lower one. For example, with two positive inferences, this situation is characterised by the identity (e) below. z denotes the place of presentation τ labelling the lower inference, the corresponding premiss of which is the conclusion of the upper inference.

$$
\cfrac{
  \text{For each } u \in |\tau| \backslash \{z\} \qquad
  \cfrac{
    \overbrace{\dots \; \psi(v) * \dot{y}v \; \dots}^{\text{For each } v \in |\tau'|}
  }{
    (\psi \triangleright \tau') * y
  } \; \mathbf{L(y, R'^+, \tau')}
}{
  \cfrac{
    \overbrace{\dots \; \varphi(u) * \dot{x}u \; \dots} \qquad \varphi(z) * \dot{x}z
  }{
    (\varphi \triangleright \tau) * x
  } \; \mathbf{L(x, R^+, \tau)}
} \; \text{(e)}
$$

$\rightsquigarrow$

$$
\cfrac{
  \text{For each } v \in |\tau'| \backslash \{t\} \qquad
  \cfrac{
    \overbrace{\dots \; \varphi(u) * \dot{x}u \; \dots}^{\text{For each } u \in |\tau| \backslash \{z\}} \quad
    \cfrac{\psi(t) * \dot{y}t}{\varphi'(z) * \dot{x}z} \text{(e1)}
  }{
    (\varphi' \triangleright \tau) * x
  } \; \mathbf{L(x, R^+, \tau)}
}{
  \cfrac{
    \overbrace{\dots \; \psi(v) * \dot{y}v \; \dots} \qquad \psi'(t) * \dot{y}t
  }{
    (\psi' \triangleright \tau') * y
  } \; \mathbf{L(y, R'^+, \tau')}
} \; \text{(e')}
$$

$$\frac{(\psi' \triangleright \tau') * y}{(\varphi \triangleright \tau) * x}$$

In the input inference tree, φ, ψ are presentation assignments and we assume that x, y are independent (condition for the commutation to be possible). Hence $\dot{x}z \in \bigcup_{v \in |\tau'|} |\psi(v)|$. Hence, there exists $t \in |\tau'|$ such that $\dot{x}z \in |\psi(t)|$. In the output inference tree, φ' and ψ' are assignments defined as follows:

$$
\begin{aligned}
\forall u \in |\tau| \quad \varphi'(u) &= \text{if } u = z \text{ then } (\psi(t) * \dot{y}t)_{\dot{x}z} \text{ else } \varphi(u) \\
\forall v \in |\tau'| \quad \psi'(v) &= \text{if } v = t \text{ then } ((\varphi' \triangleright \tau) * x)_{\dot{y}t} \text{ else } \psi(v)
\end{aligned}
$$

Identity (e1) and (e2) are direct consequences of the definitions of φ', ψ'. Identity (e) is justified as follows:

$(\psi' \rhd \tau') * y$

$= (\psi' \ddot{y} \blacktriangleright (\dot{y} \blacktriangleright \tau')) * y$ by Proposition 5.18

$= (\psi \ddot{y}|_{\dot{y}t} \blacktriangleright (\dot{y} \blacktriangleright \tau'))[\dot{y}t \rightarrow ((\varphi' \rhd \tau) * x)_{\dot{y}t}] * y$ by def. of ψ' and Lemma 7

$= (\psi \ddot{y}|_{\dot{y}t} \blacktriangleright (\dot{y} \blacktriangleright \tau'))[\dot{y}t \rightarrow ((\varphi' \ddot{x} \blacktriangleright (\dot{x} \blacktriangleright \tau)) * x)_{\dot{y}t}] * y$ by Proposition 5.18

$= (\psi \ddot{y}|_{\dot{y}t} \blacktriangleright (\dot{y} \blacktriangleright \tau'))[\dot{y}t \rightarrow (((\varphi \ddot{x}|_{\dot{x}z} \blacktriangleright (\dot{x} \blacktriangleright \tau))[\dot{x}z \rightarrow (\psi(t) * \dot{y}t)_{\dot{x}z}]) * x)_{\dot{y}t}] * y$

     by definition of φ' and Lemma 7

$=$

$\underbrace{((\psi \ddot{y}|_{\dot{y}t} \blacktriangleright (\dot{y} \blacktriangleright \tau')) * y)}_{C}[\dot{y}t \rightarrow (((\underbrace{(\varphi \ddot{x}|_{\dot{x}z} \blacktriangleright (\dot{x} \blacktriangleright \tau))}_{B} * x)[\dot{x}z \rightarrow \underbrace{(\psi(t) * \dot{y}t)_{\dot{x}z}}_{A}]))_{\dot{y}t}]$

$= C[\dot{y}t \rightarrow (B[\dot{x}z \rightarrow (A)_{\dot{x}z}])_{\dot{y}t}] = B[\dot{x}z \rightarrow (C[\dot{y}t \rightarrow (A)_{\dot{y}t}])_{\dot{x}z}]$ by Lemma 5(5.3)

$= ((\varphi \ddot{x}|_{\dot{x}z} \blacktriangleright (\dot{x} \blacktriangleright \tau)) * x)[\dot{x}z \rightarrow (((\psi \ddot{y}|_{\dot{y}t} \blacktriangleright (\dot{y} \blacktriangleright \tau')) * y)[\dot{y}t \rightarrow (\psi(t) * \dot{y}t)_{\dot{y}t}])_{\dot{x}z}]$

$= ((\varphi \ddot{x}|_{\dot{x}z} \blacktriangleright (\dot{x} \blacktriangleright \tau)) * x)[\dot{x}z \rightarrow (((\psi \ddot{y}|_{\dot{y}t} \blacktriangleright (\dot{y} \blacktriangleright \tau')))[\dot{y}t \rightarrow \psi(t)] * y)_{\dot{x}z}]$

$= ((\varphi \ddot{x}|_{\dot{x}z} \blacktriangleright (\dot{x} \blacktriangleright \tau)) * x)[\dot{x}z \rightarrow ((\psi \ddot{y} \blacktriangleright (\dot{y} \blacktriangleright \tau')) * y)_{\dot{x}z}]$ by Lemma 7

$= ((\varphi \ddot{x}|_{\dot{x}z} \blacktriangleright (\dot{x} \blacktriangleright \tau)) * x)[\dot{x}z \rightarrow ((\psi \rhd \tau') * y)_{\dot{x}z}]$ by Proposition 5.18

$= ((\varphi \ddot{x}|_{\dot{x}z} \blacktriangleright (\dot{x} \blacktriangleright \tau)) * x)[\dot{x}z \rightarrow (\varphi(z) * \dot{x}z)_{\dot{x}z}]$ by the identification (e)

$= (\varphi \ddot{x}|_{\dot{x}z} \blacktriangleright (\dot{x} \blacktriangleright \tau))[\dot{x}z \rightarrow \varphi(z)] * x$

$= (\varphi \ddot{x} \blacktriangleright (\dot{x} \blacktriangleright \tau)) * x$ by Lemma 7

$= (\varphi \rhd \tau) * x$ by Proposition 5.18

Example 5.2

In the variety-presentation framework of simple sets, we have

$$\frac{\{11, 21\} \cup A \qquad \{22\} \cup B}{\{2\} \cup ((\{11\} \cup A) \cup B)} \; \text{L}(2, \{\{1, 2\}\}^+)$$

$$\frac{\{11, 2\} \cup A \cup B \qquad\qquad\qquad \{12\} \cup C}{\{1\} \cup ((\{2\} \cup A \cup B) \cup C)} \; \text{L}(1, \{\{1, 2\}\}^+)$$

$\rightsquigarrow$

$$\frac{\{11, 21\} \cup A \qquad \{12\} \cup C}{\{1\} \cup ((\{21\} \cup A) \cup C)} \; \text{L}(1, \{\{1, 2\}\}^+)$$

$$\frac{\{1, 21\} \cup A \cup C \qquad\qquad\qquad \{22\} \cup B}{\{2\} \cup ((\{1\} \cup A \cup C) \cup B)} \; \text{L}(2, \{\{1, 2\}\}^+)$$

and we have indeed

$$\{1\} \cup ((\{2\} \cup A \cup B) \cup C) = \{2\} \cup ((\{1\} \cup A \cup C) \cup B) = \{1, 2\} \cup A \cup B \cup C$$

This equality, which is obvious with sets, requires the demonstration above to ensure that it also holds in any variety-presentation framework. □

Similarly, on can write a Commutation rule for each configuration of inferences $\frac{L(x_2, R_2^{s_2})}{L(x_1, R_1^{s_1})}$ (where $\{x_1, x_2\}$ are independent), transforming it into a configuration $\frac{L(x_1, R_1^{s_1})}{L(x_2, R_2^{s_2})}$, *except* when $s_2 = +$ and $s_1 = -$.

There are also Commutation rules transforming any configuration $\frac{L(x, R^-)}{S}$ (resp. $\frac{S}{L(x, R^+)}$) into a configuration $\frac{S}{L(x, R^-)}$ (resp. $\frac{L(x, R^+)}{S}$).

5.3.3.3 Termination of Focussing

Clearly, the Commutation rules being reversible, the Focussing procedure may never terminate in the usual sense. However, a form of termination may still be proved. The following definition characterises the expected "terminal" states:

Definition 5.24 *An inference tree is said to be* polarised *if,*

- *omitting Structural inferences, consecutive Logical inferences are always of opposite polarity,*
- *Structural inferences occur only between a positive Logical inference above and a negative Logical inference below.*

Theorem 5.25 *The Focussing procedure terminates in the following sense: if the input inference tree π has no proper axioms, no Cut inference and a conclusion α, then*

(i) *any finite sequence of Focussing steps from π terminates on an inference tree of conclusion α;*

(ii) *any infinite sequence of Focussing steps from π contains twice the same inference tree (ie. the same commutation rule has been applied back and forth);*

(iii) *there is at least one sequence of Focussing steps from π which terminates on a polarised inference tree.*

Proof The first clause (preservation of the conclusion) is obvious, since it is true of each Focussing step. To prove the second clause, note that each Focussing step either decreases the set of places appearing in the sequents of the inference trees (Composition cases), or preserves it, but then it also preserves the set of labels of the inferences (Commutation cases), and only re-arranges them. Now, for a given set of places and labels, there are only finitely many inference trees using these places and labels. The third clause is not so straightforward and its proof follows that given for pure Linear Logic in [2].   □

5.3.4 Proofs and Types

5.3.4.1 Type Maps

The Cut elimination procedure of the previous section operates on inference trees which are not typed and hence may result in error. Proofs are well-typed inference trees for which the procedure never fails. We introduce the notion of *type map*, meant to replace the traditional notion of formula (and sequent) usually associated with typing. Basically, while a traditional formula associates a connective to each path leading to a (compound) sub-formula, a type map associates a generalised polarised connective to any place.

Definition 5.26 (Type maps) *A type map is a mapping from the set of places to polarised connectives.*

Example 5.3

Let's show how to represent atom-free formulas using type-maps. Consider the formula F in Linear Logic:

$$F = ((A_1 \otimes A_2) \otimes A_3) \,\mathfrak{N}\, ((B_1 \oplus (B_2 \,\mathfrak{N}\, B_3)) \& (C_1 \otimes C_2))$$

where $A_1, A_2, A_3, B_1, B_2, B_3, C_1, C_2$ are arbitrary atom-free formulas. Let's use places (loci) to denote the paths to the sub-formulas of F, with the convention that the bias 1 (resp. 2) denotes the first (resp. second) immediate sub-formula of a formula. Thus the place 21 denotes the 1-st immediate sub-formula of the 2-nd immediate sub-formula of F, ie. the sub-formula $(B_1 \oplus (B_2 \,\mathfrak{N}\, B_3))$ of F. The connectives $\mathfrak{N}, \otimes, \&, \oplus$ are represented as the generalised connectives, respectively, $\{\{1, 2\}\}^-, \{\{1, 2\}\}^+$, $\{\{1\}, \{2\}\}^-, \{\{1\}, \{2\}\}^+$. Hence, F at place ϵ can be represented by any type map satisfying:

$\Gamma(\epsilon) = \{\{1,2\}\}^-$	$\Gamma(1) = \{\{1,2\}\}^+$	$\Gamma(11) = \{\{1,2\}\}^+$	
	$\Gamma(2) = \{\{1\}, \{2\}\}^-$	$\Gamma(21) = \{\{1\}, \{2\}\}^+$	$\Gamma(212) =$
		$\Gamma(22) = \{\{1,2\}\}^+$	$\{\{1,2\}\}^-$

assuming the sub-formulas $A_1, A_2, A_3, B_1, B_2, B_3, C_1, C_2$ are represented by Γ at places, respectively, $111, 112, 12, 211, 2121, 2122, 221, 222$. Other representations are possible, eg. using other biases than $1, 2$ or grouping together connectives (we will see below that this is legitimate is a certain sense). Thus, grouping together the connectives of same polarity, we could encode F by any type map satisfying

$\Gamma(\epsilon) = \{\{1, 21\}, \{1, 22\}\}^-$	$\Gamma(1) = \{\{11, 12, 2\}\}^+$	
	$\Gamma(21) = \{\{1\}, \{2\}\}^+$	$\Gamma(212) = \{\{1,2\}\}^-$
	$\Gamma(22) = \{\{1,2\}\}^+$	

with the same assumptions on sub-formulas $A_1, A_2, A_3, B_1, B_2, B_3, C_1, C_2$.

□

As in the example above, in a type map, we are interested only in the places that can be reached from the root.

Definition 5.27 (Equivalence on type maps) *Let* Γ *be a type map.*

- *A set* D *of places is said to be stable by* Γ *if* $\{\epsilon\} \cup \bigcup_{x \in D} |\Gamma(x)| \subset D$
- *The* extension *of* Γ *is the smallest set of places which is stable by* Γ.
- *Two type maps are said to be* equivalent *(notation* $\equiv$*) if they have the same extension and their restrictions to it are equal.*

The extension of a type map always exists. Indeed, the set of all places is always stable by Γ, and the intersection of any family of stable sets is a stable set, so that the extension of Γ can equivalently be defined as the intersection of all the sets of places which are stable by Γ. In the sequel, we will essentially work with type maps modulo equivalence.

Definition 5.28 (Dualisation, Extraction, Substitution) *Let* Γ, Δ *be type maps, and* s *a polarity. The type maps* Γ^s, $\Gamma\!\downarrow_x$ *and* $\Gamma[x : \Delta]$ *where* x *is a place are defined for all place* y *by*

$$\Gamma^s(y) = \Gamma(y)^s$$
$$\Gamma\!\downarrow_x (y) = \Gamma(\dot{x}y)$$
$$\Gamma[x : \Delta](y) = \text{if } (y = \dot{x}z) \text{ then } \Delta(z) \text{ else } \Gamma(y)$$

The first operation corresponds to dualising a formula. The second operation corresponds to extracting a sub-formula at a given place. The third operation corresponds to substituting a sub-formula at a given place by a new formula. It is easy to show that these operations are stable by equivalence of type maps, so that they are also defined on equivalence classes:

Proposition 5.29 *If* $\Gamma \equiv \Gamma'$ *then* $\Gamma^s \equiv \Gamma'^s$ *and* $\Gamma\!\downarrow_x \equiv \Gamma'\!\downarrow_x$. *If* $\Gamma \equiv \Gamma'$ *and* $\Delta \equiv \Delta'$ *then* $\Gamma[x : \Delta] \equiv \Gamma'[x : \Delta']$.

In a proof, it will be necessary to detect if two sub-formulas are dual of each other. For this, we introduce the notion of translocations.

Definition 5.30 (Translocations) *Let* Γ *be a type-map, and* s *a polarity. A* translocation *on* Γ *of polarity* s *is an independent pair* x, y *of places (notation* $\langle x; y \rangle_\Gamma^s$*) such that* $\Gamma\!\downarrow_y \equiv (\Gamma\!\downarrow_x)^s$.

Thus a translocation on Γ of polarity s is a pair of places typed by dual formulas (if $s = -$) or identical formulas (if $s = +$).

Proposition 5.31 *We have the following properties on translocations:*

$$\langle x; y \rangle_\Gamma^s \ \Rightarrow \ \langle y; x \rangle_\Gamma^s \qquad \langle x; y \rangle_\Gamma^s \ \wedge \ \langle y; z \rangle_\Gamma^{s'} \ \Rightarrow \ \langle x; z \rangle_\Gamma^{ss'}$$

5.3.4.2 Proofs

Definition 5.32 (Proofs, Types) *Let Γ be a type map and π an inference tree. π is said to be a proof of type Γ (notation $\pi \vdash \Gamma$) if one of the following conditions holds*

- *π contains no inference (ie. it is reduced to a proper axiom).*
- *π ends with an inference $\mathbf{L}(x, R^s)$ and its immediate sub-trees are of type Γ and $\Gamma(x) = R^s$.*
- *π ends with an inference $\mathbf{S}$ and its immediate sub-tree is of type Γ.*
- *π ends with an inference $\mathbf{A}(x, x')$ and $\langle x; x' \rangle_\Gamma^-$.*
- *π ends with an inference $\mathbf{C}(x)$ and its immediate sub-trees are of type, respectively, $\Gamma[x : \Delta^+]$ and $\Gamma[x : \Delta^-]$ for some type map Δ.*

Note that by proof we mean possibly open proof (ie. possibly with proper axioms), so that all the sub-trees of a proof are proofs. Furthermore proofs always have multiple types. In particular, if a type map Γ is a type for a proof, so is any type map equivalent to Γ. Hence, types can be considered modulo equivalence. More precisely,

Proposition 5.33 *Let π be a proof of conclusion α. Let Γ, Γ' be type maps*

$$\pi \vdash \Gamma \ \wedge \ \forall x \in |\alpha| \ \ \Gamma\!\downarrow_x \equiv \Gamma'\!\downarrow_x \ \Rightarrow \ \pi \vdash \Gamma'$$

This is shown by simple induction on the size of the proof.

5.3.4.3 Cut Elimination with Proofs

Theorem 5.34 *Cut elimination applied to a proof never results in error, and the output inference tree is a proof with the same types as the input one.*

Proof The demonstration is an adaptation of the traditional one. It consists in showing that in each reduction step, if the input inference tree is a proof of type Γ, then so is the output inference tree (in particular, it cannot be an error). We use the notations of Section 5.3.2 and consider each case of reduction step:

- Commutative conversions (Section 5.3.2.1):
  Since π_a is of type $\Gamma[z : \Delta]$, so is each of its own sub-proofs, and, reporting in the output inference tree, we find that it is also a proof of type Γ.

- Symmetric reduction L-L (Section 5.3.2.2):
  The sub-proofs π_a and π_b which end with Logical inferences $\mathbf{L}(z, c_a)$ and $\mathbf{L}(z, c_b)$ are of type, respectively, $\Gamma[z : \Delta^+]$ and $\Gamma[z : \Delta^-]$ for some type map Δ, so we obtain that

$$c_a = \Gamma[z : \Delta^+](z) = \Delta^+(\epsilon) \qquad c_b = \Gamma[z : \Delta^-](z) = \Delta^-(\epsilon)$$

Hence the two connectives are dual of each other, and we cannot be in the case of the reduction that results in error. In the succeeding reduction, we have:

  (i) Each input proof of $\varphi(y_k) * \dot{z}y_k$ (left premiss of the input Cut) is of type $\Gamma[z : \Delta^+]$, and by Proposition 5.33, also of type

$$\Gamma[\dot{z}y_i : \Delta_i^-]_{i=n}^{k+1}[\dot{z}y_k : \Delta_k^+]$$

  where $\Delta_i = \Delta|_{y_i}$, and the substitutions are applied downward from n to k (this is arbitrary, as the order of application is irrelevant). Indeed, the only assignment that counts is that of $\dot{z}y_k$, so the other ones can be chosen arbitrarily.

  (ii) Similarly, the input proof of $\omega * (\dot{z} \blacktriangleright \tau)$ (right premiss of the input Cut) is of type $\Gamma[z : \Delta^-]$, and by Proposition 5.33, also of type

$$\Gamma[\dot{z}y_i : \Delta_i^-]_{i=n}^{1}$$

  (iii) By induction on k (starting at 1 upwards), we obtain that each intermediate output inference tree of conclusion $\omega * (\dot{z} \blacktriangleright \tau)[\dot{z}y_i \rightarrow \varphi(y_i)]_{i=1}^{k}$ is a proof of type

$$\Gamma[\dot{z}y_i : \Delta_i^-]_{i=n}^{k+1}$$

Hence, finally (for $k = n$), the output inference tree (of conclusion $\omega * (\varphi \triangleright \tau)$) is a proof of type Γ.

- Symmetric reduction A-A (Section 5.3.2.3):
  Since π_a, π_b are of type $\Gamma[z : \Delta^+]$ and $\Gamma[z : \Delta^-]$, respectively,

we have $\langle x; z \rangle^-_{\Gamma[z:\Delta]}$ and $\langle y; z \rangle^-_{\Gamma[z:\Delta^-]}$ hence, $\langle y; z \rangle^+_{\Gamma[z:\Delta]}$ and, by Proposition 5.31, we have $\langle x; y \rangle^-_{\Gamma[z:\Delta]}$. Therefore, the output inference tree is a proof of type $\Gamma[z:\Delta]$, and by Proposition 5.33, also of type Γ.

- Asymmetric reduction (Section 5.3.2.4):
  Since π_a, π_b are of type $\Gamma[z:\Delta^+]$ and $\Gamma' = \Gamma[z:\Delta^-]$, respectively, we have $\Delta(\epsilon) = R^+$ and $\langle y; z \rangle^-_{\Gamma'}$. Hence,

$$\Gamma'(z) = \Delta^-(\epsilon) = R^- \quad \wedge \quad \Gamma'(y) = \Gamma'(z)^- = R^+$$

Furthermore, it is easy to see that for each $\tau \in R$ and each $u \in |\tau|$, we have $\langle \dot{y}u; \dot{z}u \rangle^-_{\Gamma'}$. Thus, the left sub-tree of the output inference tree is a proof of type Γ' and the output inference tree is a proof of type Γ.

$\square$

5.3.4.4 Focussing with Proofs

Definition 5.35 *Let* Γ, Γ' *be type maps. We write* $\Gamma \longrightarrow \Gamma'$ *if for all place* x, *either* $\Gamma'(x) = \Gamma(x)$ *or there exist a polarity* s, *connectives* R, R', *presentation* ω *and place* x' *such that*

$$\omega \in R \qquad x' \in |\omega|$$
$$\Gamma(x) = R^s \quad \Gamma(x') = R'^s$$
$$\Gamma'(x) = (R \triangleright_{\omega, x'} R')^s$$

Theorem 5.36 *Let* π *be an inference tree and* π' *be a polarised inference tree obtained by Focussing* π.

- *If* π *is a proof of some type* Γ, *then* π' *is a proof of some type* Γ' *such that* $\Gamma \longrightarrow^* \Gamma'$.
- *If* π' *is a proof of some type* Γ', *then* π *is a proof of some type* Γ *such that* $\Gamma \longrightarrow^* \Gamma'$.

Proof Simple induction. $\square$

5.3.5 Dealing with Exponentials

In the previous section, generalised connectives were introduced as combinations of the traditional multiplicative and additive connectives. They are now further extended to also include exponentials.

We assume given a mapping $b \mapsto \hat{b}$ from the set of biases into itself, such that $\widehat{\widehat{b}} = \widehat{b}$ for any bias. It is extended into a mapping on loci, by $\widehat{b_1 \cdots b_n} = \widehat{b_1} \cdots \widehat{b_n}$, and hence on places. The idea here is that two places such that $\widehat{x} = \widehat{y}$ are considered replicas of each-other. For any place x, the set of all its replicas is denoted

$$\overset{\frown}{x} = \{y \mid \widehat{y} = \widehat{x}\}$$

Places such that $\widehat{x} = x$ are called base places. Thus, for any place x, the place $\widehat{x}$ is a base place (master copy) and x is one of its copies. It is easy to show that:

Proposition 5.37 *For any places* x, y, z:

$$\widehat{\widehat{x}} = \widehat{x} \qquad \overset{\frown}{\widehat{x}} = \overset{\frown}{x} \qquad \overset{\frown}{xy} = \overset{\frown}{\widehat{x}\widehat{y}}$$
$$\widehat{x} \leq \widehat{y} \Rightarrow \exists y' \in \overset{\frown}{y} \; x \leq y' \wedge \exists x' \in \overset{\frown}{x} \; x' \leq y$$
$$x \leq y \wedge \widehat{x} = \widehat{y} \Rightarrow x = y$$

Now, connectives can allow the unbounded use of a resource at place x, simply by making available all the copies of that place, ie. $\overset{\frown}{x}$. From the typing point of view, the only constraint is that all the copies of the same base place behave identically.

Definition 5.38 *A (generalised)* connective *is a set* R *of pairs* $N \boxtimes \omega$ *where* N *is a finite set of places,* ω *is a presentation and* $N, |\omega|$ *are disjoint, such that* $|R| = \bigcup_{N \boxtimes \omega \in R} N \cup |\omega|$ *is independent and contains only base places.* Polarised connectives *are of the form* R^s *where* R *is a connective and* s *a polarity.*

In each component $N \boxtimes \tau$ of a connective, the set N is meant to hold all the unbounded places. The inference system of Coloured Linear Logics with exponentials is given in Figure 5.4. The nodes of the inferences are labeled not by varieties as in the multiplicative-additive case, but by pairs $M \boxtimes \alpha$ where M is a *possibly infinite* set of places and α is a variety. The set M captures the set of copies of unbounded places available at each node.

- In a negative inference $\mathbf{L}(x, R^-)$, for each $N \boxtimes \tau \in R$, the corresponding premiss inherits the unbounded places of the conclusion but also adds those in $\widehat{N}$ prefixed by x.
- In a positive inference $\mathbf{L}(x, R^+, N \boxtimes \tau)$, for each $y \in |\tau|$, the corresponding premiss inherits a piece of the unbounded places of the conclusion. Furthermore, there is also a premiss for each $y \in N$,

Let ω be a presentation, φ be a presentation assignment and P be a [set of places] assignment. Let x, x' be places, X be a set of places, and R be a connective.

- Logical rule **Negative** case: if $x \notin |\omega| \cup M$

$$\cdots \quad \frac{M \cup \dot{x}\, \widehat{N} \boxtimes \omega * (\dot{x} \triangleright \tau) \quad \cdots}{M \boxtimes \omega * x} \; \mathbf{L}(x, R^-)$$

with one premiss for each $N \boxtimes \tau \in R$.

- Logical rule **Positive** case: if φ is linear on $|\tau|$ and $x \notin |\varphi \triangleright \tau| \cup P(|\tau| \cup N)$

$$\cdots \quad \frac{\overbrace{P(y) \boxtimes \varphi(y) * \dot{x}y}^{y \in |\tau|} \quad \cdots \quad \overbrace{P(y) \boxtimes \overline{\dot{x}y}}^{y \in N} \quad \cdots}{\bigcup_{|\tau| \cup N} P \boxtimes (\varphi \triangleright \tau) * x} \; \mathbf{L}(x, R^+, N \boxtimes \tau)$$

with $N \boxtimes \tau \in R$ and one premiss for each $y \in |\tau| \cup N$.

- Identities:

  if $x \neq x'$

$$\frac{}{\emptyset \boxtimes x * x'} \; \mathbf{A}(x, x')$$

  if $|\omega| \cap |\omega'| = \emptyset$

$$\frac{M \boxtimes \omega * x \qquad M' \boxtimes \omega' * x}{M \cup M' \boxtimes \omega * \omega'} \; \mathbf{C}(x)$$

  if $\widehat{X} = \{\hat{x}\}$

$$\frac{M \cup X \boxtimes \alpha \qquad M' \boxtimes \overline{\hat{x}}}{M \cup M' \boxtimes \alpha} \; \mathbf{C}(X, x)$$

- Structural rule:

  if $\alpha \preceq \alpha'$ and $M' \subseteq M$

$$\frac{M' \boxtimes \alpha'}{M \boxtimes \alpha} \; \mathbf{s}$$

  if $x \notin M$

$$\frac{M \boxtimes \omega * x}{M \cup \{x\} \boxtimes \overline{\omega}} \; \mathbf{P}$$

Fig. 5.4. The generic locative inference system of full Coloured Linear Logics (with exponentials)

which also inherits a piece of the unbounded places of the conclusion but none of the bounded places.

- There is a new variant of the Cut rule, where the cut place is the only bounded place in the right premiss while an arbitrary number of copies of it, unbounded, appear in the left premiss.

- Finally, there is a completely new inference rule, called Promotion and denoted **P**. It enables the use of the unbounded places.

The conditions defining inference trees (Definition 5.20) are modified as follows:

- in each Cut inference of the first type, the cut place is different from and independent of each of the places occurring in the conclusion;
- in each Cut inference of the second type, the cut place as well as its copies in the left premiss are different from and independent of each of the places occurring in the conclusion;
- if the conclusion is $M \boxtimes \alpha$, then M and $|\alpha|$ are disjoint and $M \cup |\alpha|$ is independent.

Proposition 5.21 is therefore extended accordingly:

At any sequent $M \boxtimes \alpha$ in an inference tree, M and $|\alpha|$ are disjoint and $M \cup |\alpha|$ is independent.

Type maps (Definition 5.26) are modified as follows:

A type map is a mapping from the set of *base* places to polarised connectives.

The conditions for an inference tree π to be a proof of type Γ (Definition 5.32) are modified as follows:

- π contains no inference (ie. it is reduced to a proper axiom).
- π ends with an inference $\mathbf{L}(x, R^s)$ and $\Gamma(\widehat{x}) = R^s$ and each immediate sub-tree of π is of type Γ.
- π ends with an inference $\mathbf{S}$ or $\mathbf{P}$ and its immediate sub-tree is of type Γ.
- π ends with an inference $\mathbf{A}(x, x')$ and $\langle \widehat{x}; \widehat{x'} \rangle_{\Gamma}^{-}$.
- π ends with an inference $\mathbf{C}(x)$ or $\mathbf{C}(X, x)$ and its immediate sub-trees are of type, respectively, $\Gamma[\widehat{x} : \Delta^+]$ and $\Gamma[\widehat{x} : \Delta^-]$ for some type map Δ.

The main difference w.r.t. Definition 5.32 is the use of $\widehat{x}$ instead of x in the type constraints.

Definition 5.39 *Let r be a base place and σ a permutation over $\widehat{r}$ (ie. all the copies of r). Then the mapping*

$$\sigma^*(x) = \text{if } (x = \dot{u}y \wedge \widehat{u} = r) \text{ then } \sigma(u)y \text{ else } x$$

is well defined. For any tree π labeled with sequents, $\pi^{r,\sigma}$ denotes the tree obtained by replacing each sequent $M \boxtimes \alpha$ in π by $\sigma^*(M) \boxtimes \sigma^* \triangleright \alpha$.

Let's show that σ^* is well defined, ie. for any place x there is at most one pair u, y such that $x = \dot{u}y$ and $\widehat{u} = r$. Indeed, assume $x = \dot{u}y = \dot{u}'y'$ with $\widehat{u} = \widehat{u}' = r$. Hence $u \leq x$ and $u' \leq x$, and, since $\leq$ is a tree ordering, we have $u \leq u'$ or $u' \leq u$. But $\widehat{u} = \widehat{u}'$. By Proposition 5.37, we get $u = u'$ and hence $y = y'$.

Proposition 5.40 *If π is an inference tree of conclusion $M \boxtimes \alpha$ and $\forall x \in |\alpha|\ \widehat{x} \not< r$, then $\pi^{r,\sigma}$ is an inference tree. Furthermore, if π is a proof of type Γ then so is $\pi^{r,\sigma}$.*

Proof This is shown by induction. For example, if π is an inference tree of the form

$$\genfrac{}{}{}{}{\genfrac{}{}{}{}{\vdots \pi_a}{\cdots \quad M \cup \dot{x}\,\widehat{N} \boxtimes \omega * (\dot{x} \triangleright \tau) \quad \cdots}{}}{M \boxtimes \omega * x} \ \ {\scriptstyle L(x, R^-)}$$

then, $\pi^{r,\sigma}$ is of the following form, with $x' = \sigma^*(x)$:

$$\cfrac{\cfrac{\vdots \pi_a^{r,\sigma}}{\sigma^*(M) \cup \sigma^*(\dot{x}\,\widehat{N}) \boxtimes \sigma^* \triangleright (\omega * (\dot{x} \triangleright \tau))}{\quad}\ {\scriptstyle (e)}}{\cfrac{\cdots \quad \sigma^*(M) \cup \dot{x}'\,\widehat{N} \boxtimes (\sigma^* \triangleright \omega) * (\dot{x}' \triangleright \tau) \quad \cdots}{\cfrac{\sigma^*(M) \boxtimes (\sigma^* \triangleright \omega) * x'}{\sigma^* M \boxtimes \sigma^* \triangleright (\omega * x)}}\ {\scriptstyle L(x', R^-)}}$$

By the induction hypothesis, $\pi_a^{r,\sigma}$ is an inference tree. Hence, so is $\pi^{r,\sigma}$. The identity (e) is justified as follows.

- If $x = \dot{u}y$ for some $u \in \widehat{r}$, let $u' = \sigma(u)$. Hence $x' = \dot{u}'y$. Now, for any place z, we have $\dot{x}z = \dot{u}\dot{y}z$ and $\sigma^*(\dot{x}z) = \dot{u}'yz = \dot{x}'z$. Hence $\sigma^*\dot{x} = \dot{x}'$.

- If $x \neq \dot{u}y$ for any $u \in \widehat{r}$, hence, $r \not\leq \widehat{x}$ (by Proposition 5.37) and $x' = x$. By assumption $\widehat{x} \not< r$. Hence $r, \widehat{x}$ are distinct independent, and so are $r, \widehat{\dot{x}z}$ for any place z. Hence (by Proposition 5.37) $r, \widehat{\dot{x}z}$ are distinct independent, and $\sigma^*(\dot{x}z) = \dot{x}z$. Hence, $\sigma^*\dot{x} = \dot{x} = \dot{x}'$.

Therefore, in both cases, $\sigma^*\dot{x} = \dot{x}'$. Hence $\sigma^*(\dot{x}\,\widehat{N}) = (\sigma^*\dot{x})(\widehat{N}) = \dot{x}'\,\widehat{N}$. Similarly, $\sigma^* \triangleright (\dot{x} \triangleright \tau) = (\sigma^* \triangleright \dot{x}) \triangleright \tau = \dot{x}' \triangleright \tau$. $\qquad\square$

Theorem 5.41 *Theorems 5.22 and 5.34 also hold with exponentials:*

Cut elimination on inference trees terminates and, for proofs, preserves types (in particular, it does not fail).

The proof is essentially the same as in the multiplicative-additive case. There is one interesting new symmetric reduction case where $\widehat{X} = \widehat{x} = \widehat{x}' = r$ and σ denotes the transposition x, x'.

$$
\cfrac{\cfrac{M \cup X \boxtimes \omega * x}{M \cup X \cup \{x\} \boxtimes \overline{\omega}} \; \text{P} \qquad \cfrac{\vdots \pi}{M' \boxtimes \overline{x}'}}{M \cup M' \boxtimes \overline{\omega}} \; \text{C}(X \cup \{x\}, x')
$$

$\rightsquigarrow$

$$
\cfrac{\cfrac{M \cup X \boxtimes \omega * x \qquad \cfrac{\vdots \pi^{r,\sigma}}{M' \boxtimes \overline{x}}}{M \cup M' \cup X \boxtimes \overline{\omega}} \; \text{C}(x) \qquad M' \boxtimes \overline{x}'}{M \cup M' \boxtimes \overline{\omega}} \; \text{C}(X, x')
$$

Theorem 5.42 *Theorems 5.25 and 5.36 also hold with exponentials: Focussing on inference trees terminates in the sense of Theorem 5.25 and, for proofs, transforms types as stated in Theorem 5.36.*

5.4 Conclusion

This paper extends results known for Linear Logic (Cut elimination and Focussing) to Coloured Linear Logics, which include arbitrary (to a certain extent) structural rules. It is shown that a few axioms, defining the variety-presentation frameworks, are sufficient to ensure the consistency and the symmetry of the whole system. The basic intuition behind these axioms is that at any point in a sequent proof, when distinguishing the principal formula for decomposition, the view of the rest of the sequent (context) is altered by that choice (hence the need for two kinds of structures: varieties, before the choice, and presentations, after the choice). The axioms only impose loose constraints on the possible structural rules, so that all sorts of "monsters" can be imagined (non-commutative, non-associative, multi-modal logics, possibly with infinitely many non-redundant connectives etc.). This situation is not satisfactory and requires further work, first to understand the relevance of the basic intuition which led to it, and second, by taking alternative viewpoints (proof-nets, semantics, etc.), to understand what essential properties have been lost or gained compared to the pure Linear Logic case.

Acknowledgements

This paper owes a lot to Ludics. However, at the time when it was started, Jean-Yves Girard was still in the process of developing Ludics, and I preferred to adopt a more classical sequent calculus presentation of things, using only basic elements of Ludics: the use of loci instead of formulas, and the expression of proof manipulations (Cut elimination and Focussing) in an exclusively locative manner. I did not introduce the concept of designs, however essential to Ludics, although I believe that today, that step could be taken.

This paper also owes a lot to Non Commutative Logic, and in particular to the many lively discussions with one of its proponents, Paul Ruet.

Finally, I am grateful to the anonymous referee as well as the non anonymous one (Franccois Lamarche) for their helpful comments.

Bibliography

[1] M. Abrusci and P. Ruet. Non-commutative logic i: the multiplicative fragment. *Annals of Pure and Applied Logic*, 101(1):29–64, 2000.

[2] J-M. Andreoli. Logic programming with focusing proofs in linear logic. *Journal of Logic and Computation*, 2(3), 1992.

[3] J-M. Andreoli. Focussing and proof construction. *Annals of Pure and Applied Logic*, 107(1):131–163, 2001.

[4] P. de Groote and F. Lamarche. Classical non associative lambek calculus. *Studia Logica*, 71(2), 2002.

[5] J-Y. Girard. Linear logic. *Theoretical Computer Science*, 50:1–102, 1987.

[6] J-Y. Girard. Locus solum: From the rules of logic to the logic of rules. *Mathematical Structures in Computer Science*, 2001.

[7] A. Joyal. Une théorie combinatoire des séries formelles. *Advances in Mathematics*, 42:1–82, 1981.

[8] F. Lamarche. On the algebra of structural contexts, 2002. Preprint.

[9] M. Moortgat. Categorial type logic. In J. van Benthem and A. ter Meulen, editors, *Handbook of Logic and Language*, pages 93–177. Elsevier Science, 1997.

[10] P. Ruet. Non-commutative logic ii: Sequent calculus and phase semantics. *Mathematical Structures in Computer Science*, 10(2):277–304, 2000.

[11] D. Yetter. Quantales and (non-commutative) linear logic. *Journal of Symbolic Logic*, 55(1):41–64, 1990.

6

An Introduction to Uniformity in Ludics

Claudia Faggian, Marie-Renée Fleury-Donnadieu and Myriam Quatrini

Institut de Mathématiques de Luminy

Abstract

In this note we develop explicit examples to help understanding the role of uniformity in Ludics. This is the key notion which underlies the move from behaviours to *bihaviours*, and is necessary to achieve full completeness.

The research on full completeness for logic and λ-calculus has motivated a large amount of work, in particular over the past 10 years. J.-Y. Girard has presented in [1] and in [2] a new approach in a framework called *Ludics*. The central notion in Ludics is that of *design*. A design corresponds to a proof, regarded under all possible points of view (syntactical proof, λ-term, function, clique, ...). Its orthogonal (counter-proof, anti-clique, ...) is also a design. Both proofs and counter-proofs are objects of the same nature. Then one can work with a set of designs equal to its biorthogonal, the *behaviour* (logical formula, type, coherent space,...). The main novelty of this approach is to overcome the duality syntax/semantics : the way for full completeness is then open.

In Ludics there are two levels of completeness : (i) internal completeness and (ii) full correspondence between ludic objects and logic. The internal completeness is about the decomposition of a connective (in some sense a counterpart of the sub-formula property). This is an essential part but not the whole story of full completeness, that can be stated as:

If $\mathfrak{D}$ is a "good" design in the behaviour **A** *associated to a closed* Π^1 *formula A then there is a $MALL_2$-proof of A which is interpreted by $\mathfrak{D}$.* ($MALL_2$ is the second order linear propositional calculus presented in [1]).

In order to obtain this result, the notion of behaviour has to be enriched, becoming that of "bihaviour," where the central role is played by the property of uniformity. Here we are interested in this notion. Our aim is to help understanding the uniformity property. Why is it necessary? How does it contribute to the completeness result?

In this note, we will omit some details. The calculus we refer to is $MALL_2$, the second order linear propositional calculus introduced in [1]. However, for the purpose of our examples, the reader can think of a standard two-sided calculus for multiplicative-additive linear logic, augmented with the connective $\downarrow$, the "shift", which simply changes the polarity of a formula.

6.1 Why uniformity: examples

The full completeness theorem of Ludics is based on two ingredients: internal completeness and uniformity. The former allows us to decompose compound formulas (behaviours), while the latter takes care of the atoms, in a sense we shall make precise.

Internal completeness is a striking property of Ludics connectives. A connective is a way to compose new behaviours from given ones, where a behaviour $\mathbf{G}$ is a set of designs closed under biorthogonal: $\mathbf{G} = \mathbf{G}^{\perp\perp}$. Internal completeness for the connectives of Ludics means that the set of designs given by the construction is equal to its biorthogonal, hence the slogan "there is no biorthogonal".

Let us take the example of $\oplus$. Given two distinct behaviours $\mathbf{A}$ and $\mathbf{B}$, $\mathbf{A} \oplus \mathbf{B}$ is defined as $\mathbf{A} \cup \mathbf{B}$. This is a behaviour, as it turns out to be equal to $(\mathbf{A} \cup \mathbf{B})^{\perp\perp}$. The operation of biorthogonal does not add new designs: the description in terms of $\mathbf{A}$ and $\mathbf{B}$ is complete. In fact, any compound behaviour is born equal to its biorthogonal.

The consequence of internal completeness is that we have a direct description of all the designs in a compound behaviour, as the biorthogonal does not introduce new objects. A nice example is that of $\otimes$. The operation which composes two designs $\mathfrak{A}, \mathfrak{B}$ from distinct behaviours $\mathbf{A}$ and $\mathbf{B}$, is enough to describe $\mathbf{A} \otimes \mathbf{B}$, which is defined as $\{\mathfrak{A} \otimes \mathfrak{B}, \mathfrak{A} \in \mathbf{A}, \mathfrak{B} \in \mathbf{B}\}^{\perp\perp}$. Again, the biorthogonal is not needed, therefore for any $\mathfrak{D} \in \mathbf{A} \otimes \mathbf{B}$ we know we can decompose it as $\mathfrak{D}_1 \otimes \mathfrak{D}_2$, with $\mathfrak{D}_1 \in \mathbf{A}$ and $\mathfrak{D}_2 \in \mathbf{B}$.

To any closed Π_1 formula F of $\mathbf{MALL_2}$, one can associate a behaviour $\mathbf{F}$ which is its interpretation. One would like to say that any material, "winning" design in $\mathbf{F}$ corresponds to a proof of F in $\mathbf{MALL_2}$ (full completeness). As it is well explained by Girard, the proof of completeness can be resumed into a slogan: "find the last rule". Concretely, given a design $\mathfrak{D}$, the whole process consists in "producing a last syntactical rule."

If $\mathfrak{D}$ belongs to the interpretation of the formula $\mathbf{F}$, to decompose it is always immediate for the negative formulae, and exploits the "internal completeness" for the positive formulae.

In this way, we are able to find the premises of the rule, represented by the designs $\mathfrak{D}_i \in \mathbf{F}_i$, where $\mathbf{F}_i$ correspond to the subformulae of $\mathbf{F}$. The fact that each $\mathfrak{D}_i$ is again material and winning, allows us to carry on the induction.

In practice one works with sequents of behaviours rather then single formulae. We recall that the sequent of behaviours $\vdash \mathbf{G}$ is equal to $\mathbf{G}$, $\mathbf{G} \vdash$ is equal to $\mathbf{G}^\perp$ and :

$$\mathfrak{F} \in \mathbf{H} \vdash \Delta \text{ iff for all } \mathfrak{E} \in \mathbf{H} \quad [\![\mathfrak{F}, \mathfrak{E}]\!] \in\vdash \Delta,$$
$$\mathfrak{F} \in\vdash \mathbf{H}, \Delta \text{ iff for all } \mathfrak{E} \in \mathbf{H}^\perp \quad [\![\mathfrak{F}, \mathfrak{E}]\!] \in\vdash \Delta.$$

The "closure principle," which exploits associativity and separation, allows us to reduce the problem to the decomposition of a single formula at a time. As an example, consider a design $\mathfrak{F}$ in $\vdash \mathbf{P} \oplus \mathbf{Q}, \mathbf{R}$. Assuming $\mathfrak{F}$ first focus on (the address of) $\mathbf{P} \oplus \mathbf{Q}$, we cut $\mathfrak{F}$ against a design $\mathfrak{E} \in \mathbf{R}^\perp$ and produce a design $[\![\mathfrak{F}, \mathfrak{E}]\!]$ which belongs to $\mathbf{P} \oplus \mathbf{Q}$. We are then able to find a design $\mathfrak{D}$ either in $\mathbf{P}$ or in $\mathbf{Q}$. Suppose it is in $\mathbf{P}$, we can retrieve the wanted premise as the design $\mathfrak{F}$ in $\vdash \mathbf{P}, \mathbf{R}$ such that $\mathfrak{D} = [\![\mathfrak{F}, \mathfrak{E}]\!]$.

However, one need to be careful when dealing with the negative case, such as $\mathfrak{F} \in \mathbf{R} \vdash \mathbf{P} \oplus \mathbf{Q}$, where normalization entails a choice of the premises. $\mathbf{R}$ depends on the interpretation of the atoms, which can be whatever behaviour we like. When we cut $\mathfrak{F}$ against a design in $\mathbf{R}$, the result also may depend on the choice of interpretation.

Think simply of $\mathbf{R} = \mathbf{X}$, where $\mathbf{X}$ is the interpretation of an atom[†]. Even if we are able to decompose the design in $\mathbf{P} \oplus \mathbf{Q}$ for any interpretation of $\mathbf{X}$, we could not be able to put things together again, as we shall illustrate. Internal completeness is no longer enough. The key point is that the premises need to be uniquely defined.

Next examples are paradigmatic of the situations which involve uniformity, making it necessary to move from behaviours to bihaviours. In Section 6.3 we will discuss the "identity axioms."

[†] This is a convention we adopt in all the paper.

6.1.1 Example: a matter of focalization

Let us consider a behaviour $\mathbf{X} \vdash \mathbf{P}, \mathbf{Q}$, where $\mathbf{X}$ is the interpretation of an atom and $\mathbf{P}, \mathbf{Q}$ are compound behaviours, respectively located in σ and τ. Consider a design $\mathfrak{F}$ with infinite premisses, one for each finite subset of $\mathbb{N}$:

$$\cfrac{\cfrac{\vdots}{\vdash \xi.I, \sigma, \tau}(\sigma, I) \qquad \cfrac{\vdots}{\vdash \xi.J, \sigma, \tau}(\tau, J)}{\xi \vdash \sigma, \tau}(\xi, \mathcal{P}_f(\mathbb{N}))$$

In order to decompose $\mathbf{P}$ or $\mathbf{Q}$ we first need to cut $\mathfrak{F}$ with a design in $\mathbf{X}$; for any choice we should have the same proof. However, let us take $\mathbf{X} = \{\mathfrak{D}_I \cup \mathfrak{D}_J\}^{\perp\perp}$, where

$$\mathfrak{D}_I: \qquad\qquad\qquad\qquad \mathfrak{D}_J:$$

$$\cfrac{\vdots}{\vdash \xi}(\xi, I) \qquad\qquad\qquad \cfrac{\vdots}{\vdash \xi}(\xi, J)$$

The normal form $[\![\mathfrak{F}, \mathfrak{D}_I]\!]$ is

$$\cfrac{\cdots \quad \cfrac{\vdots}{\sigma.i \vdash \tau} \quad \cdots}{\vdash \sigma, \tau}(\sigma, I)$$

which first focalizes on σ, the address of $\mathbf{P}$. The normal form $[\![\mathfrak{F}, \mathfrak{D}_J]\!]$ is instead

$$\cfrac{\cdots \quad \cfrac{\vdots}{\tau.j \vdash \sigma} \quad \cdots}{\vdash \sigma, \tau}(\tau, J)$$

which first focalizes on τ. The "last rule" is therefore not uniquely defined.

6.1.2 Example: decomposing $P \oplus Q$

Let $\mathfrak{F}$ be a design of base $\xi \vdash \sigma$ belonging to the behaviour $\mathbf{X} \vdash \mathbf{P} \oplus \mathbf{Q}^\dagger$; $\mathbf{X}$ is located in ξ, $\mathbf{P}$ in $\sigma.1$ and $\mathbf{Q}$ in $\sigma.2$.

$\dagger$ We are making a slight simplification here, in fact P stands for $\downarrow P$ and Q for $\downarrow Q$.

$$\dfrac{\dfrac{\vdots}{\sigma.1 \vdash \xi.I}\,(\sigma,\{1\})}{\vdash \xi.I, \sigma} \qquad \dfrac{\dfrac{\vdots}{\sigma.2 \vdash \xi.J}\,(\sigma,\{2\})}{\vdash \xi.J, \sigma}\,(\xi, \mathcal{P}_f(\mathbb{N}))$$
$$\xi \vdash \sigma$$

We interpret the atom as in the previous example. In order to decompose $\mathbf{P} \oplus \mathbf{Q}$, we cut $\mathfrak{F}$ with a design $\mathfrak{D}$ in $\mathbf{X}$, to obtain a design in $\vdash \mathbf{P} \oplus \mathbf{Q}$.

If we cut $\mathfrak{F}$ with $\mathfrak{D}_I$, we obtain a design of the form

$$\dfrac{\dfrac{\vdots}{\sigma.1 \vdash}\,(\sigma,\{1\})}{\vdash \sigma}$$

which is in the component $\mathbf{P}$.

If instead we cut with $\mathfrak{D}_J$, we obtain a design in the component $\mathbf{Q}$:

$$\dfrac{\dfrac{\vdots}{\sigma.2 \vdash}\,(\sigma,\{2\})}{\vdash \sigma}$$

Once again, it is impossible to associate $\mathfrak{F}$ with a derivation of $X \vdash P \oplus Q$. Depending on the design we choose in the atom interpretation, we obtain once a design in $\mathbf{X} \vdash \mathbf{P}$ and once a design in $\mathbf{X} \vdash \mathbf{Q}$.

6.2 From proofs to uniformity

All the designs in the previous examples are incarnated and daimon-free, but we cannot associate a proof to them. The premises of the last rule depend on the interpretation of the atom: different choices lead to different rules. There is something "non-uniform" in this; we are going to make explicit this intuition.

Consider as working example for the discussion a design $\mathfrak{F}$ of base $\xi \vdash \sigma$ belonging to $\mathbf{X} \vdash \mathbf{P}$ for any interpretation of the atom. The first rule is necessarily $(\xi, \mathcal{P}_f(\mathbb{N}))$, because $\mathbf{X}$ could be any behaviour, and $[\![\mathfrak{F}, \mathfrak{D}]\!]$ must converge for any possible $\mathfrak{D}$. The content of uniformity is that whatever premise the normalization with $\mathfrak{D}$ selects, the proof should "continue in the same way". We need to separate the designs according to how they interact with the orthogonal. Unfortunately, by definition of the orthogonal, convergence does not allow us any discrimination. To have a finer distinction, we need to consider a larger universe. The central role played by the premises of $\xi \vdash \sigma$ leads to the notion of **partial design** of a behaviour Γ.

A partial design $\mathfrak{D}'$ is a "part of a design" $\mathfrak{D} \in \mathbf{G}$: a subtree that has the same base, but where some of the premises may be missing.

A typical example of partial design is a *slice* of a design $\mathfrak{D}$: a subtree of $\mathfrak{D}$ obtained selecting in all negative rules at most one premise. An extreme example of partial designs are the empty ones (*Fid, Skunk*). $\mathbf{G}^p$ denotes the set of all designs (total and partial) of a behaviour $\mathbf{G}$.

Now we can express the fact that all the partial designs

$$\frac{\vdots}{\xi \vdash \sigma} \, (\xi, \{I\})$$

included in

$$\frac{\vdots \quad \vdots \quad \vdots}{\xi \vdash \sigma} \, (\xi, \mathcal{P}_f(\mathbb{N}))$$

lead essentially to the same proof.

It becomes natural to introduce on $\mathbf{G}^p$ a **partial equivalence relation** (i.e. asymmetric and transitive relation) $\cong$ which separates the *partial* designs with respect to normalization. The key is that the equivalence relation identifies the closed nets normalizing into $\mathfrak{D}ai$ and those normalizing into $\mathfrak{F}id$.

Since designs and counter-designs (proofs and counter-proofs) have the same status, we need to consider the partial equivalence on $\mathbf{G}^{p\perp}$ induced by normalization, and come back by bi-orthogonal... Two partial designs are in the same class, if their reactions against equivalent partial counter-designs are the same.

As an example, the trivial equivalence relation is the one that identifies all *proper*[†] designs of $\mathbf{G}^p$ ($\cong^\perp$ distinguishes all proper designs of $\mathbf{G}^{p\perp}$).

In a sequent of behaviours $\mathbf{X} \vdash \mathbf{P}$, we want again that two partial equivalent designs (saying $\mathfrak{D}$ and $\mathfrak{D}'$) react in the same way against two equivalent designs of $\mathbf{X}$ (saying $\mathfrak{E}$ and $\mathfrak{E}'$) ; that means that $[\![\mathfrak{D}, \mathfrak{E}]\!]$ and $[\![\mathfrak{D}', \mathfrak{E}']\!]$ produce two equivalent designs in $\mathbf{P}$.

On compound behaviours, the equivalence relations we are interested in must conserve the properties of the connective. For example, the behaviour $\mathbf{P} \oplus \mathbf{Q}$ is the union of the two distinct behaviours $\mathbf{P}, \mathbf{Q}$. The equivalence relation keeps distincts designs coming from distinct be-

---

† Any positive design distinct from $\mathfrak{D}ai$ and $\mathfrak{F}id$.

haviours, while two designs are equivalent if they are equivalent either in **P** or **Q**.

A design $\mathfrak{D}$ candidat to be a proof has to be equivalent to itself; it is called *uniform*.

Let us summarize a few definitions:

(i) **A bihaviour** is a couple $(\mathbf{G}, \cong)$ equal to its biorthogonal.

(ii) **Sequents of bihaviours.** Consider the sequent $\mathbf{G}_0 \vdash \mathbf{G}_1, \cdots, \mathbf{G}_n$. We obtain a bihaviour by considering the partial equivalence defined by: $\mathfrak{E} \cong_{\mathbf{G}_0 \vdash \mathbf{G}_1, \cdots, \mathbf{G}_n} \mathfrak{E}'$ iff $\forall \mathfrak{D}_0 \cong \mathfrak{D}'_0 \in \mathbf{G}_0^p$ $\forall \mathfrak{D}_i \cong \mathfrak{D}'_i \in$
$\mathbf{G}_i^{\perp p}$ $[\![\mathfrak{E}, \mathfrak{D}_0, \mathfrak{D}_1, \cdots, \mathfrak{D}_n]\!] = [\![\mathfrak{E}', \mathfrak{D}'_0, \mathfrak{D}'_1, \cdots, \mathfrak{D}'_n]\!]$.

(iii) **Compound bihaviours.** Let $(\mathbf{G}_1, \cong_1)$ and $(\mathbf{G}_2, \cong_2)$ be two disjoint bihaviours on the same base.

$\cong_{\mathbf{C}_1 \oplus \mathbf{C}_2}$ is defined by: for all $\mathfrak{D}, \mathfrak{D}' \in (\mathbf{G}_1 \oplus \mathbf{G}_2)^p$ $\mathfrak{D} \cong_{\mathbf{C}_1 \oplus \mathbf{C}_2}$
$\mathfrak{D}'$ iff $\exists i \in \{1, 2\}$ such that $\mathfrak{D}$ and $\mathfrak{D}' \in \mathbf{C}_i$ and $\mathfrak{D} \cong_{\mathbf{C}_i} \mathfrak{D}'$.

$\cong_{\mathbf{C}_1 \otimes \mathbf{C}_2}$ is defined by: for all $\mathfrak{D}_1 \otimes \mathfrak{D}_2, \mathfrak{D}'_1 \otimes \mathfrak{D}'_2 \in (\mathbf{C}_1 \otimes \mathbf{C}_2)^p$
$\mathfrak{D}_1 \otimes \mathfrak{D}_2 \cong_{\mathbf{C}_1 \otimes \mathbf{C}_2} \mathfrak{D}'_1 \otimes \mathfrak{D}'_2$ iff $\mathfrak{D}_1 \cong_{\mathbf{G}_1} \mathfrak{D}'_1$ and $\mathfrak{D}_2 \cong_{\mathbf{G}_2} \mathfrak{D}'_2$.

(iv) **Uniform designs.** Let $\mathfrak{D}$ be a partial design in the bihaviour $(\mathbf{G}, \cong)$,

$$\mathfrak{D} \text{ is uniform iff } \mathfrak{D} \cong \mathfrak{D}.$$

6.2.1 Back to the examples

We are now able to make precise the intuition that the designs in the starting examples are *not uniform*.

6.2.1.1 On example 6.1.1 (Focalization)

Let us consider the example of Section 6.1.1. Both $[\![\mathfrak{F}, \mathfrak{D}_I]\!]$ and $[\![\mathfrak{F}, \mathfrak{D}_J]\!]$ are located on the base $\vdash \sigma, \tau$. We close each of these nets, cutting with $\mathfrak{D}ai^-$ on the base $\sigma \vdash$ and with $\mathfrak{Sk}$ on the base $\tau \vdash$. We know that such two partial designs belong to *any* negative (partial) behaviour.

$$\mathfrak{D}ai_\sigma^- = \frac{\overline{\cdots \quad \dfrac{}{\vdash \sigma.I} ^\dagger \quad \cdots}}{\sigma \vdash} \, (\sigma, \mathcal{P}_f(\mathbb{N}))$$

$$\mathfrak{Sk}_\tau = \frac{}{\tau \vdash} \, (\tau, \emptyset)$$

Since $[\![\mathfrak{F}, \mathfrak{D}_I]\!]$ first focus on σ, while $[\![\mathfrak{F}, \mathfrak{D}_J]\!]$ first focus on τ, it is immediate that $[\![\;]\!] \mathfrak{F}, \mathfrak{D}_I [\![, \mathfrak{D}ai_\sigma^-, \mathfrak{Sk}_\tau]\!] = \mathfrak{D}ai$, while $[\![\;]\!] \mathfrak{F}, \mathfrak{D}_J [\![, \mathfrak{D}ai_\sigma^-, \mathfrak{Sk}_\tau]\!] = \mathfrak{Fid}$. We thus have that $[\![\mathfrak{F}, \mathfrak{D}_I]\!] \not\cong [\![\mathfrak{F}, \mathfrak{D}_J]\!]$. On the other hand, if we take on

X the trivial relation that quotients all *proper* designs, we have $\mathfrak{D}_I \cong \mathfrak{D}_J$. The definition of sequent of bihaviours implies that $\mathfrak{F} \not\approx \mathfrak{F}$: $\mathfrak{F}$ is not uniform.

6.2.1.2 On example 6.1.2 (Plus)

Let us look at the example of section 6.1.2.

The two premises do not behave in the same way: one works in the left component of $\mathbf{P} \oplus \mathbf{Q}$, the other works with the right one. We already observed that $[\![\mathfrak{F}, \mathfrak{D}_I]\!] \in \mathbf{P}$ and $[\![\mathfrak{F}, \mathfrak{D}_J]\!] \in \mathbf{Q}$. The equivalence relation on a disjoint union of behaviours $(\mathbf{P} \oplus \mathbf{Q})$ distinguishes the element coming from distinct components. Hence $[\![\mathfrak{F}, \mathfrak{D}_I]\!] \not\approx [\![\mathfrak{F}, \mathfrak{D}_J]\!]$. As in the previous example, it is enough to consider on **X** the trivial equivalence to realize that $\mathfrak{F}$ is not uniform.

To show *an uniform design* in the same behaviour, let assume that $\mathbf{P} = \Phi(\mathbf{X})$ and $\mathbf{Q} = \Psi(\mathbf{X})$ are distinct delocations of $\mathbf{X}^\dagger$.

Consider $\mathfrak{E} =$

$$\mathfrak{F}ax$$

$$\vdots$$

$$\forall I \in \mathcal{P}_f(\mathbb{N}) \quad \dfrac{\dfrac{\sigma.1 \vdash \xi.I}{\vdash \xi.I, \sigma}}{\xi \vdash \sigma} (\xi, \mathcal{P}_f(\mathbb{N}))$$

For any design $\mathfrak{D}$, $[\![\mathfrak{E}, \mathfrak{D}]\!] = \Phi(\mathfrak{D})^\ddagger$. Hence, as soon as $\mathfrak{D}, \mathfrak{D}' \in \mathbf{X}$ are equivalent, so are $[\![\mathfrak{E}, \mathfrak{D}]\!]$ and $[\![\mathfrak{E}, \mathfrak{D}']\!]$.

6.3 Uniformity and Fax

The proof of completeness goes on decomposing the positive formulas of the sequent, and accumulating atoms on the left-hand side. This process stops when it reaches a positive atom on the right-hand side: $\Gamma \vdash \mathbf{X}, \mathbf{\Delta}$. Ideally, we should have reached (the interpretation of) an "identity axiom" $X \vdash X$. The only good inhabitant should be the design that interpret the identity, or rather the infinite η-expansion of it: the $\mathfrak{F}ax$.

In the case of $\mathbf{X} \vdash \mathbf{X}$, $\mathfrak{F}ax$ is the only incarnated design which does not make use of daimon. We do not need anything else to prove it. However, to deal with the general case, uniformity become necessary to establish the following central result ([1]):

---

† To be precise, we whould write $\mathbf{P} = \downarrow\uparrow \Phi(\mathbf{X})$ and $\mathbf{Q} = \downarrow\uparrow \Psi(\mathbf{X})$.
‡ Precisely, $\downarrow\uparrow \Phi(D)$.

Proposition 6.1 (Polymorphic Lemma) *If $\mathfrak{F} \in \Gamma \vdash \mathbf{X}$, Δ is uniform, incarnated and daimon-free then $\mathbf{X} \in \Gamma$ and $\mathfrak{F}$ (essentially) behaves as a $\mathfrak{F}ax$*

The essential case is $\mathbf{X}, \mathbf{X} \vdash \mathbf{X}$.

Designs in $\mathbf{X} \vdash \mathbf{X}$

Uniformity is not required to prove that

$\mathfrak{F}ax$ is the only design $\mathfrak{F} \in \mathbf{X} \vdash \mathbf{X}$ which is stubborn and incarnated.

Proof: Let $\mathfrak{F} \in \mathbf{X} \vdash \mathbf{X}$ be a design based on $1 \vdash \sigma$. We first observe that the first negative rule must be $(1, \mathcal{P}_f(\mathbb{N}))$, to allow $\mathfrak{F}$ to converge with any possible design.

Now we fix a ramification I and observe that the rule above $\vdash 1 * I, \sigma$ cannot focalize on a $1 * i$: only σ is available as focus, as one can check choosing a convenient designs $\mathfrak{D}$ which makes the cut $[\![\mathfrak{F}, \mathfrak{D}]\!]$ fail when the condition is not realized. Moreover, for the same argument, the only possible rule is (σ, I).

The last step is to check that the repartition of the addresses is one-to-one:

$$\frac{\cdots \quad \sigma * i \vdash 1 * i \quad \cdots}{\vdash 1 * I, \sigma} \, (\sigma, I)$$

This also can be checked choosing a convenient $\mathfrak{D}$ and applying combinatory arguments.

Designs in $\mathbf{X}, \mathbf{X} \vdash \mathbf{X}$

Let now make explicite the **two uniform designs in $\mathbf{X}, \mathbf{X} \vdash \mathbf{X}$** : $\mathfrak{F}ax_1$ and $\mathfrak{F}ax_2$. We can think of $\mathfrak{F}ax_1$ as the first projection, i.e. that $\mathfrak{F}ax_1$ maps a pair of designs of $\mathbf{X}$ (saying $\mathfrak{D}_1 \otimes \mathfrak{D}_2$) on a delocation of $\mathfrak{D}_1$.
Suppose that the behaviours $\mathbf{X}$ in the left side of the sequent are located on 1 and that they are disjoint (say we applied two delocations, the first mapping all the bias on even biases, the second mapping the biases on odd biases). The behaviour $\mathbf{X}$ in the right side of the sequent is located on σ by the delocation θ. Let $\mathfrak{F}ax_1=$

$$\frac{\cfrac{\mathfrak{F}ax_{\sigma*i,1*i}}{\cdots \quad \cfrac{\sigma * i \vdash 1 * i}{\vdash 1 * (I \cup J), \sigma} \, (\sigma, I) \quad \cdots}}{1 \vdash \sigma} \, (*)$$

$(*) : (1, \{I \cup J : I \in \mathcal{P}_f(2\mathbf{N}), J \in \mathcal{P}_f(2\mathbf{N}+1)\}).$

$\mathfrak{Fax}_2$ is obtained by exchanging (σ, I) with (σ, J). Observe that $[\![\mathfrak{Fax}_1, \mathfrak{D}_1 \otimes \mathfrak{D}_2]\!] = \theta(\mathfrak{D}_1)$ and $[\![\mathfrak{Fax}_2, \mathfrak{D}_1 \otimes \mathfrak{D}_2]\!] = \theta(\mathfrak{D}_2)$. The core of the Polymorphic Lemma really consists in showing that

> $\mathfrak{Fax}_1$ *and* $\mathfrak{Fax}_2$ *are the only uniform incarnated and daimon-free designs in* $\mathbf{X}, \mathbf{X} \vdash \mathbf{X}$.

The argument relies on the following points:
- the normalization between uniform designs produces an uniform design.
- the only uniform incarnated design in the behaviour $(\mathbf{D} = \{\mathfrak{D}\}^{\perp\perp})$ is $\mathfrak{D}$ itself.

To sketch a case, if $\mathfrak{D}_1 \neq \mathfrak{D}_2$ one considers the bihaviour $\mathbf{X} = \mathbf{D}_1 \oplus \mathbf{D}_2$, and since $[\![\mathfrak{F}, \mathfrak{D}_1 \otimes \mathfrak{D}_2]\!] \in \mathbf{X}$, its incarnation must be either $\mathfrak{D}_1$ or $\mathfrak{D}_2$. Monotonicity of normalization then allows one to complete the argument.

Observe that there are also plenty of **non uniform designs in** $\mathbf{X}, \mathbf{X} \vdash \mathbf{X}$. To have one we can build an $\mathfrak{F}$ such that the premises above $\vdash 1 * (I \cup J), \sigma$ depend on $I \cup J$. For example we set that above $I_0 \cup J_0$ we have the same rules as in $\mathfrak{Fax}_2$ and for all the others $I \cup J$ we have the same rules as in $\mathfrak{Fax}_1$. We then obtain the following design, where again I contains only even bias, and J only odd bias.

$$
\mathfrak{Fax}_{\sigma*i,1*i} \qquad\qquad\qquad\qquad \mathfrak{Fax}_{\sigma*j,1*j}
$$

$$
\cfrac{\cfrac{\sigma*i \vdash 1*i \quad \cdots}{\vdash 1 * \{I \cup J\}, \sigma}(\sigma, I) \qquad \cdots \qquad \cfrac{\cdots \quad \sigma*j \vdash 1*j}{\vdash 1 * \{I_0 \cup J_0\}, \sigma}(\sigma, J_0)}{1 \vdash \sigma}(*)
$$

$(*) : (1, \{I \cup J : I \in \mathcal{P}_f(2\mathbf{N}), J \in \mathcal{P}_f(2\mathbf{N}+1)\}).$

Observe that $[\![\mathfrak{F}, \mathfrak{D}_1 \otimes \mathfrak{D}_2]\!] = \theta(\mathfrak{D}_1)$ or $\theta(\mathfrak{D}_2)$ depending on the first actions in $\mathfrak{D}_1$ and $\mathfrak{D}_2$.

Consider now any bihaviour $\mathbf{X}$ containing four designs $\mathfrak{D}_i$ and $\mathfrak{D}'_i$ ($i = 1, 2$) such that $\mathfrak{D}_i \cong \mathfrak{D}'_i$ but $\mathfrak{D}_1 \not\cong \mathfrak{D}'_2$. We then have $\mathfrak{D}_1 \otimes \mathfrak{D}_2 \cong \mathfrak{D}'_1 \otimes \mathfrak{D}'_2$, but $[\![\mathfrak{F}, \mathfrak{D}_1 \otimes \mathfrak{D}_2]\!] = \mathfrak{D}_1 \not\cong \mathfrak{D}'_2 = [\![\mathfrak{F}, \mathfrak{D}'_1 \otimes \mathfrak{D}'_2]\!]$. Then $\mathfrak{F}$ is not uniform.

Bibliography

[1] J.-Y. Girard. Locus Solum. *Mathematical Structures in Computer Sciences* 11/301 - 506, 2001

[2] J.-Y. Girard. From foundation to ludics. *This book*

[3] C. Faggian. Sur la dynamique de la ludique : une étude de l'interaction. *Thèse, Université de la Méditerranée,* 2002

[4] M-R Fleury-Donnadieu, Myriam Quatrini. First Order in Ludics. *To appear in Mathematical Structures in Computer Sciences*

7

Slicing Polarized Additive Normalization

Olivier Laurent[a], Lorenzo Tortora de Falco[b]

[a] *IML-CNRS Marseille*
[b] *Roma III*

Abstract

To attack the problem of "computing with the additives", we introduce a notion of sliced proof-net for the polarized fragment of linear logic. We prove that this notion yields computational objects, sequentializable in the absence of cuts. We then show how the injectivity property of denotational semantics guarantees the "canonicity" of sliced proof-nets, and prove injectivity for the fragment of polarized linear logic corresponding to the simply typed λ-calculus with pairing.

7.1 Introduction

The question of equality of proofs is an important one in the "proofs-as-programs" paradigm. Traditional syntaxes (sequent calculus, natural deduction, ...) distinguish proofs which are clearly the same as computational processes. On the other hand, denotational semantics identifies "too many" proofs (two different stages of the same computation are always identified). The seek of an object sticking as much as possible to the computational nature of proofs led to the introduction of a new syntax for logic: proof-nets, a graph-theoretic presentation which gives a more geometric account of proofs (see [5]). This discovery was achieved by a sharp (syntactical and semantical) analysis of the cut-elimination procedure.

Any person with a little knowledge of the multiplicative framework of linear logic (LL), has no doubt that proof-nets are the canonical representation of proofs. But as soon as one moves from such a fragment, the notion of proof-net appears "less pure". A reasonable solution for the multiplicative and exponential fragment of LL (with quantifiers) does exist (combining [2] and [7], like in [17]). Turning to multiplicative and additive LL (MALL), the situation radically changes: since the introduction of proof-nets [5], the additives were treated in an unsatisfactory way, by means of "boxes". Better solutions have been proposed in [9] and [19], until the paper [11] introduced "the good notion" of proof-net for cut-free MALL. But still, trying to deal with the full propositional fragment means entering a true jungle. Of course, it is possible to survive (i.e. to compute) in this jungle, as shown in [5, 17]. So what? The problem is that the objects (the proof-nets) used are definitely not *canonical*[†].

Recently, a new fragment of LL appeared to have a great interest: in [6] and [3] the *polarized fragment* of LL is shown to be enough to translate faithfully classical logic. A study of proof-nets for such a fragment was undertaken in [12], and the notion of [9] drastically simplified. In [15] a proof of strong normalization and confluence of the cut-elimination procedure is given for polarized LL, using the syntax of [5] (notice that for full LL confluence is wrong and strong normalization is still not completely proven). Despite these positive results, the notion of proof-net still appears as (more or less desperately, depending on the cases) non canonical.

The first contribution of the present paper is the proposal of a mathematical counterpart for the term "canonical". And here is where denotational semantics comes into the picture: in [18], the question of *injectivity* of denotational semantics is addressed for proof-nets. Roughly speaking, denotational semantics is said to be injective when the equivalence relation it defines on proofs coincides with the one defined by the cut-elimination procedure. Our proposal is to let semantics decide on the canonicity of some notion of proof-net: this is canonical when there exists a (non contrived, obviously!) denotational semantics which is injective with respect to the would-be canonical notion of proof-net.

Notice that this is a rather severe notion of canonicity. Indeed, proof-nets for multiplicative LL are canonical (and this is probably true also for MALL using [11]), but the previously mentioned extension to multiplicative and exponential LL is not guaranteed to be canonical: the time being we only know that coherent (set and multiset based) semantics *is not* injective for such proof-nets (see [18]). Finally, the known syntaxes

---

[†] We will use the term canonical in an intuitive way, following the idea that a canonical representation of a proof is not sensitive to inessential commutations of rules.

for full LL (with additives) are obviously not canonical for the usual semantics of linear logic.

The notion of *slice* was first introduced in [5]. The idea is very simple: instead of dealing with both the components of an additive box "at the same time", what about working with these two components separately? This attitude is tempting because it ignores the superimposition notion underlying the connective & (which is precisely the difficult point to understand). It is shown in [9] that the correctness of the slices of a proof-structure does not imply the correctness of the proof-structure itself (see also [11]). However, this turns out to be true in a polarized and cut-free framework (theorem 7.32).

In section 7.2, we give some intuitions on the original notion of slice for MALL coming from [5].

We then define, in section 7.3, a notion of *sliced proof-structure* for polarized LL (definition 7.5), and we show how to translate sequent calculus proofs into sliced proof-structures. To obtain canonical objects, we deal with atomic axioms and proof-structures in the style of the "nouvelle syntaxe" of Danos and Regnier [16]. For this purpose, we introduce ♭-formulas which do not occur in sequent calculus, but are very useful in our framework: a formula ♭A is necessarily the premise of a ?-link. The notation (and the meaning) of ♭A is clearly very much inspired from Girard's works on ludics [9] and on light linear logic [8].

We introduce in section 7.4 the relational semantics. We adapt the definition of experiment of [5] to our framework, and we define the interpretation of a sliced proof-structure (definition 7.13). Particular experiments coming from [18] are also introduced (injective 1-experiments), to be used later in section 7.8.

Section 7.5 is devoted to define and to study the notion of "correct" sliced proof-structure (or sliced *proof-net*). The polarization constraints allow to apply to our framework the *correctness criterion* of [12]. We define a *sliced* cut-elimination procedure (definition 7.21), we prove that correctness is preserved by our sliced cut-elimination steps (theorem 7.24) and that our semantical interpretation is sound (theorem 7.26). Our sliced proof-nets are thus proven to be computational objects.

In section 7.6, we prove that in the absence of cuts, the correctness criterion (plus some obviously necessary conditions on sets of slices) is enough to "glue" in a unique way different slices: a sliced proof-net comes from a sequent calculus proof (theorem 7.32). This result follows [12] (where the &-jumps of [9] are removed) and [14] (where the remaining jumps for weakenings are also removed).

Section 7.7 explains and justifies in details our method: the use of injective denotational semantics as a witness of canonicity of our sliced proof-nets.

The reader should notice that this is the very first time a notion of proof-net containing the additives and the exponentials can really pretend to be canonical.

Finally, section 7.8 shows that our method makes sense: there exist interesting fragments of polarized LL for which denotational semantics is injective (and thus the corresponding proof-nets are canonical), like the λ-calculus with pairing. The result that we prove is an extension of the result of [18]. Thanks to a remark of L. Regnier on the λ-calculus (expressed by proposition 7.48), we could avoid to reproduce the entire proof. We thus get injectivity only for relational semantics, but in a quick and simple way.

Let us conclude by stressing the fact that the last section is simply an example to illustrate the method explained in section 7.7, and it is (very) likely that injectivity for coherent and relational semantics holds for the whole polarized fragment. This would give canonical proof-nets for polarized LL, that is for classical logic (see [15]).

7.2 A little history of slices

Slices were first introduced in [5], and the following examples come directly from the ideas of that work.

In this section, we only want to give some hints of what will be developed in the following ones. In particular, all the notions used here simply have an intuitive meaning, and will be formally defined later.

Intuitively, a slice of a proof is obtained by choosing, for every occurrence of the rule &, one of the two premises. With the sequent calculus proof obtained by adding a cut between

$$\cfrac{\cfrac{\overline{\vdash A^{\perp}, A}\ ax}{\vdash A^{\perp} \oplus B^{\perp}, A}\ \oplus_1 \qquad \cfrac{\overline{\vdash B^{\perp}, B}\ ax}{\vdash A^{\perp} \oplus B^{\perp}, B}\ \oplus_2}{\cfrac{\vdash A^{\perp} \oplus B^{\perp}, A\ \&\ B}{\vdash (A^{\perp} \oplus B^{\perp})\ \invamp\ (A\ \&\ B)}\ \invamp}\ \&$$

and

$$\frac{\vdash A \,\&\, B, A^\perp \oplus B^\perp \quad \dfrac{\dfrac{}{\vdash A^\perp, A} \, ax}{\vdash A^\perp \oplus B^\perp, A} \, \oplus_1}{\vdash (A \,\&\, B) \otimes (A^\perp \oplus B^\perp), A, A^\perp \oplus B^\perp} \, ax \, \otimes$$

one would like to associate a graph, like:

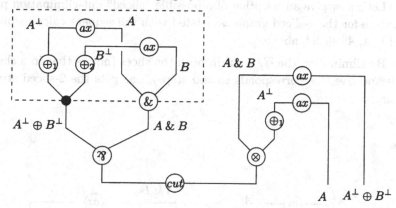

where the dashed box is an attempt to express some kind of "superimposition" of two subgraphs. Choosing to work separately with each of these two subgraphs means "slicing" the proof-net into the two following slices (where the binary &-link is replaced by two unary &-links):

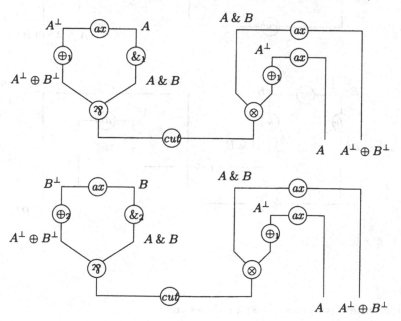

252   O. Laurent, L. Tortora de Falco

In [9], Girard shows that the correctness of slices is not enough to ensure the correctness of the whole graph: it is easy to see that there exists a proof-structure with conclusion $A \otimes (B \& C), (A^\perp \mathbin{⅋} B^\perp) \oplus (A^\perp \mathbin{⅋} C^\perp)$, with two correct slices, which is itself not correct. We will come back to this point with our theorem 7.32.

Let's now give an intuition of a possible "sliced" cut-elimination procedure for the 2-sliced graph associated with the sequent calculus proof of $\vdash A, A^\perp \oplus B^\perp$ above.

By eliminating the $\mathbin{⅋}/\otimes$ cut in both the slices (notice that in a sliced perspective this corresponds to *two* steps), one gets the 2-sliced structure:

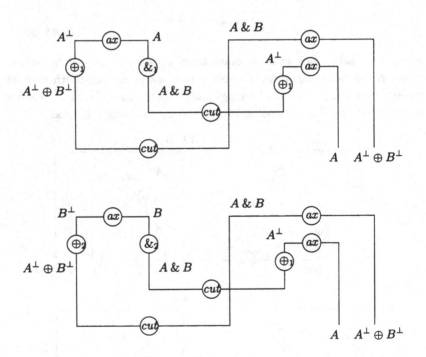

which after (two) axiom steps reduces to:

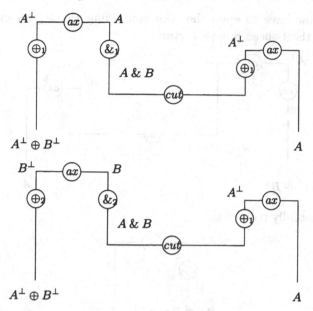

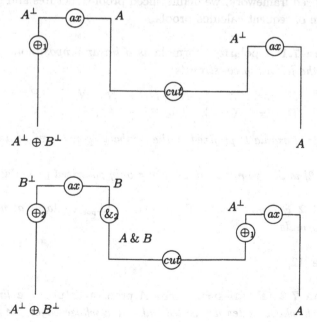

We meet here an important point: in one of the slices we have a $\&_1/\oplus_1$ cut which can be easily reduced, but in the second one we have a $\&_2/\oplus_1$ cut and no way of reducing it. By performing one step of cut-elimination (the only possible one), we obtain the 2-sliced structure:

and we now have to erase the slice containing the $\&_2/\oplus_1$ cut, thus obtaining the 1-sliced proof-structure:

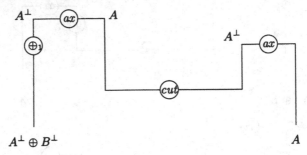

which eventually reduces to:

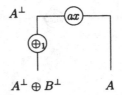

7.3 Sliced proof-structures

In a polarized framework, we define sliced proof-structures and give the translation of sequent calculus proofs.

Definition 7.1 *A polarized formula is a linear propositional formula verifying the following constraints:*

$$N \ ::= \ X \ \mid \ N \,\invamp\, N \ \mid \ N \,\&\, N \ \mid \ ?P$$
$$P \ ::= \ X^{\perp} \ \mid \ P \otimes P \ \mid \ P \oplus P \ \mid \ !N$$

or a positive formula P prefixed by the symbol $\flat$ (considered as a negative formula).

    $\mathsf{LL_{pol}}$ *[12] is the fragment of LL using only polarized formulas.*

Lemma 7.2 *Every sequent $\vdash \Gamma$ provable in $\mathsf{LL_{pol}}$ contains at most one positive formula.*

Proof See [12]. $\square$

Definition 7.3 (Proof-structure) *A proof-structure is a finite oriented graph whose nodes are called links, and whose edges are typed by* formulas of LL_{pol}. *When drawing a proof-structure we represent edges oriented up-down so that we may speak of moving upwardly or downwardly in the graph, and of links or edges "above" or "under" a given link/edge. Links are defined together with an arity and a coarity, i.e. a given number of incident edges called the premises of the link and a given number of emergent edges called the conclusions of the link.*

- *an* axiom *link or* ax-link *has no premise and two conclusions typed by dual atomic formulas,*
- *a* cut *link has two premises typed by dual formulas (which are also called the active formulas of the cut link) and no conclusion,*
- *a* $\mathcal{P}$- *(resp. $\otimes$-) link has two premises and one conclusion. If the left premise is typed by the formula A and the right premise is typed by the formula B, then the conclusion is typed by the formula $A \mathcal{P} B$ (resp. $A \otimes B$),*
- *an* !-link *has no premise, exactly one conclusion of type* $!A$ *and some conclusions of* $\flat$-types,
- *a* $\flat$-link *has one premise of type A and one conclusion of type $\flat A$,*
- *a* ?-link *has $k \geq 0$ premises of type $\flat A$ and one conclusion of type $?A$.*

Let G be a set of links such that:

(α) *every edge of G is the conclusion of a unique link;*
(β) *every edge of G is the premise of at most one link.*

We say that the edges which are not premise of a link are the conclusions of G.

We say that G is a proof-structure if with every !-link with conclusions $!A, \flat\Gamma$ is associated a proof-structure with conclusions $A, \flat\Gamma$ (called its box).

The links of the graph G are called the links with depth 0 of the proof-structure G. If a link n has depth k in a box associated with an !-link of G, it has depth $k + 1$ in G. The depth of an edge a is the depth of the link of which a is conclusion. The depth of G is the maximal depth of its links.

Convention: In the sequel, proof-structures will always have a finite depth.

Remark 7.4 *Notice that, by definition, the boxes of a proof-structure satisfy a* nesting *condition: two boxes are either disjoint or contained one in the other.*

Notice also that the type of every conclusion of a box is a negative formula.

Definition 7.5 (Sliced proof-structure) *A sliced proof-structure is a finite set S of slices such that all the slices have the same conclusions, up to the ones of type $\flat$.*

If S contains n slices, and if $\Gamma, \flat\Delta_i$ are the conclusions of the slice s_i of S, then $\Gamma, \flat\Delta_1, \ldots, \flat\Delta_n$ are the conclusions of S.

A slice s is a proof-structure possibly containing some unary $\&_1$-, $\&_2$- (resp. $\oplus_1$-, $\oplus_2$-) links, whose premise has type A, B and whose conclusion has type $A \& B$ (resp. $A \oplus B$). With every $!$-link n of s with main conclusion $!C$ is now associated a sliced proof-structure S_n (which is still called the box associated with n). This means, in particular, that C appears in every slice of S_n, while every $\flat$-conclusion of n appears in exactly one slice of S_n.

Definition 7.6 (Single-threaded slice) *A single-threaded slice is a slice s such that the sliced proof-structures associated with the $!$-links of s contain only one slice, which is itself a single-threaded slice.*

The notions of *depth* in a single-threaded slice, in a slice, and in a sliced proof-structure are the straightforward generalizations of the same notions for proof-structures given in definition 7.3.

Remark 7.7 *With every sliced proof-structure S is naturally associated a set of single-threaded slices, to which we will refer as the set of the "single-threaded slices of S (or associated with S)" denoted by sgth(S).*

Remark 7.8 *Every formula A of a sliced proof-structure is a conclusion of a unique link introducing A. (Notice that this is of course not the case in any version of proof-nets for the full propositional fragment of LL).*

We are now going to associate with every linear sequent calculus proof a sliced proof-structure.

Definition 7.9 (Translation of the sequent calculus) *Let R be the last rule of the (η-expanded) linear sequent calculus proof π. We define the sliced proof-structure S_π (with the same conclusions as π) by induction on π.*

- *If R is an axiom with conclusions $X, X^\perp$, then the unique slice of S_π is an axiom link with conclusions $X, X^\perp$.*

- If R is a $\mathcal{B}$- or a $\oplus$-rule, having as premise the subproof π', then S_π is obtained by adding to every slice of $S_{\pi'}$ the link corresponding to R.

- If R is a $\otimes$- or a cut rule with premises the subproofs π_1 and π_2, then S_π is obtained by connecting every slice of S_{π_1} and every slice of S_{π_2} by means of the link corresponding to R. Notice that if S_{π_1} (resp. S_{π_2}) contains k_1 (resp. k_2) slices, then S_π contains $k_1 \times k_2$ slices.

- If R is a &-rule with premises the subproofs π_1 and π_2, then S_π is obtained by adding a $\&_1$- (resp. $\&_2$-) link to every slice of S_{π_1} (resp. S_{π_2}) and by taking the union of these two sliced proof-structures.

- If R is a dereliction rule on A having as premise the subproof π', then S_π is obtained by adding to each slice of $S_{\pi'}$ a $\flat$-link with premise A and conclusion $\flat A$ and a unary ?-link with premise $\flat A$ and conclusion $?A$.

- If R is a weakening rule on $?A$, then S_π is obtained by adding a ?-link with arity 0 and conclusion $?A$.

- If R is a contraction rule on $?A$ having as premise the subproof π', then by induction hypothesis, every slice of $S_{\pi'}$ has two formulas $?A$ among its conclusions. By remark 7.8, these two formulas are both conclusions of a ?-link. We replace the two ?-links by a unique ?-link with the required arity, and thus obtain the slices of S_π.

- If R is a promotion rule with conclusions $!C, ?A_1, \ldots, ?A_n$ having as premise the subproof π', then let s'_i be one of the $p \geq 1$ slices of $S_{\pi'}$. For every slice s'_i of $S_{\pi'}$ with conclusions $C, ?A_1, \ldots, ?A_n$, we call s_i the graph obtained by erasing the ?-links with conclusions $?A_1, \ldots, ?A_n$. s_i is a slice with conclusions:

$$C, \flat A_{1,i}^1, \ldots, \flat A_{1,i}^{q_1,i}, \ldots, \flat A_{n,i}^1, \ldots, \flat A_{n,i}^{q_n,i}$$

with $q_{j,i} \geq 0$. The unique slice of S_π is an !-link with conclusions $!C, \flat A_{1,1}^1, \ldots, \flat A_{n,1}^{q_n,1}, \ldots, \flat A_{1,p}^1, \ldots, \flat A_{n,p}^{q_n,p}$, to which we add for every $1 \leq j \leq n$ a ?-link having as premises $\flat A_{j,i}^k$ ($1 \leq i \leq p$ and $1 \leq k \leq q_{j,i}$) and as conclusion $?A_j$. The sliced proof-structure associated with the unique !-link of S_π is the set of the s_i ($1 \leq i \leq p$).

Remark 7.10 *Let's try to give a more informal (but, hopefully clearer) description of the last case of the previous definition. For every formula*

$?A_j$, we replace the ?-link introducing it in each slice by a unique ?-link in the (unique) slice of S_π.

Let us conclude the section by giving an example of the accuracy of our sliced structures. The following sequent calculus proof:

$$
\cfrac{
  \cfrac{
    \cfrac{
      \cfrac{\overline{\quad}}{\vdash A, A^\perp}\; ax
      }{\vdash A, ?A^\perp}\; ?d
    }{\vdash A, ?A^\perp, ?B^\perp}\; ?w
  \qquad
  \cfrac{
    \cfrac{
      \cfrac{\overline{\quad}}{\vdash B, B^\perp}\; ax
      }{\vdash B, ?A^\perp, B^\perp}\; ?w
    }{\vdash B, ?A^\perp, ?B^\perp}\; ?d
  }{\vdash A \,\&\, B, ?A^\perp, ?B^\perp}\; \&
}{\vdash\, !(A \,\&\, B), ?A^\perp, ?B^\perp}\; !
$$

is translated as the sliced structure:

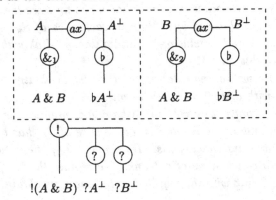

The previous structure is built inductively with respect to the depth: with the sequent calculus proof one associates the graph consisting in the !-link and in the two ?-links, and with the !-link are associated two slices (the ones inside the two dashed rectangles).

Notice that following the Danos-Regnier representation of proof-nets called "nouvelle syntaxe", consisting in "pulling down" the structural rules, the two weakenings of the sequent calculus proof simply vanished.

7.4 Semantics

We consider the concrete semantics of experiments introduced in [5]. We develop here only the case of relational semantics but the notion of experiment suits also very well coherent set-based and multiset-based semantics (see [17]).

Our results (like the existence of an injective 1-experiment used in the proof of lemma 7.49) will be completely proven only in the relational case, but the extension to the coherent semantics is just a matter of checking some minor details, consisting in the extension to our framework of the results proven in [18] without the additives.

Definition 7.11 (Relational interpretation of formulas) *The space interpreting a formula A will be denoted in the sequel by $\mathcal{A}$. It is a set, defined by induction on the complexity of A:*

- $\mathcal{X} = \mathcal{X}^{\perp}$ *is any set;*
- $\mathcal{A \otimes B} = \mathcal{A} \,\mathscr{B}\, \mathcal{B}$ *is the cartesian product of the sets $\mathcal{A}$ and $\mathcal{B}$;*
- $\mathcal{A \,\&\, B} = \mathcal{A} \oplus \mathcal{B}$ *is the disjoint union of the sets $\mathcal{A}$ and $\mathcal{B}$;*
- $!\mathcal{A} = ?\mathcal{A} = \flat\mathcal{A}$ *is the set of finite multisets of elements of $\mathcal{A}$.*

Definition 7.12 (Experiment) *If S is a sliced proof-structure, an experiment of S is an experiment of one of the slices of S.*

An experiment e of a slice s of S is an application which associates with every edge a of type A with depth 0 of s an element $e(a)$ of $\mathcal{A}$, called the label *of a. We define such an application by induction on the depth p of s.*

If $p = 0$, then:

- *If $a = a_1$ is the conclusion of an axiom link with conclusions the edges a_1 and a_2 of type X and $X^{\perp}$ respectively, then $e(a_1) = e(a_2)$.*
- *If a is the conclusion of a $\mathscr{B}$- (resp. $\otimes$-) link with premises a_1 and a_2, then $e(a) = (e(a_1), e(a_2))$.*
- *If a is the conclusion of a link $\oplus_i$ (resp. $\&_i$), $i \in \{1, 2\}$ with premise a_1, then $e(a) = (i, e(a_1))$.*
- *If a is the conclusion of a dereliction link with premise a_1, then $e(a) = \{e(a_1)\}$.*
- *If a is the conclusion of a ?-link of arity $k \geq 0$, with premises $a_1, \ldots, a_k$, then $e(a) = e(a_1) \cup \cdots \cup e(a_k)$, and $e(a) \in ?\mathcal{C}$ (if $k = 0$ we have $e(a) = \emptyset$).*
- *If a is the premise of a cut link with premises a and b, then $e(a) = e(b)$.*

If the conclusions of S are the edges $a_1, \ldots, a_l$ of type, respectively, $A_1, \ldots, A_l$, and e is an experiment of S such that $\forall i \in \{1, \ldots, l\}\ e(a_i) = x_i$, then we shall say that $(x_1, \ldots, x_l) \in \mathcal{A}_1 \,\mathscr{B} \ldots \mathscr{B}\, \mathcal{A}_l$ is the result *of the experiment e of S. We shall also denote it by $x_1, \ldots, x_l$.*

If $p > 0$, then e satisfies the same conditions as in case $p = 0$, and for every !-link n with depth 0 in s and with conclusions c of type $!C$

and $a_1, \ldots, a_l$ of type, respectively, $\flat A_1, \ldots, \flat A_l$, there exist $k \geq 0$ experiments $e_1, \ldots, e_k$ of the sliced proof-structure S' associated with n such that

- $e(c) = \{x_1, \ldots, x_k\}$, where x_j is the label associated with the edge of type C by e_j,
- If s' is the (unique!) slice of S' containing the edge a'_j with the same type as a_j, then $e(a_j)$ is the union of the labels associated with a'_j by the k experiments of s'. Notice that it might be the case that none of the k experiments is defined on a'_j: in this case one has $e(a_j) = \emptyset$.

Of course, we have that $e(c) \in \,!C$, and $e(a_j) \in \flat A_j$ (this would be an extra requirement in the coherent case).

Definition 7.13 (Interpretation) *The interpretation or the semantics of a sliced proof-structure S with conclusions Γ is the set:*

$$[\![S]\!] := \{\gamma \in \mathbin{\text{⅋}} \Gamma : \text{there exists an experiment } e \text{ of } S \text{ with result } \gamma\},$$

where $\mathbin{\text{⅋}} \Gamma$ is the space interpreting the $\mathbin{\text{⅋}}$ of the formulas of Γ.

Remark 7.14 *The interpretation of a sliced proof-structure S depends on the interpretation chosen for the atoms of the formulas of S. Once this choice is made, $[\![S]\!]$ is (by definition) the union of the interpretations of the slices of S.*

The reader should notice that the union of the interpretations of the single-threaded slices of S is not enough to recover $[\![S]\!]$ (except in some particular cases, for example when S is a cut-free proof-net, see section 7.8). This is a crucial point (behind which hide the complex relations between the additive and multiplicative worlds) showing the impossibility of working only with single-threaded slices.

Indeed, were we working with a "single-threaded semantics", by cutting the single-threaded version of the example at the end of section 7.3 (on the formula $!(A\&B)$) with the proof-net corresponding to the following proof (which is a single-threaded slice since there is no $\&$-rule):

$$
\cfrac{
  \cfrac{
    \cfrac{
      \cfrac{\overline{\vdash A, A^\perp}\ ax}{\vdash A, A^\perp \oplus B^\perp}\ \oplus_1
    }{\vdash A, ?(A^\perp \oplus B^\perp)}\ ?d
  }{\vdash\, !A, ?(A^\perp \oplus B^\perp)}\ !
  \qquad
  \cfrac{
    \cfrac{
      \cfrac{\overline{\vdash B, B^\perp}\ ax}{\vdash B, A^\perp \oplus B^\perp}\ \oplus_2
    }{\vdash B, ?(A^\perp \oplus B^\perp)}\ ?d
  }{\vdash\, !B, ?(A^\perp \oplus B^\perp)}\ !
}{
  \cfrac{\vdash\, !A \otimes\, !B, ?(A^\perp \oplus B^\perp), ?(A^\perp \oplus B^\perp)}{\vdash\, !A \otimes\, !B, ?(A^\perp \oplus B^\perp)}\ ?c
}\ \otimes
$$

we would get a proof-net with an empty semantics. Moreover, applying the cut-elimination procedure described in the next section (to the set of single-threaded slices associated with that same net) would lead to an empty set of slices.

The following notion of 1-experiment is a particular case of the more general notion of n-obsessional experiment introduced in [18].

Definition 7.15 (1-experiment) *An experiment e of a sliced proof-structure S is a 1-experiment, when with every !-link of S one has (using the notations of definition 7.12) $k = 1$, and e_1 is a 1-experiment.*

Remark 7.16 *Let S be a sliced proof-structure.*

 (i) *Let e be a 1-experiment of S. If a is any edge of S of type A, then with a the experiment e associates at most one element of A, whatever the depth of a is. In case e is not a 1-experiment, this is (in general) the case only for the edges with depth 0.*

 (ii) *The 1-experiments of S are exactly the 1-experiments of the single-threaded slices of S.*

 (iii) *We say that a 1-experiment e of a single-threaded slice s of S is injective when for every pair of (different) axiom links n_1 and n_2 of s, if x_1 (resp. x_2) is the (unique) label associated by e with the conclusions of n_1 (resp. n_2), then $x_1 \neq x_2$.*

 (iv) *If S contains no cut links, then there always exists an injective experiment of any single-threaded slice of S (just associate distinct labels with the axiom links and "propagate" them downwardly). This is not that obvious in the coherent case (due to the presence of ?-links): it is actually wrong in a non polarized framework, even for single-threaded slices coming from sequent calculus proofs (see [18]).*

7.5 Proof-nets and cut-elimination

We now define a notion of correct sliced proof-structure: a *proof-net* is a sliced proof-structure satisfying some geometrical condition. For these sliced proof-nets, a "sliced" cut-elimination procedure is given: a cut-elimination step is a step in *one* of the slices.

We show that the cut-elimination steps preserve the correctness of the structures, and that the interpretation given by definition 7.13 is sound (i.e. invariant with respect to these steps).

7.5.1 Definitions

Definition 7.17 (Acyclic sliced proof-structure) *The* correction
graph *(see [12]) of a slice s is the directed graph obtained by erasing
the edges conclusions of s, forgetting the sliced proof-structure associated with every !-link with depth 0 in s and by orienting negative (resp.
positive) edges downwardly (resp. upwardly).*

A single-threaded slice satisfies (AC) *when its correction graph, so as
the correction graph of all its boxes, is acyclic.*

A sliced proof-structure S *is* acyclic, *when every single-threaded slice
associated with* S *satisfies* (AC).

Definition 7.18 (Proof-net) *Let* S *be an* acyclic *sliced proof-structure
without any* ♭-conclusion. S *is a* proof-net *if every slice of* S *has exactly one* ♭-link *or one positive conclusion (at depth 0). Moreover, we
require that the sliced proof-structures (the boxes)* $S_1, \dots, S_k$, *recursively
associated with the !-links of* S *also satisfy these properties.*

Remark 7.19 *More geometrically, notice that this only* ♭-link *(or link
above the positive conclusion) is the only non-weakening initial node
(without incident edge) of the correction graph.*

Remark 7.20 *It is easy (and standard) to show, by induction on the
sequent calculus proof, that the sliced proof-structure associated by definition 7.9 with a sequent calculus proof is a proof-net.*

*Notice that the condition given by definition 7.18 is nothing but the
proof-net version of lemma 7.2.*

We come now to the definition of the cut-elimination procedure. If the
cut link c has depth n in the sliced proof-structure S, the cut-elimination
step associated with c will be a step for the sliced proof-structure associated with the !-link (of depth $n - 1$) the box of which contains c.

Definition 7.21 (Cut-elimination) *Let* S *be an* acyclic *sliced proof-structure without* ♭-*conclusions. We define a one step reduct* S' *of* S. *Let
$s \in S$ and c be a cut link of s. We define* $\{s_i'\}_{i \in I}$, *obtained by applying
some transformations to* s. S' *is the set of the slices obtained from* S *by
substituting* $\{s_i'\}_{i \in I}$ *for s.*

- *If c is a cut link of type ax, then $\{s'\}$ is obtained, as usual, by
  erasing the axiom link and the cut link.*

- If c is a cut link of type $\mathrm{\gamma}/\otimes$, let A and B (resp. $A^\perp$ and $B^\perp$) be the premises of the $\mathrm{\gamma}$-link (resp. $\otimes$-link). $\{s'\}$ is obtained by erasing the $\mathrm{\gamma}$-link, the $\otimes$-link and the cut link and by putting two new cut links between A and $A^\perp$, and B and $B^\perp$.

- If c is a cut link of type $\&_i/\oplus_i$, then $\{s'\}$ is obtained by erasing the two links and by moving up the cut link to their premises.

- If c is a cut link of type $\&_1/\oplus_2$ (or $\&_2/\oplus_1$), then $I = \emptyset$ (we simply erase s). Moreover, if s is the unique slice of the sliced proof-structure S_n associated with the !-link n, we also erase the slice containing n (and so on recursively...).

- If c is a cut link of type $!/?$ with a 0-ary $?$-link, then the !-link (together with its box) and its conclusion edges are erased. We then erase the 0-ary $?$-link (and the cut) thus obtaining $\{s'\}$ (notice that some $?$-links have lost some premises).

- If c is a cut link of type $!/?$ with a 1-ary $?$-link under a $\flat$-link, let T be the sliced proof-structure associated with the !-link. With each slice t_i of T, we associate the slice s'_i defined by erasing the $?$-link and the $\flat$-link, by replacing in s the !-link by t_i and by cutting the main conclusion of t_i with the premise of the $\flat$-link.

- If c is a cut link of type $!/?$ with a 1-ary $?$-link whose premise is a $\flat$-conclusion of an !-link l', let T be the sliced proof-structure associated with l' and l be the cut !-link. Let $?A/!A^\perp$ be the cut formula. $\{s'\}$ is obtained by erasing l and its conclusions and by replacing the conclusion $\flat A$ of l' by all the $\flat$-conclusions of l. And with this new !-link (which we still denote by l') is associated a sliced proof-structure T' obtained by replacing the (unique) slice t of T having $\flat A$ among its conclusions by the slice obtained by adding to the conclusion of type $\flat A$ of t a unary $?$-link and cutting its conclusion (of type $?A$) with the conclusion of type $!A^\perp$ of l. (The sliced proof-structure associated with l remains unchanged).

- If c is a cut link of type $!/?$ with a n-ary $?$-link l with $n > 1$, then $\{s'\}$ is obtained by creating a new unary $?$-link l' having as premise one of the premises of l (and erasing the corresponding edge above l), by duplicating the !-link and by cutting the copy with the conclusion of l', every $\flat$-conclusion of the copy of the !-link is premise of the same links as the edge it is a copy of (namely, they are intuitively premise of the same $?$-link). The sliced proof-structures associated with the two copies of the !-link are the same.

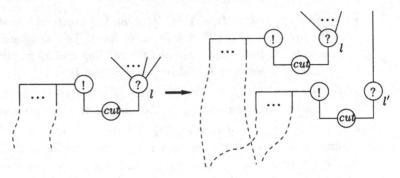

Remark 7.22 *The attentive reader certainly noticed that there are exactly two cases in which the previous definition requires the acyclicity condition:*

(i) *when the two premises of a cut link are both conclusions of the same axiom link,*

(ii) *when the two premises of a cut link are both conclusions of the same !-link (in fact, in our framework, this means that the premise of type ? of the cut link is the conclusion of a ?-link whose premise is a conclusion of type ♭ of the !-link).*

In these two cases the cut-elimination procedure is not defined. By the following section, the acyclicity of a sliced proof-structure is a sufficient condition to ensure that cut-elimination never yields to these configurations.

7.5.2 Preservation of correctness

Proposition 7.23 (Preservation of acyclicity) *If S' is a sliced proof-structure obtained from the acyclic sliced proof-structure S (without ♭ conclusions) by performing some steps of cut-elimination, then S' is acyclic.*

Proof We study every cut-elimination step, using the notations of definition 7.21:

- For the $\&_i/\oplus_j$ $(i \neq j)$ and $!/0$-ary ? steps, we erase a part of the graphs, such an operation cannot create cycles.
- For the ax and $\&_i/\oplus_i$ steps, some paths are replaced by shorter ones changing nothing to cycles.

- For the $\mathscr{R}/\otimes$ step, if p is a path containing a cycle in S', it must use one of the two new cut links starting from the premise A of the $\mathscr{R}$-link and going to the premise $A^\perp$ of the $\otimes$-link, for example. If p exists, then replacing in S the part from A to $A^\perp$ by the path going from A through the $\mathscr{R}$-link, the cut link and the $\otimes$-link to $A^\perp$ would give a cycle in S.

- For the $!/1$-ary ? step with a b-link just above the ?-link, if s'_i contains a cycle p, either it is inside t_i and thus comes from a cycle in S or it goes outside t_i, but due to the orientation, it is impossible for a path to go outside t_i and to come inside t_i since t_i has only emergent edges (since it has only negative conclusions from remark 7.4).

- For the $!/1$-ary ? step with an !-link just above the ?-link: at the depth p of the cut link, some paths are just replaced by shorter ones, and this cannot create any cycle. At depth $p + 1$, adding a cut and an !-link to an acyclic graph cannot create any cycle.

- For the $!/n$-ary ? step $(n > 1)$, if p is a cycle in S', it has to cross one of the two residues of the cut link of S. But identifying the two ?-links, the two cut links and the two !-links in p would give a cycle in S thus p doesn't exist.

$\square$

Theorem 7.24 (Preservation of correctness) *If S' is a sliced proof-structure obtained from the proof-net S by performing some steps of cut-elimination, then S' is a proof-net.*

Proof S' is acyclic by proposition 7.23. To conclude, we now prove that if S' is a one-step reduct of S, then (whatever reduction step has been performed) S' has exactly one positive conclusion or one b-link at depth 0 and in every slice of every box (assuming that the reduced cut has depth 0 in S):

- The multiplicative and additive steps are straightforward and the $!/0$-ary ? step, too.

- For the $!/1$-ary ? step with a b-link just above the ?-link, the b-link at depth 0 is erased and replaced by the one coming from every slice of the box of the !-link (which necessarily exists by remark 7.4 and definition 7.18).

- For the $!/1$-ary ? step with an !-link just above the ?-link, the b-links and the positive conclusions at depth 0 are not modified and at depth 1, we just add an !-node to a slice.

- For the $!/n$-ary ? step ($n > 1$), some links are duplicated but the b-links (and the positive conclusions) are unchanged.

$\square$

7.5.3 Soundness of the interpretation

We are going to prove that the cut-elimination procedure previously defined preserves the semantical interpretation. We use exactly the same technique as in [5], and give the details of the proof only in the most relevant cases. The proof is given for the relational semantics, and it can be straightforwardly extended to both the set and multiset based coherent semantics (see remark 7.27).

Remark 7.25 *By induction on the sequent calculus proof π, one can check that the semantics of π (as defined for example in [5]) is the semantics of the sliced proof-structure S_π of definition 7.9.*

Theorem 7.26 (Semantical soundness) *If S' is a sliced proof-structure obtained from the acyclic proof-structure S without b-conclusions by performing some steps of cut-elimination. Then $[\![S]\!] = [\![S']\!]$.*

Proof Let Γ be the conclusions of S and S' and γ an element of $\mathfrak{R}\,\Gamma$. We show that there exist a slice s of S and an experiment e of s with result γ, iff there exist a slice s' of S' and an experiment e' of s' with result γ.

One has to check this is the case for every cut-elimination step defined in definition 7.21. We will use for these steps the notations of definition 7.21. Let c be a cut link of a slice s of S. Notice that our claim is obvious for the slices which are not concerned by the cut-elimination step that we consider, and we then restrict to the other ones: we prove that there exists an experiment e of s with result γ, iff there exist a slice s' of $\{s_i'\}_{i \in I}$ and an experiment e' of s' with result γ.

By induction on the depth of c in s, we can restrict to the case where c has depth 0. The steps associated with the ax and $\mathfrak{R}/\otimes$ cut links are the same as in [5].

- If c is a cut link of type $\&_i/\oplus_i$, and e is an experiment of s, let (i, x) be the element of $\mathcal{A} \& \mathcal{B} = \mathcal{A}^\perp \oplus \mathcal{B}^\perp$ associated by e with the two edges premises of c. Then the experiment e' of s' we look for is the "restriction" of e to s': the label associated by e with the two premises of the unary $\&_i$ and $\oplus_i$ links of s is x, and x is also the label associated by e' with the two premises of the "residue" of c in s'. For the converse, one clearly proceeds in the same way.

- If c is a cut link of type $\&_1/\oplus_2$ (or $\&_2/\oplus_1$), then there exists no experiment of s (remember the condition of definition 7.12 on the label of the premises of a cut link), and no experiment of $\{s'_i\}_{i \in I}$ (remember $I = \emptyset$).

- If c is a cut link of type $!/?$ with a 0-ary ?-link, then we are simply applying the weakening step of [5].

- If c is a cut link of type $!/?$ with a 1-ary ?-link whose premise is the conclusion of a $\flat$-link, let T be the sliced proof-structure associated with the !-link. With each slice t_i of T, this step associates a slice s'_i.

  Let e be an experiment of s, let $\{x\}$ be the element of $!\mathcal{A} = ?\mathcal{A}^\perp$ associated by e with the two edges premises of c. By definition of experiment, because the label of the conclusion of the !-link is a singleton, there is a unique slice t_i of the sliced proof-structure T (associated with the !-link), and a unique experiment e_i of t_i from which e is built. The label associated with the conclusion of type A of t_i will be $x \in \mathcal{A}$. Again by definition of experiment, the label associated by e with the premise of type $A^\perp$ of the $\flat$-link is $x \in \mathcal{A}^\perp$. We can then build (from e_i) an experiment e'_i of s'_i with the same result as e. For the converse, one proceeds in the same way: an experiment e' of some slice s'_i induces an experiment e_i of t_i, and an experiment e of s.

- If c is a cut link of type $!/?$ with a 1-ary ?-link whose premise is a conclusion (of type $\flat$) of the box associated with the !-link l' (different from l), then there is nothing new with respect to the commutative step of [5].

- If c is a cut link of type $!/?$ with a n-ary ?-link l with $n > 1$, then let e be an experiment of s, let $\{x_1, \ldots, x_k\} = a_1 \cup \cdots \cup a_n$ be the element of $!\mathcal{A} = ?\mathcal{A}^\perp$ associated by e with the two edges premises of c. Suppose that a_1 is the label of the one among the premises of the ?-link of arity n, which becomes the conclusion of the new unary ?-link. We have $\{x_1, \ldots, x_k\} = a_1 \cup \{y_1, \ldots, y_h\}$. This splitting is actually a splitting of the k experiments of the sliced proof-structure associated with the !-link. This remark is enough to conclude the existence of an experiment e' of s' with the same result as e. Conversely, let e' be an experiment of s'. Because the sliced proof-structure associated with the two !-links is the same we can build an experiment e of s with the same result as e'.

$\square$

Remark 7.27 *To prove the soundness of the (set and multiset based) coherent semantics, one first needs to generalize the following result of [5] to* $\mathsf{LL_{pol}}$: *"if S is an acyclic sliced proof-structure with conclusions Γ (where Γ contains no $\flat$ formula), then $[\![S]\!]$ is a clique of the coherent space $\mathscr{F} \Gamma$."*

This result has to be used in the proof of the previous theorem in the cases of $!/?$ *cuts.*

7.6 Sequentialization for (cut-free) slices

We show that the conditions on sliced proof-structures given in definitions 7.17 and 7.18 yield a *correctness criterion* for cut-free proof-structures (theorem 7.32): they allow to characterize exactly those proof-structures coming from sequent calculus proofs.

A novelty due to our sliced presentation is that we have to be able to glue together slices. Thanks to the polarization constraint this will be possible, provided one restricts to cut-free proof-structures. In the whole section, all our proof-structures will be cut-free.

Definition 7.28 (Equivalence of links) *Let s_1, s_2 be two slices of a sliced proof-structure S. Let n_1 and n_2 be two links of s_1 and s_2 at depth 0 having the same negative non-$\flat$ conclusion A. We define, by induction on the number of links under A in s_1, the meaning of n_1 and n_2 are equivalent links denoted by $n_1 \equiv n_2$.*

If A is a conclusion of s_1 then it is also a conclusion of s_2 and $n_1 \equiv n_2$ if they are the links introducing A in s_1 and s_2.

Let A be the premise of the unary link m_1 (resp. m_2) of s_1 (resp. s_2) and the conclusion of n_1 (resp. n_2): if $m_1 \equiv m_2$, then $n_1 \equiv n_2$.

Let A be the left or right premise of the binary link m_1 (resp. m_2) of s_1 (resp. s_2) and the conclusion of n_1 (resp. n_2): if $m_1 \equiv m_2$, then $n_1 \equiv n_2$.

It is clear that $\equiv$ is an equivalence relation on the negative links at depth 0 of S.

Remark 7.29 *If $n_1 \equiv n_2$ then n_1 and n_2 are links of the same kind except if $n_1 = \&_1$ and $n_2 = \&_2$.*

Definition 7.30 (Weights) *Let S be a sliced proof-structure and let $\&^1, \ldots, \&^k$ be the equivalence classes for $\equiv$ of the $\&$-links at depth 0 of S. We associate with each $\&^i$ an eigen weight p_i that is a boolean variable (in the spirit of [9]). The weight of a slice s of S is (with an empty product equal to 1 by convention):*

$$w(s) = \prod_{\&^i_1 \in s} p_i \prod_{\&^i_2 \in s} \bar{p}_i$$

and the weight of the set S is:

$$w(S) = \sum_{s \in S} w(s)$$

The sliced proof-structure S is full *if $w(S) = 1$ and* compatible *if we have $w(s)w(t) = 0$ for $s \neq t$.*

Remark 7.31 *We can now be more precise than in remark 7.20: the sliced proof-structure associated by definition 7.9 with a cut-free sequent calculus proof is a (cut-free) proof-net, which is full and compatible.*

Theorem 7.32 (Sequentialization) *If S is a cut-free sliced proof-structure, S is the translation of an $\mathsf{LL_{pol}}$† sequent calculus proof if and only if S is a full and compatible proof-net.*

Proof We prove the second implication by induction on the size of S (the first one is remark 7.31). Since S has no b-conclusions, the conclusions of the slices of S are the same. The size of a slice s is the triple (depth(s),number of ?-links with arity at least 2 and depth 0,number of links with depth 0), lexicographically ordered, and the size of S is the sum (component by component) of the sizes of the slices of S.

Let s be a slice of S. We shall say that a link of s is terminal when its conclusion is a conclusion of s.

- If s has a terminal $\invamp$-link, a corresponding link appears in each slice since they have the same conclusions. We can remove these links in each slice and we obtain a sliced proof-structure S' verifying the hypothesis of the theorem.

---

† The extension of this result to the multiplicative units is straightforward. The case of $\top$ presents no real difficulty but requires a heavier treatment (see [13]).

- If s has a terminal &-link, a corresponding link appears in each slice. For some slices this link will be a $\&_1$-link (we call S_1 the set of slices obtained by erasing the $\&_1$-links in these slices) and for some others a $\&_2$-link (we call S_2 the corresponding set without the $\&_2$-links). We have to show that S_1 and S_2 are full and compatible. The weight of S_1 (resp. S_2) is obtained by taking $p = 1$ (resp. $p = 0$) in $p.w(S)$ (resp. $\bar{p}.w(S)$) thus this weight is 1. Let s and t be two slices of S_1 with weights $w_1(s)$ and $w_1(t)$, their weights in S are $p.w_1(s)$ and $p.w_1(t)$ thus $w_1(s)w_1(t) = 0$ (idem for S_2). We can now apply the induction hypothesis to S_1 and S_2.

Now, s has no terminal $\Re$-links and no terminal &-links thus it has no such links at depth 0 by *polarization*. This entails that s is the only slice of S by compatibility.

- If s has a terminal 0-ary ?-link, we can remove it: this corresponds to a weakening rule.
- If s has a terminal n-ary ?-link with $n \geq 2$, we break it into n unary links, we apply the induction hypothesis and perform $n-1$ contraction rules in the sequent calculus proof thus obtained.
- If s has a unary ?-link under a b-link, we remove both of them, and this corresponds to a dereliction rule. (Notice that we can apply the induction hypothesis, because when removing the two links we replace a b-link at depth 0 by a positive conclusion).
- If none of the previous conditions is satisfied then s has no $\Re$-, &-, ?-links at depth 0 (except unary ?-links under !-links). This means that if s has a terminal $\otimes$-link, it is the unique one and it is splitting: we can apply induction hypothesis to the two sub-proof-structures.
- If s has a terminal $\oplus$-link, we just remove it and apply the induction hypothesis.

If s doesn't correspond to any of the cases above, either it is an axiom link (straightforward) or it is reduced to an !-link with a unary ?-link under each b-conclusion. Let S' be the box associated with the !-link. By adding to the slices of S' some 0-ary ?-links (like in example page 258) and a 1-ary ?-link under each b-conclusion, one gets a sliced proof-structure S''. Let π'' be the proof obtained by sequentializing S'', the sequentialization π of S is obtained by adding a promotion rule to π''. (As an exercise, the reader can apply this sequentialization method to the sliced proof-structure of page 258). □

Remark 7.33 *Notice that (according to remark 7.8) every negative con-clusion $M \parr N$ (resp. $M \& N$) of a proof-net S is the conclusion of a $\parr$ (resp. $\&$) link. The previous proof shows that there exists a sequential-ization of S whose last rule introduces this formula. The reader might have recognized a proof-net version of the reversibility of the connectives $\parr$ and $\&$.*

Remark 7.34 *In fact, a(n apparently) stronger version of theorem 7.32 could be given: the reader certainly noticed that nowhere in the proof of the theorem we have used the acyclicity property of our proof-nets. This is simply due to the fact that every cut-free sliced proof-structure S is acyclic. Indeed, a path starting from a positive edge of S upwardly goes to an axiom link or an !-link and then goes down to a conclusion stopping there; while a path starting from a negative edge goes directly down to a conclusion and stops.*

7.7 Computing with slices

We now introduce a general method, allowing to use denotational se-mantics in order to guarantee the "canonicity" of our proof-nets. More precisely, we introduce the notion of *injective* semantics (which comes from [17]), and show how the existence of such a semantics is a witness of the canonicity of our sliced proof-nets as computational objects.

Remark 7.35 *We will use in the sequel the strong normalization prop-erty for proof-nets with respect to the cut-elimination procedure. We do not give the proof of such a result, which is proven in [15] (for LL_{pol}) in the framework of polarized proof-nets with additive boxes.*

Let F be a subsystem of our sliced proof-structures, and let $[\![.]\!]$ be an interpretation of the sliced proof-structures of F (satisfying theorem 7.26 and) injective: if S_1 and S_2 are two cut-free sliced proof-nets such that (for every interpretation of the atomic formulas) $[\![S_1]\!] = [\![S_2]\!]$, then $S_1 = S_2$.

Another way to speak of injectivity is the following: $[\![.]\!]$ is injec-tive when the semantical equivalence class of every proof-net contains a unique cut-free proof-net. In this (strong) sense our objects are canon-ical. In particular, such a property entails confluence: if S_1^0 and S_2^0 are two normal forms of the proof-net S, then by theorem 7.26 and injec-tivity $S_1^0 = S_2^0$.

Another crucial point is that injectivity allows to compute with the sliced proof-structures of F coming from sequent calculus proofs. Indeed, let π be any linear propositional sequent calculus proof, let S_π be the sliced proof-structure associated with π by definition 7.9, and let S_0 be the normal form of S_π. Now compute a normal form π_0 of π semantically correct (i.e. satisfying $[\![\pi]\!] = [\![\pi_0]\!]$), which can be done by performing cut-elimination directly in sequent calculus in several different ways. By remark 7.25, $[\![\pi]\!] = [\![S_\pi]\!]$ and $[\![\pi_0]\!] = [\![S_{\pi_0}]\!]$, by theorem 7.26, $[\![S_\pi]\!] = [\![S_0]\!]$, and we know that $[\![\pi]\!] = [\![\pi_0]\!]$. By injectivity, we can then conclude that $S_{\pi_0} = S_0$. In fact, our approach to injectivity (in section 7.8, and more generally in [18]) is "to rebuild" a cut-free proof from its semantics: on the one hand the injectivity property guarantees that any reasonable way of computing with sliced proof-structures coming from sequent calculus proof is sound ($S_{\pi_0} = S_0$), and on the other hand the technique used to prove injectivity suggests the possibility of semantically computing the normal form (S_0) of a proof (π_0). This last approach is very close to the so-called "normalization by evaluation" (see [1, 4]).

Summing up, one has:

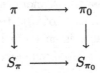

This diagram expresses a simulation property of the cut-elimination (in sequent calculus) by proof-net reductions. The injectivity property of the semantics allows to obtain such a result by *semantical* means.

Notice that the mentioned argument holds for *any existing syntax for* $\mathsf{LL_{pol}}$ instead of sequent calculus (like proof-nets with additive boxes see [5] and [15], multiboxes see [19], proof-nets with weights see [9] and [12]): let R be a proof in such a system, it will always be possible to translate R as a sliced proof-structure S_R with the same semantics as R (in the previously mentioned syntaxes, this is straightforward). Let S_0 be the normal form of S_R. Let R_0 be a normal form of R and let S_{R_0} be the sliced proof-structure associated with R_0. As before, we have $S_0 = S_{R_0}$.

We are claiming that our proof-nets are canonical computational objects: they are actually the first example of such objects in presence of the additive and exponential connectives. Indeed, (sliced) proof-nets are computational objects by theorem 7.24, and they are canonical by the injectivity property (as we already explained).

Notice that none of the previously mentioned polarized syntaxes can really claim to yield a canonical representation of proofs: denotational semantics is not injective for proof-nets with boxes nor multiboxes (even though this last syntax realizes a much greater quotient on proofs), and it is well-known that with a sequent calculus proof can be associated several proof-nets with weights (and the cut-elimination procedure is not always defined for such proof-nets).

We then have a new canonical syntax, independent from sequent calculus, allowing to make correct computations. Despite the fact that we don't have a procedure to sequentialize proof-nets with cuts, we know that if we start from a sequentializable proof-net S, we eventually reach a normal form S_0 which is itself sequentializable. This means on the one hand that nothing is lost, and on the other hand that the new objects which naturally appear (and which are not necessarily sequentializable) have a clear and well-structured computational behaviour. Actually, this is precisely the point where our approach differs from the one of [11]: we mainly focus on the computational behaviour of our objects (cut-elimination), while [11]'s main issue is correctness. Indeed, the "proof-nets" (i.e. the correct proof-structures) introduced by Hughes and Van Glabbeek are all sequentializable and this is not the case of ours. However, the translation of sequent calculus into sliced proof-structures is a function (this is not the case for [11]'s nets), and our cut-elimination procedure is local (just perform it, separately, in each slice) while Hughes and Van Glabbeek have to reduce all the slices at the same time. The non-sequentializable sliced proof-structures naturally appearing during (sliced) cut-elimination have a perfectly well-understood computational behaviour, and we do not see any reason to reject them.

The equivalence relation on sequent calculus proofs defined by our (sliced) proof-nets can be very well compared to the one defined by ordinary proof-nets in the multiplicative fragment of linear logic.

But do there exist some (interesting) subsystems F of sliced proof-structures with an injective semantics?

Such systems and semantics certainly exist in the absence of the additives (see [18]), it is very likely also the case for [11]. The next section gives a positive answer to the previous question in presence of both additive and exponential connectives. We want to mention here that this is just a first (limited) result, and it is very likely that it can be extended to full LL_{pol}.

7.8 An application: λ-calculus with pairing

We prove that (relational) semantics is injective for the fragment λLL_{pol} of LL_{pol}, which corresponds to the simply typed λ-calculus with pairing.

Definition 7.36 (λ-calculus with pairing)

$$t \ ::= \ x \ | \ \lambda x.t \ | \ (t)t \ | \ \pi_1 t \ | \ \pi_2 t \ | \ <t,t>$$

Definition 7.37 (Girard's translation) *The types of the λ-calculus with pairing are translated as negative formulas as follows:*

$$
\begin{aligned}
X \ &\rightsquigarrow \ X \\
A \rightarrow B \ &\rightsquigarrow \ ?A^{\perp} \ \mathscr{P} \ B \\
A \wedge B \ &\rightsquigarrow \ A \ \& \ B
\end{aligned}
$$

and terms are translated by the straightforward extension of Girard's translation [5, 2] for the λ-calculus.

Let λLL_{pol} be the sub-system of LL_{pol} containing only the following formulas:

$$
\begin{aligned}
N \ &::= \ X \ | \ N \ \& \ N \ | \ ?P \ \mathscr{P} \ N \\
P \ &::= \ X^{\perp} \ | \ P \oplus P \ | \ !N \otimes P
\end{aligned}
$$

(and their sub-formulas) together with the $\flat P$-formulas, and such that all the conclusions of proofs are negative formulas.

Terms are translated by proof-nets of λLL_{pol}. The constraint that axiom links introduce only atomic formulas entails that the translation contains an implicit η-expansion of terms.

In the present section, in order to prove injectivity for λLL_{pol}, we restrict to proof-structures, slices and sliced proof-structures of λLL_{pol} *without cut links* (corresponding to normal terms).

Definition 7.38 *Let s be a single-threaded slice. We denote by $L(s)$ (the "linearization" of s) the graph obtained by replacing every !-link n by the associated slice. More precisely, if n is an !-link having a conclusion of type $!A$ with an associated slice s_n, we replace n by a modified unary !-link with as premise the conclusion A of s_n; the $\flat$-conclusions of n are replaced by the corresponding $\flat$-conclusions of s_n.*

Remark 7.39 *If e is a 1-experiment of s, then with every edge a of type A of L(s) is associated a unique label e(a) of A.*

For the 1-experiment e, we will denote by $e|_{L(s)}$ the labeling of the edges of L(s) associated with e.

Lemma 7.40 *Let s and s′ be two single-threaded slices. Let e (resp. e′) be an injective 1-experiment of s (resp. s′) with result γ (resp. γ').*

If $\gamma = \gamma'$, then $L(s) = L(s')$ and $e|_{L(s)} = e'|_{L(s')}$.

Proof Our claim is that the graph $L(s)$ so as the labels of its edges are completely determined by the types of the conclusions of s and by the result of an injective 1-experiment of s. Indeed, let's start from some edge a of $L(s)$, with its type A and its label $x \in A$. There are exactly three cases in which either the type A of a is not enough to determine the link of $L(s)$ having a as conclusion or the link is known but the bottom-up propagation of the labels is not obviously deterministic:

(i) $A = C \,\&\, D$: then a might be conclusion of a $\&_1$- or of a $\&_2$-link. But the label of a tells us which of these two cases holds, and which is the label of the premise of the $\&$-link.

(ii) $A = C \oplus D$: exactly like in the previous case.

(iii) $A = ?C$: then, *because e is a 1-experiment*, the cardinality of the label of a is the arity of the ?-link with conclusion a. This also implies that there is a unique way to determine the labels of the premises of the ?-link.

To conclude, notice that the fact that e is injective allows to uniquely determine the axiom links of $L(s)$. $\qquad\square$

7.8.1 Recovering boxes in λLL_{pol}

We are now going to use in a strong way the particular shape of the (sliced) proof-nets of λLL_{pol}. We show that for a single-threaded slice s of this fragment, the graph $L(s)$ contains as much information as s. In other terms, once $L(s)$ is known, the fact that s is a single-threaded slice of a sliced proof-structure which is the translation of a term, uniquely determines the way to "put" the boxes on the graph $L(s)$.

Lemma 7.41 *If s is a slice of a λLL_{pol} proof-net, there is exactly one b-link with depth 0 in every slice of every box of s.*

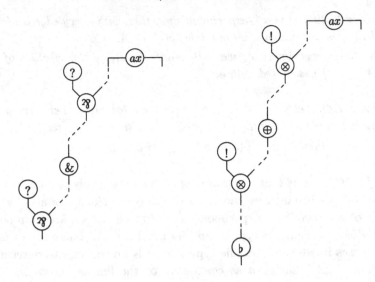

Fig. 7.1. Combs

Proof Just the correctness criterion (theorem 7.32).   □

Lemma 7.42 *Let s be a single-threaded slice, and a an edge of type A of s.*

If A is a negative (resp. positive) formula, then the graph above a is a comb (see figure 7.1):

- *the teeth of the "negative comb" are edges of type ?, while the backbone is made of unary &-nodes and of ⅋-nodes and moving upwards along it one necessarily ends in the unique (negative) atomic edge of the comb.*

- *dually, the teeth of the "positive comb" are edges of type !, while the backbone is made of ⊕-nodes and of ⊗-nodes and moving upwards along it one necessarily ends in the unique (positive) atomic edge of the comb.*

We will speak of the comb associated with a. *Notice that a is considered as an edge of the comb.*

Proof Immediate consequence of the definition of $\lambda\mathsf{LL}_{\mathrm{pol}}$ and of the definition of single-threaded slice.   □

Remark 7.43 *As a consequence of the previous lemma, with every neg-ative edge α of a single-threaded slice s, is associated an oriented path Φ_α of s (see figure 7.2): it is the path with starting edge α (oriented up-wardly), following the backbone of the negative comb up to the negative atomic edge X of the comb, crossing the axiom link and its positive con-clusion $X^\perp$ (oriented now downwardly) and moving downwardly along the backbone of a positive comb (crossing $\oplus$- and $\otimes$-links) until a b-link is reached (there are no other possibilities).*

We will refer to Φ_α (in the sequel of the paragraph) as the oriented path associated with the negative edge α.

Until the end of this section 7.8.1, we will fix the following notations according to figure 7.2:

- α is the (negative) edge premise of an !-link l of the single-threaded slice s
- B_l is the box associated with l
- n is the last link of Φ_α which is a b-link (by remark 7.43)
- c is the (positive) premise of n
- $l_1, \ldots, l_k$ are the $k \geq 0$!-links of s whose conclusions are the teeth of the positive comb associated with c
- $B_1, \ldots, B_k$ are the boxes associated with $l_1, \ldots, l_k$

Remark 7.44 *Every edge of s "above α" is contained in B_l. Moreover, all the links of Φ_α (including n) are contained in B_l.*

Lemma 7.45 *Every edge with depth 0 of B_l is either an edge of Φ_α, or the conclusion of the b-link n, or a tooth of one of the two combs of Φ_α, or the b-conclusion of a !-link.*

Proof See figure 7.2.

Let G be the correction graph of B_l (see definition 7.17). The initial nodes of G are n and the 0-ary ?-links. Every link in G is accessible by an oriented path from an initial link, but any 0-ary ?-link is the premise of a $\mathcal{P}$-link (in λLL_{pol}) that must be also accessible through its other premise. By induction on the number of links above this $\mathcal{P}$-link we easily show that it is accessible from a non ?-link. So that every link (except 0-ary ?-links) at depth 0 in B_l is accessible from n. We said that every conclusion of a 0-ary ?-link is the premise of a $\mathcal{P}$-link, and we just

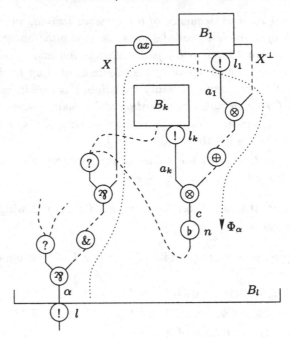

Fig. 7.2. Lemma 7.45 (dashed lines are given as examples)

proved that this γ-link is accessible from n: the conclusions of the 0-ary ?-links must then be teeth of the negative comb of Φ_α. $\qquad\square$

We are now going to define a (partial) order relation on the !-links of $L(s)$, for every single-threaded slice s. We then show that this relation coincides with the nesting of boxes and it is enough to recover the boxes of s.

Definition 7.46 *Let s be a single-threaded slice, let l and m be two !-links of s and let α be the premise of l in $L(s)$. We define the relation $<_1$ on the !-links of s as follows: $m <_1 l$ iff the oriented path Φ_α crosses the $\otimes$-link of s having m as premise. We define the relation $\leq$ as the reflexive and transitive closure of $<_1$.*

Lemma 7.47 *Let s be a single-threaded slice. If l (resp. m) is an !-link of s and B_l (resp. B_m) is the box associated with l (resp. m), then $m \leq l$ iff B_l contains B_m.*

In particular, this implies that the relation $\leq$ is indeed a partial order relation.

Proof Suppose that $m \leq l$. From the nesting condition, it is clearly enough to show that if $m <_1 l$, then B_l contains m, and this is a consequence of remark 7.44.

Conversely, suppose that B_l contains B_m. It is again enough to consider the case in which B_l is the smallest box containing B_m. By lemma 7.45, the conclusion of m is one of the teeth of the positive comb of Φ_α. We just proved that $m <_1 l$. $\qquad\square$

Proposition 7.48 *Let s and s' be two single-threaded slices of $\lambda\mathsf{LL}_{\mathrm{pol}}$. If $L(s) = L(s')$, then $s = s'$.*

Proof The reason why this holds is that the paths of $L(s)$ are the same as the paths of s.

We still use figure 7.2 and show, by induction on the number of !-links of $L(s)$ smaller (with respect to $\leq$) than l (noting that it is a finite number by lemma 7.47, since our graphs are finite), that once $L(s)$ is known, we know how to recover B_l. By induction hypothesis, we know how to recover $B_1, \ldots, B_k$. By lemma 7.45, every edge with depth 0 of B_l is either an edge of one of the two combs of Φ_α, or the conclusion of n. By remark 7.44, all the just mentioned edges are edges of B_l. Then B_l can only be the graph containing $B_1, \ldots, B_k$, the two combs of Φ_α (including the ?-links) and the conclusion of n. $\qquad\square$

7.8.2 Injectivity for $\lambda\mathsf{LL}_{\mathrm{pol}}$

We prove the following lemma for relational semantics. It certainly holds in the coherent case too, but a detailed proof would require some more intermediate results.

In the sequel, we will write $[\![S]\!] = [\![S']\!]$, always meaning that the equality holds *for every interpretation of the atoms* of the formulas of S and S'.

Lemma 7.49 *Let S and S' be two sliced proof-structures with the same conclusions.*
If $[\![S]\!] = [\![S']\!]$, then $sgth(S) = sgth(S')$, where $sgth(S)$ (resp. $sgth(S')$) is the set of the single-threaded slices of S (resp. S').

Proof By contradiction, suppose that $sgth(S) \neq sgth(S')$. There exists a single-threaded slice s of S which is different from all the single-threaded slices of S'. Let e be an injective 1-experiment of s with result γ. Such an experiment obviously exists, at least in the case of relational semantics. From $[\![S]\!] = [\![S']\!]$, there exists an experiment e' of S' with result γ. It is easy to convince oneself that e' is a 1-experiment of S' (see [18] for a proof without additives). From remark 7.16, e' is then a 1-experiment of a single-threaded slice s' of S'.

By lemma 7.40, we obtain that $L(s) = L(s')$, and then by proposition 7.48, $s = s'$ which is a contradiction. $\square$

Definition 7.50 (♭-free subgraph) *The ♭-free subgraph of a slice s is the graph obtained by keeping only the part of s at depth 0 and by replacing every !-link by an !-link without any ♭-conclusion. This erases some ♭-edges that are premises of ?-links.*

Definition 7.51 (Non-contradiction of slices) *Let s and s' be two single-threaded slices with the same non-♭ conclusions, the fact that s and s' are non-contradictory is defined by induction on the depth of s. s and s' are non-contradictory if either there exists a $\&_i$ (resp. $\&_j$) link n (resp. n') of s (resp. s') at depth 0 such that $n \equiv n'$ and $i \neq j$, or s and s' have the same ♭-free subgraph and the boxes of s and s' are non-contradictory.*

A sliced proof-structure S is non-contradictory if for every pair of single-threaded slices s and s' of S, s and s' are non-contradictory.

Theorem 7.52 (Injectivity) *Let S and S' be two non-contradictory proof-nets with the same conclusions.*
If $[\![S]\!] = [\![S']\!]$, then $S = S'$.

Proof By lemma 7.49, we have $sgth(S) = sgth(S')$. For a given set of non-contradictory single-threaded slices, there is only one way to reconstruct a sliced proof-structure: to glue (recursively with respect to the depth) the single-threaded slices with the same part at depth 0. $\square$

Remark 7.53 *The reader should not think that the hypothesis of "non-contradiction" of proof-nets weakens our injectivity theorem: it is the opposite! Indeed, our requirement for a sliced proof-structure to deserve the name of proof-net is just that "it contains only correct slices" (see definition 7.18). This (minimal) requirement is already enough to make* *correct computations (theorem 7.24), which are also semantically sound (theorem 7.26). But it is obvious that a set of correct slices is not sequentializable (in general), and we could prove theorem 7.32 only by adding the "compatibility" and "fullness" conditions. A full and compatible proof-net is always non-contradictory, and the non-contradiction hypothesis (weaker than the compatibility and fullness one) is already enough to prove theorem 7.52.*

7.8.3 Computing with the λ-calculus slices

To apply the content of section 7.7 to λLL_{pol}, just notice that if S_{π_0} is the sliced proof-structure associated with the cut-free sequent calculus proof π_0, then by remark 7.31, S_{π_0} is full and compatible (thus non-contradictory). If π is a sequent calculus proof of λLL_{pol} and S_0 is a normal form of S_π, then $[\![S_{\pi_0}]\!] = [\![S_0]\!]$ and by theorem 7.52, $S_0 = S_{\pi_0}$.

Acknowledgments: We thank Laurent Regnier who suggested us the property of the λ-calculus expressed by proposition 7.48, thus allowing us to (drastically) simplify the proof of theorem 7.52.

Bibliography

[1] Ulrich Berger, Matthias Eberl, and Helmut Schwichtenberg. Normalization by evaluation. In *Prospects for hardware foundations (NADA)*, volume 1546 of *Lecture Notes in Computer Science*, pages 117–137. Springer, 1998.

[2] Vincent Danos. *La Logique Linéaire appliquée à l'étude de divers processus de normalisation (principalement du λ-calcul)*. Thèse de doctorat, Université Paris VII, 1990.

[3] Vincent Danos, Jean-Baptiste Joinet, and Harold Schellinx. A new deconstructive logic: linear logic. *Journal of Symbolic Logic*, 62(3):755–807, September 1997.

[4] Olivier Danvy, Morten Rhiger, and Kristoffer H. Rose. Normalization by evaluation with typed abstract syntax. *Journal of Functional Programming*, 11(6):673–680, November 2001.

[5] Jean-Yves Girard. Linear logic. *Theoretical Computer Science*, 50:1–102, 1987.

[6] Jean-Yves Girard. A new constructive logic: classical logic. *Mathematical Structures in Computer Science*, 1(3):255–296, 1991.

[7] Jean-Yves Girard. Quantifiers in linear logic II. In Corsi and Sambin, editors, *Nuovi problemi della logica e della filosofia della scienza*, pages 79–90, Bologna, 1991. CLUEB.

[8] Jean-Yves Girard. Light linear logic. *Information and Computation*, 143(2):175–204, 1995.

[9] Jean-Yves Girard. Proof-nets: the parallel syntax for proof-theory. In Ursini and Agliano, editors, *Logic and Algebra*, New York, 1996. Marcel Dekker.

[10] Jean-Yves Girard. Locus solum: From the rules of logic to the logic of rules. *Mathematical Structures in Computer Science*, 11(3):301–506, June 2001.

[11] Dominic Hughes and Rob van Glabbeek. Proof nets for unit-free multiplicative-additive linear logic. In *Proceedings of the eighteenth annual symposium on Logic In Computer Science*, pages 1–10. IEEE, IEEE Computer Society Press, June 2003.

[12] Olivier Laurent. Polarized proof-nets: proof-nets for LC (extended abstract). In Jean-Yves Girard, editor, *Typed Lambda Calculi and Applications '99*, volume 1581 of *Lecture Notes in Computer Science*, pages 213–227. Springer, April 1999.

[13] Olivier Laurent. *Étude de la polarisation en logique*. Thèse de doctorat, Université Aix-Marseille II, March 2002.

[14] Olivier Laurent. Polarized proof-nets and $\lambda\mu$-calculus. *Theoretical Computer Science*, 290(1):161–188, January 2003.

[15] Olivier Laurent, Myriam Quatrini, and Lorenzo Tortora de Falco. Polarized and focalized linear and classical proofs. Prépublication 24, Institut de Mathématiques de Luminy, Marseille, France, September 2000. To appear in *Annals of Pure and Applied Logic*.

[16] Laurent Regnier. *Lambda-Calcul et Réseaux*. Thèse de doctorat, Université Paris VII, 1992.

[17] Lorenzo Tortora de Falco. *Réseaux, cohérence et expériences obsessionnelles*. Thèse de doctorat, Université Paris VII, January 2000. Available at: http://www.logique.jussieu.fr/www.tortora/index.html.

[18] Lorenzo Tortora de Falco. Obsessional experiments for linear logic proof-nets. Quaderno 11, Istituto per le Applicazioni del Calcolo, Roma, Italy, June 2001. To appear in *Mathematical Structures in Computer Science*.

[19] Lorenzo Tortora de Falco. The additive multiboxes. *Annals of Pure and Applied Logic*, 120(1–3):65–102, January 2003.

8

A Topological Correctness Criterion for Multiplicative Non-Commutative Logic

Paul-André Melliès

CNRS, Université Paris 7

Abstract

We formulate Girard's long trip criterion for multiplicative linear logic (MLL) in a topological way, by associating a ribbon diagram to every switching, and requiring that it is homeomorphic to the disk. Then, we extend the well-known planarity criterion for multiplicative cyclic linear logic (McyLL) to multiplicative non-commutative logic (MNL) and show that the resulting planarity criterion is equivalent to Abrusci and Ruet's original long trip criterion for MNL.

8.1 Introduction

In his seminal article [7] on linear logic, Jean-Yves Girard develops two alternative notations for proofs:

- a *sequential* syntax where proofs are expressed as derivation trees in a sequent calculus,
- a *parallel* syntax where proofs are expressed as bipartite graphs called *proof-nets*.

The proof-net notation plays the role of natural deduction in intuitionistic logic. It exhibits more of the intrinsic structure of proofs than the derivation tree notation, and is closer to denotational semantics. Typically, a derivation tree defines a unique proof-net, while a proof-net may represent several derivation trees, each derivation tree witnessing a particular order of *sequentialization* of the proof-net.

The parallel notation requires to separate "real proofs" (proof-nets) from "proof alikes" (called *proof-structures*) using a *correctness criterion*. Intuitively, the criterion reveals the "geometric" essence of the logic, beyond its "grammatical" presentation as a sequent calculus. In the case of MLL, the (unit-free) multiplicative fragment of (commutative) linear logic, Girard introduces a "long trip condition" which characterizes proof-nets among proof-structures. The criterion is then extended to full linear logic in [9].

The article is divided in two parts. In part one, we recall Girard's long trip criterion (section 8.2) reformulate the criterion topologically (section 8.3) and relate it to an alternative formulation by Vincent Danos and Laurent Regnier (section 8.4). In part two, we shift from commutative to non-commutative logic. So, we start by reformulating carefully the well-known planarity criterion for multiplicative cyclic logic (McyLL) (section 8.5). And we recall multiplicative non-commutative logic (MNL) (section 8.6) as well as the long trip criterion devised for MNL by V. Michele Abrusci and Paul Ruet [3] (section 8.7). Finally, we generalize to MNL the "planarity" criterion for McyLL (section 8.8) and show that the criterion is equivalent to Abrusci-Ruet "long trip" criterion (section 8.9). We conclude the article with an appendix discussing the topological status of logics like MLL, McyLL or MNL (section 8.10).

8.2 Girard's long trip correctness criterion

We recall below the long trip correctness criterion, which appears in [7], and characterizes the proofs of the (unit-free) multiplicative fragment of linear logic (MLL).

MLL formulas and negation. — An MLL *formula* is a tree with leaves p, q, r, ... and $p^\perp$, $q^\perp$, $r^\perp$,... called *atoms*, and binary connectives $\otimes$ and $\mathscr{B}$. The *negation* $A^\perp$ of a formula A is the formula defined inductively by so-called de Morgan laws:

$$(A \otimes B)^\perp = B^\perp \mathscr{B} A^\perp, \quad (A \mathscr{B} B)^\perp = B^\perp \otimes A^\perp, \quad (p)^\perp = p^\perp, \quad (p^\perp)^\perp = p.$$

It follows that $(A^\perp)^\perp = A$ for every formula A.

MLL sequent calculus. — An MLL sequent is a finite sequence of formulas, noted $\vdash A_0, ..., A_{k-1}$. We usually write formulas as latin letters A, B, C, and finite sequences of formulas as greek letters Γ, Δ. A *derivation tree* is a tree with a sequent at each node, constructed inductively by the five rules below.

$$(\text{Ax}) \ \frac{}{\vdash A^{\perp}, A} \qquad (\text{Cut}) \ \frac{\vdash \Gamma, A \qquad \vdash A^{\perp}, \Delta}{\vdash \Gamma, \Delta}$$

$$(\otimes) \ \frac{\vdash \Gamma, A \qquad \vdash B, \Delta}{\vdash \Gamma, A \otimes B, \Delta} \qquad (\mathfrak{N}) \ \frac{\vdash \Gamma, A, B}{\vdash \Gamma, A \,\mathfrak{N}\, B}$$

$$(\text{Exch}) \ \frac{\vdash \Gamma, A, B, \Delta}{\vdash \Gamma, B, A, \Delta}$$

MLL links. — An MLL *link* is a graph of the following form, whose vertices are labelled with MLL formulas:

1. Axiom link

with two conclusions A and $A^{\perp}$, and no premise,

2. Cut link

with two premises A and $A^{\perp}$, and no conclusion,

3. $\otimes$ and $\mathfrak{N}$ links

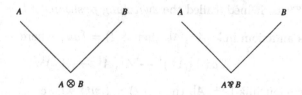

where the formula A is the *first* premise, the formula B is the *second* premise, and $A \otimes B$ (or $A \,\mathfrak{N}\, B$) is the conclusion.

MLL proof-structures. — A *proof-structure* Θ is a graph constructed with links such that every (occurrence of) formula is the conclusion of one link, and the premise of at most one link. We define a *conclusion* of Θ as a formula which is not the premise of any link. A link of Θ is *terminal* when its conclusion is a conclusion of Θ.

Every derivation tree defines a proof-structure, but conversely, not

every proof-structure is deduced from a derivation tree. The simplest
example is the proof-structure:

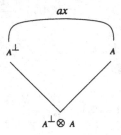

So, which proof-structures exactly are obtained from a derivation tree?
Here follows Girard's remarkable answer, the so-called *long trip* criterion.

Decorated formulas. — Call decorated formula a couple $(A, \uparrow)$ or
$(A, \downarrow)$ where A is an MLL formula and $\uparrow$ or $\downarrow$ is a tag. We write $A^\uparrow$ and
$A^\downarrow$ for the decorated formulas $(A, \uparrow)$ and $(A, \downarrow)$. Now, for each axiom,
cut, $\otimes$ or $\mathscr{V}$ link l, we define two sets l^{in} and l^{out} of decorated formulas,
as follows:

- l^{in} is the set of all decorated formulas $A^\downarrow$ where A is a premise of l,
  and all decorated formulas $A^\uparrow$ where A is a conclusion of l;
- l^{out} is the set of all decorated formulas $A^\uparrow$ where A is a premise of l,
  and all decorated formulas $A^\downarrow$ where A is a conclusion of l.

Switching positions. — For every link l, a set $\mathcal{S}(l)$ of functions from
l^{in} to l^{out} is defined, called the *switching positions* of l:

- if l is an axiom link $[A^\perp, A]$, then $\mathcal{S}(l) = \{ax\}$ where

$$ax : (A^\perp)^\uparrow \mapsto A^\downarrow, A^\uparrow \mapsto (A^\perp)^\downarrow;$$

- if l is a cut link $[A^\perp, A]$, then $\mathcal{S}(l) = \{cut\}$ where

$$cut : (A^\perp)^\downarrow \mapsto A^\uparrow, A^\downarrow \mapsto (A^\perp)^\uparrow;$$

- if l is a $\otimes$-link $[A, A \otimes B, B]$, then $\mathcal{S}(l) = \{\otimes_R, \otimes_L\}$ where

$$\otimes_R : A^\downarrow \mapsto B^\uparrow, B^\downarrow \mapsto (A \otimes B)^\downarrow, (A \otimes B)^\uparrow \mapsto A^\uparrow,$$

$$\otimes_L : A^\downarrow \mapsto (A \otimes B)^\downarrow, B^\downarrow \mapsto A^\uparrow, (A \otimes B)^\uparrow \mapsto B^\uparrow;$$

- if l is a $\mathscr{V}$-link $[A, A \mathscr{V} B, B]$, then $\mathcal{S}(l) = \{\mathscr{V}_R, \mathscr{V}_L\}$ where

$$\mathscr{V}_R : A^\downarrow \mapsto A^\uparrow, B^\downarrow \mapsto (A \otimes B)^\downarrow, (A \otimes B)^\uparrow \mapsto B^\uparrow,$$

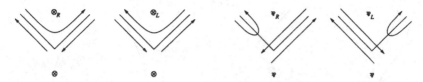

Fig. 8.1. Girard switching positions for tensor and par

$$\mathscr{R}_L\colon A^{\downarrow} \mapsto (A \otimes B)^{\downarrow}, B^{\downarrow} \mapsto B^{\uparrow}, (A \otimes B)^{\uparrow} \mapsto A^{\uparrow}.$$

Long trip criterion. — A *switching s* of an MLL proof-structure Θ is a function which associates a switching position $s(l) \in \mathcal{S}(l)$ to every link l of Θ. The *switched proof-structure* **trip**(Θ, s) is the oriented graph with vertices the decorated formulas labelling Θ, and with an edge from A^x to B^y iff $B^y = s(l)A^x$, for some link l in Θ, or $A^x = C^{\downarrow}$ and $B^y = C^{\uparrow}$, for some conclusion C of Θ.

Definition 8.1 (Girard) *A* Girard proof-net *is a proof-structure* Θ *such that every switched proof-structure* **trip**(Θ, s) *contains a unique cycle. This unique cycle is called* the long trip.

Intuitively, every switching s defines a trajectory for a particle visiting the proof. Each $\otimes$ and $\mathscr{R}$ link is visited according to one switching position of figure 8.1; the particle rebounces on axioms, cuts and conclusions. A proof-structure is a proof-net when the particle visits every part, without being captured into a cycle, this for every switching.

Three important properties are established in [7].

(i) *soundness:* every MLL derivation tree translates as a Girard proof-net.

(ii) *sequentialization:* every Girard proof-net is the translation of an MLL derivation tree. The proof is based on the notions of (maximal) empire, and splitting tensor.

(iii) *cut-elimination:* MLL enjoys cut-elimination.

8.3 Our topological reformulation

The characterization of proofs provided by Girard's criterion is not only "geometric", it is also "computational". Expressed in game semantics, the criterion characterizes proofs as uniform strategies which do not deadlock during communication, and which interact with every part of

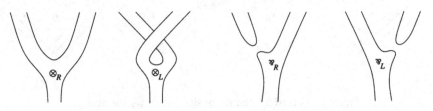

Fig. 8.2. Ribbon version of figure 8.1

the formula, see [1]. In fact, switchings should be understood as counter-proofs in an extended "para-logic", see [10].

One technical point is that long trips are oriented in Girard's criterion. However, the orientation may be avoided by reformulating the criterion topologically. The idea is to replace oriented edges by *ribbons*, and to apply the convention below.

Convention. —

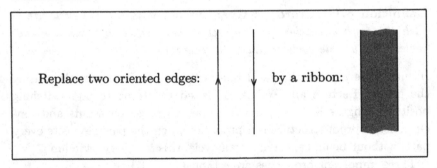

According to the convention, the ⊗ and ⅋ switching positions of figure 8.1 are replaced by the ribbon diagrams of figure 8.2, while the (switching position of) axiom and cut links are replaced by simple ribbons:

Similarly, each conclusion C is replaced by a 2-dimensional "cul-de-sac":

Now, every proof-structure Θ and every switching s induces a surface **ribbon**(Θ, s) obtained by replacing every switched link and conclusion of Θ by its ribbon diagram, and pasting all diagrams together. This enables us to reformulate Girard's long trip criterion below, see also section 8.4 for a proof that the two formulations are equivalent (lemma 8.4.)

Definition 8.2 (topological proof-net) *A topological proof-net is a proof-structure Θ such that the surface* **ribbon**(Θ, s) *is homeomorphic to the disk, for every Girard switching s.*

8.4 Danos and Regnier correctness criterion

Many alternative formulations of Girard's long trip criterion are possible. We recall here the "tree" criterion formulated by Vincent Danos and Laurent Regnier in [5]. A Danos-Regnier switching for an MLL proof-structure Θ is the data for every $\mathscr{V}$-link of a switching position chosen among $\mathscr{V}_R$ and $\mathscr{V}_L$:

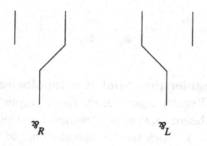

Given a Danos-Regnier switching s, the switched graph **graph**(Θ, s) is defined by replacing every $\mathscr{V}$-link in Θ by the corresponding switching position. Danos and Regnier's formulation of the criterion follows.

Definition 8.3 (Danos-Regnier) *A Danos-Regnier proof-net is a proof-structure whose all switching graphs are trees, ie. connected and acyclic graphs.*

Herebelow, we establish that the three formulations of proof-net (Girard, Danos-Regnier, topological) are equivalent. The proof is not really difficult, but informative enough to appear here. We will consider the "shrink" operation contracting ribbons into one-dimensional edges, like this:

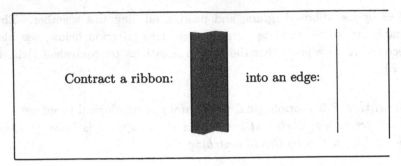

It is worth observing that the operation "shrinks" the ribbon diagrams of figure 8.2 into the Danos-Regnier switching positions above. In particular, the operation contracts the two $\otimes_R$ and $\otimes_L$ positions into the "invisible" Danos-Regnier switching position $\otimes_R = \otimes_L$:

Every Danos-Regnier proof-net is a topological proof-net. — Consider a Danos-Regnier proof-net Θ. Every (topological) switching s defines a surface $\textbf{ribbon}(\Theta, s)$ which "retracts" as the tree $\textbf{graph}(\Theta, s)$. Thus, the surface is a "thick tree" homeomorphic to the disk. We conclude.

Every topological proof-net is a Girard proof-net. — Consider a topological proof-net Θ. Every (topological = Girard) switching s defines a surface $\textbf{ribbon}(\Theta, s)$ homeomorphic to the disk. Its border $\textbf{trip}(\Theta, s)$ is unique, therefore a long trip. We conclude.

Every Girard proof-net is a Danos-Regnier proof-net. — This is the only delicate step of our series of equivalence. We proceed by contradiction. Suppose that Θ is a Girard proof-net, and not a Danos-Regnier proof-net. By definition, there exists a Danos-Regnier switching s such that $\textbf{graph}(\Theta, s)$ is not a tree. The difficult point is to define a (topological = Girard) switching s' inducing a surface $\textbf{ribbon}(\Theta, s')$ with two borders at least. When $\textbf{graph}(\Theta, s)$ is not connected, we take $s' = s$.

When **graph**(Θ, s) contains a cycle $\mathcal{C}$, it is always possible to alter the switching positions of the $\otimes$-links visited by $\mathcal{C}$ in **ribbon**(Θ, s) in such a way that the altered switching s' verifies **graph**$(\Theta, s) = $ **graph**(Θ, s') and that the cycle $\mathcal{C}$ "lifts" to a border of **ribbon**(Θ, s'). Note that the resulting surface **ribbon**(Θ, s') has two borders at least. Each such border induces a cycle in **trip**(Θ, s'). It follows that **trip**(Θ, s') is not a long trip, and we conclude.

Lemma 8.4 *The three formulations of MLL proof-net are equivalent.*

Intuitively, the topological criterion stands halfway between Girard and Danos-Regnier criteria, keeping the best of both worlds. For instance, the switching position $\otimes_L$ is necessary to test a proof-structure in the long trip criterion; but not in the Danos-Regnier and topological formulations.

Lemma 8.5 *In definition 8.2, switchings may be replaced by $\otimes_L$-free switchings.*

This point is best illustrated by the proof-structure (8.1) pointed out by Abrusci and Ruet [3]. Switching every $\otimes$-link as $\otimes_R$ is enough to show that Θ is not a topological proof-net — since the induced switching surface is not planar. On the other hand, the surface has a unique border... So, it takes one switching position $\otimes_L$ at least to detect that (8.1) is not a Girard proof-net.

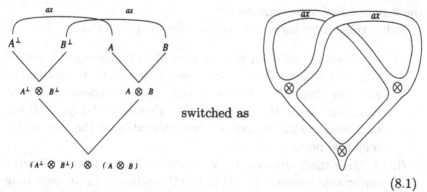

$$(8.1)$$

This is the advantage of thinking topologically: the long trip criterion counts the number of borders of **ribbon**(Θ, s) while the topological criterion takes also into account its planarity and genus.

8.5 A planarity correctness criterion for cyclic linear logic

Suggested by Girard in [8] expounded by Yetter in [22] cyclic linear logic (cyLL) is the variant of linear logic obtained by limiting the exchange rule EXCH to *cyclic* permutations:

$$(\text{CYEXCH}) \quad \frac{\vdash A_0, ..., A_{k-1}}{\vdash A_{\xi(0)}, ..., A_{\xi(k-1)}} \quad \text{where } \xi \text{ is a cyclic permutation.}$$

In this section, we consider McyLL, the multiplicative (unit-free) fragment of cyclic linear logic. As in [3], we use the notations $\odot$ for "next" and $\triangledown$ for "sequential" to distinguish the cyclic connectives from their commutative counterparts $\otimes$ and $\invamp$. The definitions of formula, sequent and proof-structure are the same in McyLL as in MLL, with the only difference that the connectives $\odot$ and $\triangledown$ replace $\otimes$ and $\invamp$ everywhere, respectively. Negation is defined as in MLL:

$$(A \odot B)^{\perp} = B^{\perp} \triangledown A^{\perp}, \qquad (A \triangledown B)^{\perp} = B^{\perp} \odot A^{\perp}.$$

Except for the restriction on the exchange rule, the rules of McyLL are the same as in MLL:

$$(\text{AX}) \quad \frac{}{\vdash A^{\perp}, A} \qquad\qquad (\text{CUT}) \quad \frac{\vdash \Gamma, A \qquad \vdash A^{\perp}, \Delta}{\vdash \Gamma, \Delta}$$

$$(\odot) \quad \frac{\vdash \Gamma, A \qquad \vdash B, \Delta}{\vdash \Gamma, A \odot B, \Delta} \qquad\qquad (\triangledown) \quad \frac{\vdash \Gamma, A, B}{\vdash \Gamma, A \triangledown B}$$

It is worth noting that the formula $(A \odot B) \multimap (B \odot A)$ is not provable in McyLL, where $A \multimap B$ is notation for $A^{\perp} \triangledown B$. This is the reason why the logic is called non-commutative.

Today, three correctness criteria are available for McyLL.

(i) A "planarity" criterion characterizes McyLL proof-nets as *planar* MLL proof-nets. This criterion was observed by Girard at the very first days of cyclic linear logic, and is well-known today. It appears explicitly in [4, 16, 17]. Franccois Métayer delivers an alternative but equivalent characterization of the logic in his simplicial presentation [14].

(ii) A "long trip" criterion by V. Michele Abrusci adapts Girard's correctness criterion for MLL, by (1) limiting $\odot$ to the switching position $\otimes_R$ and (2) adding a new position $\triangledown_3$ to the switching positions of $\triangledown$. The criterion is formulated for McyLL in [2] and extended to non-commutative logic (MNL) in [3]. The criterion is exposed in section 8.7 where we also discuss a recent version of the criterion by Virgil Mogbil and Quentijn Puite [15].

(iii) A recent "seaweed" criterion by Roberto Maieli [12] formulates a criterion for McyLL and MNL in the fashion of Danos and Regnier criterion for MLL. The idea is to replace trees by series-parallel order varieties (seaweed).

We formulate very carefully the "planarity" criterion for McyLL, which is not as straightforward as it seems. The first part of the criterion requires that an McyLL proof-net Θ translates as an MLL proof-net Θ^*.

Definition 8.6 (commutative translation) *The commutative translation Θ^* of an McyLL proof-structure Θ is the MLL proof-structure obtained as the result of replacing every $\odot$ and ∇ link by $\otimes$ and $\invamp$, respectively.*

The second part of the criterion requires "planarity" of Θ, or more precisely planarity of the (orientable) surface **ribbon**(Θ) obtained as in section 8.3, by replacing every $\{\odot, \nabla, \text{axiom}, \text{cut}\}$-link and conclusion in Θ by the associated ribbon diagram

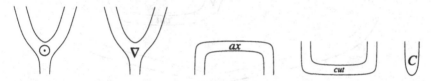

The unexpected point is that planarity of **ribbon**(Θ) is *not* sufficient to characterize McyLL proofs among McyLL proof-structures. Typically, the McyLL proof-structure Θ of conclusion

$$\vdash (A^\perp \nabla B^\perp), (A \odot B)$$

is not sequentializable in McyLL, but its surface **ribbon**(Θ) is planar:

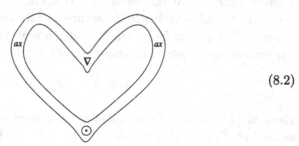

$$(8.2)$$

So, how should one characterize McyLL proof-nets? One possible answer is to require that all conclusions of Θ lie on the same border of **ribbon**(Θ). It is not very complicated to prove that this requirement added to planarity characterizes all *cut-free* proofs among *cut-free* proof-structures. Unfortunately, the criterion is too weak to characterize proofs with cuts, as witnessed by the example below of a non-sequentializable McyLL proof-structure, with a unique conclusion.

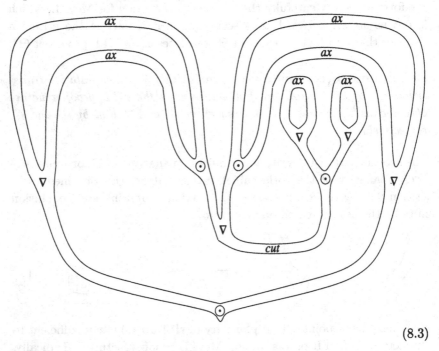

$$(8.3)$$

Remark. — The proof-structure (8.3) is interpreted as a disk in Métayer's simplicial presentation. This explains why Métayer's sequentialization theorem for McyLL [14] is limited to *cut-free* proof-nets.

Planar logic. — At this point, it is tempting to define a conservative logic over McyLL, which would capture exactly the idea of "planarity". Let us call it *planar logic*. Its formulas are McyLL formulas, and its sequents are finite sets of (occurrences of) McyLL sequents, written

$$\vdash \Gamma_1 \mid \cdots \mid \Gamma_n$$

Each McyLL sequent Γ_i is called a *component* of the sequent. Two sequents $\vdash \Gamma_1 \mid \ldots \mid \Gamma_n \mid \Delta$ and $\vdash \Gamma_1 \mid \ldots \mid \Gamma_n$ of the logic are generally identified when Δ is the empty component. Planar logic enables general exchange between components:

$$(\text{Exch}) \quad \frac{\vdash \cdots \mid \Gamma \mid \Delta \mid \cdots}{\vdash \cdots \mid \Delta \mid \Gamma \mid \cdots}$$

and cyclic permutations ξ inside a component:

$$(\text{cyExch}) \quad \frac{\vdash \cdots \mid A_0, ..., A_{k-1}}{\vdash \cdots \mid A_{\xi(0)}, ..., A_{\xi(k-1)}}$$

The remaining rules of planar logic follow:

$$(\text{Ax}) \quad \frac{}{\vdash A^\perp, A} \qquad\qquad (\text{Cut}) \quad \frac{\vdash \cdots \mid \Gamma, A \qquad \vdash A^\perp, \Delta \mid \cdots}{\vdash \cdots \mid \Gamma, \Delta \mid \cdots}$$

$$(\triangledown) \quad \frac{\vdash \cdots \mid \Gamma, A, \Delta, B}{\vdash \cdots \mid \Gamma, A\triangledown B \mid \Delta} \qquad (\odot) \quad \frac{\vdash \cdots \mid \Gamma, A \qquad \vdash B, \Delta \mid \cdots}{\vdash \cdots \mid \Gamma, A\odot B, \Delta \mid \cdots}$$

Every proof π of $\vdash \Gamma_1 \mid \cdots \mid \Gamma_m$ of planar logic defines a McyLL proof-structure Θ whose translation Θ^* is a MLL proof-net, and whose surface **ribbon**(Θ) is planar with $m + n$ borders $\sigma_1, ..., \sigma_m$ and $\tau_1, ..., \tau_n$; each border σ_i visits the formulas of Γ_i in the order in which they appear in the component; none of the remaining borders τ_j visits a conclusion of Θ.

Conversely, every McyLL proof-structure Θ whose translation Θ^* is a MLL proof-net, and whose surface **ribbon**(Θ) is planar, sequentializes as a proof π of planar logic. Typically, the "twist" proof-structure (8.2) sequentializes as the proof

$$\frac{\dfrac{\vdash A^\perp, A \qquad \vdash B, B^\perp}{\vdash A^\perp, A\odot B, B^\perp}\;\odot}{\vdash A^\perp\triangledown B^\perp \mid A\odot B}\;\triangledown$$

But (8.2) does not sequentialize as a proof of $\vdash A^\perp\triangledown B^\perp, A\otimes B$. In a similar way, the proof-structure (8.3) sequentializes as a derivation tree of planar logic:

$$\frac{\dfrac{\dfrac{\dfrac{\dfrac{\vdash A, A^\perp \qquad \vdash A, A^\perp}{\vdash A, A^\perp \odot A, A^\perp}\;\odot}{\vdash A^\perp \odot A, A^\perp\triangledown A}\;\triangledown \qquad \dfrac{\dfrac{\vdash B, B^\perp \qquad \vdash B, B^\perp}{\vdash B, B^\perp \odot B, B^\perp}\;\odot}{\vdash B^\perp\triangledown B, B^\perp \odot B}\;\triangledown}{\vdash A^\perp \odot A, (A^\perp\triangledown A)\odot (B^\perp\triangledown B), B^\perp \odot B}\;\odot}{\vdash (A^\perp\triangledown A)\odot (B^\perp\triangledown B)\mid(A^\perp \odot A)\triangledown(B^\perp \odot B)}\;\triangledown \qquad \dfrac{\dfrac{\vdash B^\perp, B}{\vdash B^\perp\triangledown B}\;\triangledown \qquad \dfrac{\vdash A^\perp, A}{\vdash A^\perp\triangledown A}\;\triangledown}{(B^\perp\triangledown B)\odot (A^\perp\triangledown A)}\;\odot}{\vdash (A^\perp\triangledown A)\odot (B^\perp\triangledown B)}\;\text{cut}$$

It is worth noting that cut-elimination preserves the planarity of proof-structures, but generally reduces the number of borders of the surface. Typically:

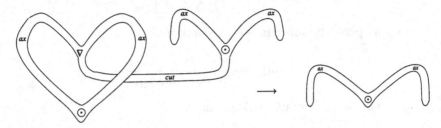

Accordingly, planar logic enjoys the following cut-elimination property: if π is a proof of $\vdash \Gamma_1 \mid \cdots \mid \Gamma_m$ in planar logic, and π' is a proof obtained after a series of cut-elimination steps applied to π, then π' is a proof of a sequent $\vdash \Delta_1 \mid \cdots \mid \Delta_n$ which reduces to the sequent $\vdash \Gamma_1 \mid \cdots \mid \Gamma_m$ by applying a series of "divide" rules:

$$(\text{DIVIDE}) \quad \frac{\vdash \cdots \mid \Gamma, \Delta \mid \cdots}{\vdash \cdots \mid \Gamma \mid \Delta \mid \cdots}$$

Conservativity of planar logic over McyLL follows from this and the cut-elimination property of McyLL, established in corollary 8.10. Indeed, the *cut-free* proofs of a McyLL sequent $\vdash \Gamma$ are the same in McyLL and in planar logic.

Planar logic seems interesting for itself. But from now on, we stick to cyclic linear logic, and characterize its sequentializable proof-structures, notwithstanding the difficulties.

Index. Internal and external borders. — Given an McyLL proof-structure Θ, and a border σ of **ribbon**(Θ), we shall count the number of ∇-links visited by the border σ on their thick side, see (8.4). We call this number the *index* of σ. A border of index 0 is called *external*, and a border of index more than 1 is called *internal*.

(8.4)

Conversely, the border of **ribbon**(Θ) which visits the thick side of a given ∇-link of Θ, is called the *internal border* of this link.

The correctness criterion. — Example (8.2) and (8.3) suggest to reinforce the definition of McyLL proof-net as follows.

Definition 8.7 (McyLL proof-net) *An* McyLL *proof-net is an* McyLL *proof-structure* Θ *such that,*

(i) *its commutative translation* Θ* *is an MLL proof-net,*

(ii) *its surface* **ribbon**(Θ) *is planar with a unique external border* σ,

(iii) σ *contains all the conclusions.*

The criterion rejects the proof-structures (8.2) and (8.3) because one of their conclusions lies on an internal border. The criterion rejects the proof-structure (8.5) of conclusion ⊢ $(B \odot A) \multimap (A \odot B)$ as well, because it is not planar.

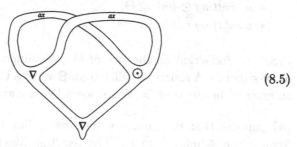

$$(8.5)$$

Remark. — The criterion implies that every internal border is of index exactly one in **ribbon**(Θ), when Θ is a McyLL proof-net. Indeed, by condition 1, the surface **ribbon**(Θ) defines a surface homeomorphic to the disk, when every ∇-link is replaced by a switching position $\mathscr{R}_L$ or $\mathscr{R}_R$. Consequently, the planar surface **ribbon**(Θ) has $n + 1$ borders, where n is the number of ∇-links appearing in Θ. Since there exists only one external border, each of the remaining n internal borders of **ribbon**(Θ) visits exactly one ∇-link.

Soundness. — It is not difficult to show by induction that the criterion is *sound*. At each step, one proves that the McyLL derivation tree of ⊢ $A_1, ..., A_k$ translates as an McyLL proof-net whose external border visits the conclusions $A_1, ..., A_k$ in the clockwise order (here, one assumes implicitly that the surface is oriented.)

Planarity 1. — We recall one elementary property of planar surfaces, which we shall use in our proof of sequentialization. If one pastes (with glue) the two borders σ_1 and σ_2 of a planar surface S, on disjoints segments A of σ_1 and B of σ_2, in such a way that orientation of S is preserved, one obtains a surface S' which is:

- planar when $\sigma_1 = \sigma_2$,
- not planar when $\sigma_1 \neq \sigma_2$.

In the next lemma, the concept of splitting $\odot$-link or cut-link is adapted from [7, 9].

Lemma 8.8 *Suppose that Θ is an McyLL proof-structure whose MLL translation Θ^* is an MLL proof-net, and whose surface* **ribbon**(Θ) *is planar. Then, either Θ is the axiom link, or every external border of* **ribbon**(Θ) *visits one of the following:*

- *the conclusion of a terminal ∇-link of Θ,*
- *a splitting $\odot$-link of Θ,*
- *a splitting cut-link of Θ.*

Proof By induction on the size of Θ. We suppose that every McyLL proof-structure Λ strictly smaller than Θ verifies the property. Consider an external border σ of Θ. We proceed by case analysis.

[A] Suppose that Θ contains a terminal ∇-link l of conclusion $A\nabla B$. Remove the ∇-link l from Θ. The resulting McyLL proof-structure Λ translates as an MLL proof-net Λ^* and has a planar surface **ribbon**(Λ). We proceed by case analysis.

1. either the external border σ visits the conclusion of the terminal ∇-link l of Θ, and we are done,

2. or the external border σ does not visit the conclusion of the ∇-link l. Since σ is not the internal border of l either, σ is the residual of an external border σ' of **ribbon**(Λ) which does not visit the conclusions A and B of Λ. This shows already that Λ is not the axiom-link. By induction hypothesis on Λ, two cases may occur. Either the external border σ' visits the conclusion of a terminal ∇-link m of Λ. In that case, the ∇-link remains terminal in Θ, and σ visits the conclusion of m: we are done. Or the external border σ' visits a splitting $\odot$-link (or cut-link) m of Λ, which splits Λ in two McyLL proof-structures Λ_1 and

Λ_2. Since the border σ visits the link m in Θ, the proof reduces to showing that m is splitting in Θ. The surface **ribbon**(Θ) is the result of glueing together the two conclusions A and B of **ribbon**(Λ). Planarity of **ribbon**(Θ) implies that the two conclusions A and B appear on the same border σ'' of **ribbon**(Λ). This border σ'' cannot be σ' because σ' does not visit the formulas A and B. Since the border σ' is the unique border of **ribbon**(Λ) visiting both Λ_1 and Λ_2, the border σ'' is either a border of **ribbon**(Λ_1) or border of **ribbon**(Λ_2). In the former case, A and B are conclusions of Λ_1, in the latter case, A and B are conclusions of Λ_2. In both cases, the link m remains splitting in Θ, and we are done.

[B] Suppose that Θ does not contain any terminal ∇-link. In that case, Θ^* is an MLL proof-net with no terminal γ-link, and it follows that the proof-structure Θ contains a splitting $\odot$-link or cut-link l, see [7, 9]. Remove the link l from Θ. The two resulting McyLL proof-structures Λ_1 and Λ_2 translate as MLL proof-nets Λ_1^* and Λ_2^* and define planar surfaces **ribbon**(Λ_1) and **ribbon**(Λ_2). Either σ visits both Λ_1 and Λ_2: in that case, we are done, because σ visits the splitting link l. Or the border σ visits Λ_1 only, or Λ_2 only. Suppose that we are in the first situation. It follows by induction hypothesis on Λ_1, which cannot be the axiom-link, that the external border σ visits either the conclusion of a terminal ∇-link m of Λ_1, or a splitting $\odot$-link m of Λ_1, or a splitting cut-link m of of Λ_1. In the two last cases, we are done, because the link m remains splitting in Λ. In the first case, note that the conclusion of m is not the premise in Θ of the splitting link l. Thus, the ∇-link m is a terminal link in Θ, whose conclusion is visited by σ. We conclude. $\square$

Sequentialization. — We prove that every McyLL proof-net sequentializes as an McyLL derivation tree, theorem 8.9. The proof is not really complicated, except for the cut-link case, which requires the preliminary lemma 8.8.

Theorem 8.9 (McyLL sequentialization) *Every McyLL proof-net is the translation of an McyLL derivation tree.*

Proof We show by induction on the number of connectives in Θ, that there exists an McyLL derivation tree π sequentializing the McyLL proof-net Θ.

Suppose that Θ contains a terminal ∇-link of conclusion $A \nabla B$. Remove this ∇-link l from Θ. The resulting McyLL proof-structure Λ is an McyLL proof-net. By induction hypothesis, there exists an McyLL derivation tree π sequentializing Λ of, say, conclusion $\vdash A_0, ..., A_{n-1}$. Let i and j be the two indices $0 \leq i, j \leq n-1$ such that $A = A_i$ and $B = A_j$. We claim that $j = i+1$ modulo n. Suppose not. Then, the conclusions $A_{i+1}, ..., A_{j-1}$ appear on the segment of border between A and B in Λ, thus on the internal border of a ∇-link l in Θ. This contradicts the hypothesis that Θ is a McyLL proof-net. We conclude that $j = i+1$ modulo n. The McyLL derivation tree π' sequentializing Θ follows immediately from π, and we are done.

Suppose now that Θ contains no terminal ∇-link. We are done when Θ is an axiom link. Otherwise, Θ^* is an MLL proof-net without terminal $\mathscr{R}$-link, and thus, there exists a splitting $\odot$-link or cut-link l in Θ, see [7, 9]. Obviously, when l is a $\odot$-link, it connects two McyLL proof-nets Λ_1 and Λ_2, and we conclude by a simple induction argument.

The remaining case, when there are only splitting cut-links, and no splitting $\odot$-link, is more delicate. Indeed, removing an arbitrary splitting cut-link l from Θ induces two McyLL proof-structures Λ_1 and Λ_2; and one of them, say Λ_1, may not be a McyLL proof-net. This case happens when Λ_2 has a unique conclusion A, whose dual formula $A^\perp$ appears on an internal border of the surface **ribbon**(Λ_1). Note that in this "pathological" case, the cut-link l is visited by an *internal* border of **ribbon**(Θ). The situation is illustrated by the cut-link number 2 in the McyLL proof-net below:

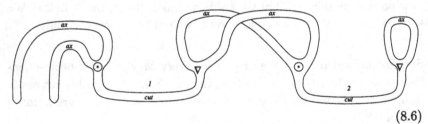

$$(8.6)$$

In other words, we need to *choose* which splitting cut-link should be removed first from a McyLL proof-net, if we want to sequentialize it. Typically, the cut-link number 1 must be removed before the cut-link number 2 in the McyLL proof-net (8.6). Fortunately, there is always a correct choice, induced by lemma 8.8. By hypothesis, the proof-net Θ does not contain any terminal ∇-link, nor splitting $\odot$-link; moreover, by definition of a McyLL proof-net, its translation Θ^* is planar. It follows by lemma 8.8 that the unique external border of Θ visits one splitting cut-link l at least. We choose to remove this cut-link l from Θ first, and avoid in this way the "pathological" case. So, we obtain two McyLL proof-nets Λ_1 and Λ_2 and conclude by a simple induction argument. □

Planarity 2. — We recall another elementary property of planar surfaces, that we shall use in our proof of cut-elimination. If one cuts (with scisors) a planar surface S which is connected, from a border σ_1 to a border σ_2 of S, one obtains a surface S' with:

- two connected components when $\sigma_1 = \sigma_2$,
- one connected component and one border less than S, when $\sigma_1 \neq \sigma_2$.

Cut-elimination. — The planarity criterion, definition 8.7, enables to prove cut-elimination of McyLL in a simple and intuitive way.

Corollary 8.10 *McyLL enjoys cut-elimination.*

Proof We prove that McyLL proof-nets are preserved by cut-elimination. Let Θ be an McyLL proof-net containing a cut-elimination pattern R. We prove that the McyLL proof-structure Λ obtained after rewriting the pattern R, is an McyLL proof-net. Cut-elimination in MLL ensures already that Λ translates as an MLL proof-net Λ^*. There remains to show that **ribbon**(Λ) is planar, and has a unique external border visiting all conclusions of Λ.

Topologically, cut-elimination consists in cutting (with scisors) the surface separating two borders σ_1 and σ_2 of **ribbon**(Θ). One border, say σ_1, visits the internal border of the ∇-link l of R, while the other border σ_2 visits the $\odot$-link. Planarity of **ribbon**(Λ) follows. Besides, the surface **ribbon**(Λ) is connected because Λ translates as an MLL proof-net Λ^*. We conclude that the two borders σ_1 and σ_2 are different in **ribbon**(Θ).

Let σ_3 denote the border of **ribbon**(Λ) obtained by "merging" the two borders σ_1 and σ_2 of **ribbon**(Θ). We mentioned that every internal border of **ribbon**(Θ) has index one, for a McyLL proof-net like Θ, see the remark after definition 8.7. In particular, the ∇-link l is the unique ∇-link visited internally by σ_1. Since cut-elimination removes this ∇-link l, the index of σ_2 and σ_3 are equal.

It follows that **ribbon**(Λ) has a unique external border σ. This border σ is the border σ_3 when the border σ_2 is external, and the residual of the external border of **ribbon**(Θ) when the border σ_2 is internal. In each case, the border σ visits all conclusions of Λ. We conclude that the proof-structure Λ is a McyLL proof-net.

We have just proved that McyLL proof-nets are preserved by cut-elimination. The end of the proof is easy. Suppose that π_1 and π_2 are McyLL derivation trees of conclusion $\vdash \Gamma, A$ and $\vdash A^\perp, \Delta$. By soundness, the derivation trees π_1 and π_2 define McyLL proof-nets Λ_1 and Λ_2, respectively. Now, connect Λ_1 to Λ_2 by a cut-link between the conclusions A and $A^\perp$. This defines a McyLL proof-net Θ which reduces by cut-elimination to a cut-free McyLL proof-net Θ'. The proof-net Θ' sequentializes to a cut-free McyLL derivation tree π', by theorem 8.9. The derivation tree π' has conclusion $\vdash \Gamma, \Delta$. We conclude that McyLL enjoys cut-elimination. $\qquad \square$

Remark. — The proof-structure (8.3) appears independently in Robert Schneck's work on non-symmetric linearly distributive categories [21]. Motivated by this example, Schneck strengthens the planarity criterion for negation-free multiplicative linear logic, and formulates a new criterion, in a similar way as we do above.

8.6 Non commutative logic

Non commutative logic (NL) was introduced by Paul Ruet in his PhD thesis [19] and developped with collaborators in a series of articles [3, 20, 13]. It is a conservative extension of both commutative linear logic (LL) and cyclic linear logic (cyLL). The idea is to equip every sequent $\vdash A_0, ..., A_{k-1}$ with additional information on the relative positions of the conclusions, provided by an *order variety* on the set of (occurrences of) formulas $A_1, ..., A_k$.

Order varieties. — An order variety α on a set X is a ternary relation which is:

(i) cyclic: $\forall x, y, z \in E, \alpha(x, y, z) \Rightarrow \alpha(y, z, x)$,

(ii) anti-reflexive: $\forall x, y \in E, \neg\alpha(x, x, y)$,

(iii) transitive: $\forall x, y, z, t \in E, \alpha(x, y, z) \wedge \alpha(x, z, t) \Rightarrow \alpha(x, y, t)$,

(iv) spreading: $\forall x, y, z, t \in E, \alpha(x, y, z) \Rightarrow \alpha(t, y, z) \vee \alpha(x, t, z) \vee \alpha(x, y, t)$.

The three first properties define a *cyclic order*, as introduced by Novak in [18]. A cyclic order is *total* when it verifies the additional property:

$$\forall x, y, z \in E, x \neq y \neq z \neq x \Rightarrow \alpha(x, y, z) \vee \alpha(x, z, y)$$

A total cyclic order is often called *oriented cycle* on X, because, at least when X is finite, it can be described by a graph $(X, \rightarrow)$ which relates $x \rightarrow y$ when there exists no $z \in X$ such that $\alpha(x, z, y)$. This graph contains a unique cycle, and $\alpha(x, y, z)$ simply means in that case that "y stands between x and z".

Order varieties generalize total cyclic orders, like partial orders generalize total orders. Every order variety on X becomes a *partial* order on $X - \{x\}$ once an origin x is fixed in X — in a reversible way, in the sense that the order variety on X may be reconstructed from the partial order on $X - \{x\}$. The following properties are established in [3, 20].

Focusing. — Given an order variety α on X and an element $x \in X$, define the partial order α_x on $X - \{x\}$, called *focus* of α on x, by:

$$\forall y, z \in X - \{x\}, \qquad \alpha_x(y, z) \iff \alpha(x, y, z)$$

Conversely, given a partial order $\omega = (X, <)$ on X and an element $z \in X$, define the binary relation on X:

$$x \overset{z}{<} y \iff x < y \text{ and } z \text{ is comparable with neither } x \text{ nor } y.$$

Then, the order variety $\overline{\omega}$ on X, the *closure* of ω on X, is defined as the ternary relation $\overline{\omega}(x, y, z)$ on X:

$$x < y < z \text{ or } y < z < x \text{ or } z < x < y \text{ or } x \overset{z}{<} y \text{ or } y \overset{x}{<} z \text{ or } z \overset{y}{<} x.$$

Parallel and series. — Given two partial orders ω on X and ω' on Y, define the partial orders $\omega \mid \omega'$ (called ω *parallel* ω') and $\omega < \omega'$ (called ω *series* ω') on $X + Y$.

$$x(\omega \mid \omega')y \iff \begin{cases} x \in X \text{ and } y \in X \text{ and } x\omega y \\ x \in Y \text{ and } y \in Y \text{ and } x\omega'y \end{cases}$$

$$x(\omega < \omega')y \iff \begin{cases} x \in X \text{ and } y \in Y \\ x \in X \text{ and } y \in X \text{ and } x\omega y \\ x \in Y \text{ and } y \in Y \text{ and } x\omega'y \end{cases}$$

Glueing. — If ω and ω' are two partial orders on disjoint sets X and

Y, then the following equality holds:

$$\overline{\omega < \omega'} \;=\; \overline{\omega \mid \omega'} \;=\; \overline{\omega' < \omega}$$

This enables to *glue* two partial orders ω on X and ω' on Y, and obtain an order variety $\omega * \omega' = \overline{\omega \mid \omega'}$ on $X + Y$. The two main properties of glueing are:

$$(\alpha_x) * x = \alpha \qquad\qquad\qquad (\omega * x)_x = \omega$$

for α an order variety on X, x an element of X, and ω a partial order on $X - \{x\}$.

Next and tensor. — Given two order varieties α on X and β on Y, and two elements $x \in X$ and $y \in Y$, one glues α and β together on x and y, in a *series* or *parallel* fashion, to obtain an order variety on $(X - \{x\}) + (Y - \{y\}) + \{z\}$:

$$\alpha \odot_{x,y}^z \beta \;=\; \overline{\alpha_x < z < \beta_y} \;=\; (\beta_y < \alpha_x) * z$$

$$\alpha \otimes_{x,y}^z \beta \;=\; \overline{\alpha_x \mid z \mid \beta_y} \;=\; (\beta_y \mid \alpha_x) * z$$

Interior. — Every cyclic order α on X contains a *largest* order variety $\natural\alpha$. The order variety $\natural\alpha$ is called the *interior* of the cyclic order α, and defined as

$$\natural\alpha \;=\; \bigcap_{x \in X} \alpha_x * x$$

Notation. — Consider an order variety α on X, and a subset Y of X. We write $\alpha \restriction_Y$ the order variety obtained by restricting the ternary relation α to the subset Y of X. Given an element x of X, the order variety $\alpha[z/x]$ is the order variety on $(X - \{x\}) + \{z\}$ obtained by replacing x by z in X.

Par. — Given an order variety α on X and two different elements $x, y \in X$, one defines the order variety $\alpha[z/x, y]$ on $(X - \{x, y\}) + \{z\}$ as

$$\alpha[z/x, y] \;=\; \natural\, (\alpha \restriction_{X - \{y\}} [z/x] \cap \alpha \restriction_{X - \{x\}} [z/y])$$

We write $\alpha[x, y]$ when x and y are two different elements of X.

MNL. — The multiplicative fragment (without units) of non commutative logic (MNL) extends both MLL and McyLL. Its formulas are

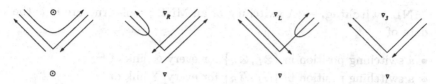

Fig. 8.3. Abrusci-Ruet switching positions for next and sequential

constructed using the connectives $\otimes$, $\bindnasrepma$ (from MLL) and $\odot$, ∇ (from McyLL). Negation in MNL simply extends negation in MLL and McyLL. An MNL sequent $\vdash \omega$ is an order variety on a finite set of (occurrences of) MNL formulas. An MNL derivation tree is a tree of MNL sequents constructed according to the following rules.

$$(\text{Ax}) \; \frac{}{\vdash A^{\perp} * A} \qquad\qquad (\text{Cut}) \; \frac{\vdash \omega * A \qquad \vdash A^{\perp} * \omega'}{\vdash \omega * \omega'}$$

$$(\otimes) \; \frac{\vdash \omega * A \qquad \vdash \omega' * B}{\vdash (\omega \mid \omega') * A \otimes B} \qquad (\bindnasrepma) \; \frac{\vdash \alpha[A, B]}{\vdash \alpha[A \bindnasrepma B / A, B]}$$

$$(\odot) \; \frac{\vdash \omega * A \qquad \vdash \omega' * B}{\vdash (\omega < \omega') * A \odot B} \qquad (\nabla) \; \frac{\vdash \omega * (A < B)}{\vdash \omega * A \nabla B}$$

8.7 Abrusci and Ruet's long trip criterion for MNL

In this section, we recall the correctness criterion for McyLL and MNL developped by V. Michele Abrusci and then Paul Ruet in [2, 3]. The criterion adapts Girard long trip condition for MLL, by:

- keeping the switching positions of MLL for $\otimes$ and $\bindnasrepma$ links,
- considering $\odot$-links as $\otimes$-links limited to the unique switching position $\odot = \otimes_R$,
- considering ∇-links as $\bindnasrepma$-links with the usual switching positions $\nabla_L = \bindnasrepma_L$ and $\nabla_R = \bindnasrepma_R$, and an additional switching position ∇_3.

Abrusci-Ruet switching positions appear in figures 8.1 and 8.3. Contrarily to the other switching positions, the position ∇_3 is not *total*: a ∇-link in position ∇_3 does not necessarily reemit a particle which enters it! Accordingly, Abrusci and Ruet weaken Girard's long trip condition in definition 8.12, and require only that, for a given proof-net Θ and switching s, there exists a unique cycle in $\mathbf{trip}(\Theta, s)$ which visits all the conclusions, *but not necessarily all the proof-net Θ*.

MNL switching. — A switching of an MNL proof-structure Θ is the data of

- a switching position in $\{\otimes_L, \otimes_R\}$ for every $\otimes$-link of Θ,
- a switching position in $\{\mathfrak{N}_L, \mathfrak{N}_R\}$ for every $\mathfrak{N}$-link of Θ,
- a switching position in $\{\nabla_L, \nabla_R, \nabla_3\}$ for every ∇-link of Θ.

Every MNL switching s defines a switched proof-structure $\mathbf{trip}(\Theta, s)$ as in section 8.2.

Bilaterality. — An additional (and technical) condition of "bilaterality" is required on the cycle. The condition ensures for instance that the proof-structure illustrated in (8.1) with $\otimes$-links replaced by $\odot$-links, is not a proof-net.

Definition 8.11 (bilateral) *Let Θ be an MNL proof-structure, and s an MNL switching of Θ. A trip σ in $\mathbf{trip}(\Theta, s)$ is bilateral if σ is not of the form*

$$A^x, ..., B^y, ..., A^{\overline{x}}, ..., B^{\overline{y}}$$

where A and B are occurrences of formulas in Θ, and $\overline{\uparrow} = \downarrow$, $\overline{\downarrow} = \uparrow$.

Abrusci-Ruet long trip criterion. —

Definition 8.12 (Abrusci-Ruet proof-net) *An* Abrusci-Ruet proof-net *is an MNL proof-structure Θ such that, for every MNL switching s:*

  (i) *there is exactly one cycle σ in $\mathbf{trip}(\Theta, s)$, called the* long trip,

  (ii) *σ contains all the conclusions,*

  (iii) *σ is bilateral.*

Three important properties are established in [3].

  (i) *soundness:* every MNL derivation tree of conclusion $\vdash \alpha$ translates as an Abrusci-Ruet proof-net Θ, in such a way that α is the largest order variety contained in each α_s, where α_s denotes the total cyclic order (or oriented cycle) on the conclusions of Θ defined by the long trip of $\mathbf{trip}(\Theta, s)$, for s an MNL switching. It is worth noting for section 8.8 that the characterization of α still works when the switchings s are restricted to the $\{\nabla_L, \nabla_R\}$-free ones,

(ii) *sequentialization:* every *cut-free* Abrusci-Ruet proof-net sequentializes as an MNL derivation tree,

(iii) *cut-elimination:* MNL enjoys cut-elimination.

In fact, points 2. and 3. are proved using an alternative characterization of Abrusci-Ruet proof-nets, rather than the original definition 8.12. — see theorem 2.20 in [3], or the discussion in section 8.9.

Remark. — Virgil Mogbil and Quintijn Puite observe in [15] that the bilaterality condition of definition 8.12 (point (iii)) may be replaced by the condition that the MNL proof-structure Θ translates as a MLL proof-net Θ*. Obviously, this condition also rejects the proof-structure illustrated in (8.1).

8.8 A planarity correctness criterion for MNL

In this section, we extend to MNL the well-known planarity criterion for McyLL, discussed at length in section 8.5. We will see in section 8.9 that the resulting planarity criterion for MNL reformulates topologically Abrusci-Ruet long trip criterion. Thus, just as in the commutative case of MLL, the topological point of view federates seemingly different correctness criteria (eg. planarity vs. long trip).

Topological switching. — A topological switching of an MNL proof-structure Θ is simply defined as a $\{\nabla_L, \nabla_R\}$-free MNL switching of Θ. Alternatively, it is the data of

- a switching position in $\{\otimes_L, \otimes_R\}$ for every $\otimes$-link of Θ,
- a switching position in $\{\invamp_L, \invamp_R\}$ for every $\invamp$-link of Θ.

Switched surface. — To every MNL proof-structure Θ and topological switching s, we associate the surface **ribbon**(Θ, s) by replacing every $\otimes$ and $\invamp$-link by the ribbon diagram corresponding to its MNL switching

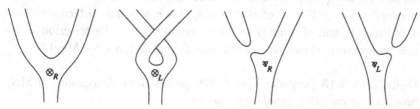

and every $\odot$ or ∇ or axiom or cut-link and conclusion by the ribbon

diagram

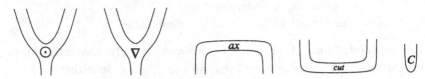

Planarity criterion for MNL. — Just as for McyLL in section 8.5, requiring planarity of **ribbon**(Θ, s) for every switching s is not sufficient to characterize MNL proofs. We have seen that requiring in addition that all conclusions lie on the same border of **ribbon**(Θ) is sufficient to characterize *cut-free* McyLL proofs. Note that this is not even the case in MNL. For instance, the *cut-free* proof-structure of conclusion $\vdash (B \odot A) \multimap (A \odot B)$ which is not sequentializable in MNL, has its two switched surfaces planar, with all conclusions (= one conclusion in each case) on the same border.

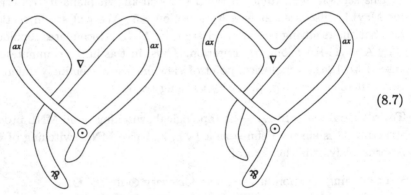

$$(8.7)$$

Fortunately, proof-structures like (8.7) may be rejected in the same way as in McyLL: by considering *external* and *internal* borders. These notions are adapted to MNL in the obvious way: given an MNL proof-structure Θ and a topological switching s, the *index* of a border b of the surface **ribbon**(Θ, s), is the number of *internal sides* of ∇-link of Θ the border b visits; A border of **ribbon**(Θ, s) is *external* or *internal* when it is of index 0, and of index 1 or more, respectively. The criterion below is a "conservative" extension to MNL of definition 8.7 for McyLL.

Definition 8.13 (topological MNL proof-net) *A topological MNL proof-net is an MNL proof-structure* Θ

1. whose commutative translation Θ^* *is an MLL proof-net,* *and such that, for every topological switching s:*

2. *the switched surface* **ribbon**(Θ, s) *is planar and has a unique external border* σ,

3. σ *contains all the conclusions.*

Obviously, the proof-structure (8.7) is rejected by the criterion: its unique conclusion lies on an internal border when $\gamma\!\!\!8$ is switched in position $\gamma\!\!\!8_R$.

Remark. — For the same reasons as in section 8.5, definition 8.7, it follows from definition 8.13 that every internal border of **ribbon**(Θ, s) is of index 1, when Θ is an MNL proof-net, and s is a topological switching.

Soundness. — Given a proof derivation π, its associated proof-structure Θ in MNL, and a topological switching s, one proves by structural induction on π that the long trip in the proof-structure **trip**(Θ, s) *is* precisely the external border of the switched surface **ribbon**(Θ, s). It follows that the long trip of **trip**(Θ, s) visits the conclusions of Θ in the same order as the external border of **ribbon**(Θ, s). By property of soundness, in section 8.7, the order variety $\vdash \alpha$ is the maximal order variety on the conclusions of Θ included in all oriented cycles induced by the external border of **ribbon**(Θ, s), for s a topological switching of Θ. Soundness follows easily.

Sequentialization. — Just as in [3, 12] we limit our sequentialization theorem to *cut-free* MNL proof-nets.

Theorem 8.14 (MNL sequentialization) *Every* cut-free *MNL proof-net is the translation of an MNL derivation tree.*

Proof The proof proceeds as in theorem 8.9 for $\odot$ and ∇-links. $\otimes$-links can be treated as $\odot$-link, and $\gamma\!\!\!8$-links are treated as follows. Suppose that l is a terminal $\gamma\!\!\!8$-link of conclusion $A \gamma\!\!\!8 B$ in a cut-free MNL proof-net Θ. Let Λ be the proof-structure obtained by removing l from Θ. Its MLL translation is a proof-net. There remains to check on Λ conditions 2 and 3 of definition 8.13. Let s be a topological switching of Λ, and $s_L = s + \{l \mapsto \gamma\!\!\!8_L\}$ and $s_R = s + \{l \mapsto \gamma\!\!\!8_R\}$ the two associated topological switchings on Θ. Obviously, **ribbon**(Λ, s), **ribbon**(Θ, s_L) and **ribbon**(Θ, s_R) denote the same surface S. Planarity of **ribbon**(Λ, s) follows. Moreover, the unique external border of **ribbon**(Θ, s_L) (on which A lies in **ribbon**(Λ, s)) and the unique external border of **ribbon**(Θ, s_R)

(on which B lies in $\mathbf{ribbon}(\Lambda, s)$) are necessarily the same border of S. It follows that $\mathbf{ribbon}(\Lambda, s)$ has a unique external border, on which A and B lie. We conclude that Λ is an MNL proof-net. $\qquad\square$

Cut-elimination. — The proof of cut-elimination for MNL follows a purely topological argument, instead of the algebraic one presented by Abrusci and Ruet in [3].

Corollary 8.15 *MNL enjoys cut-elimination.*

Proof Follows from soundness and sequentialization of MNL proof-nets in the same way as corollary 8.10 follows from soundness and sequentialization of McyLL proof-nets. The only difficulty is to establish that MNL proof-nets are preserved by cut-elimination.

Consider a topological MNL proof-net Θ containing a cut-elimination pattern, and the MNL proof-structure Λ obtained after cut-elimination of the pattern. We prove that Λ is a proof-net. Two cases may occur: either the cut-elimination pattern is "non-commutative", that is, involves a $\odot$ and a ∇ link, in which case we proceed as in corollary 8.10, with an obvious adaptation regarding preservation of uniqueness of the external border; or the cut-elimination pattern is "commutative", that is, involves a $\otimes$ link $l_\otimes$ and a γ link l_γ, with respective conclusions $A \otimes B$ and $B^\perp \gamma A^\perp$, in which case we proceed as follows. We fix a topological switching s of Λ, and consider the four topological switchings of Θ

$$s_{XY} = s + (l_\gamma \mapsto \gamma_X) + (l_\otimes \mapsto \otimes_Y)$$

for $X, Y \in \{L, R\}$. From now on, we call S the surface obtained by cutting (with scisors) the branch A of the $\otimes$-link $l_\otimes$ in $\mathbf{ribbon}(\Theta, s_{LR})$. Like $\mathbf{ribbon}(\Theta, s_{LR})$, the surface S is planar. The cut-link between $l_\otimes$ and l_γ has two borders σ and τ in S, which may be distinguished by indicating that the surfaces $\mathbf{ribbon}(\Theta, s_{LR})$ and $\mathbf{ribbon}(\Theta, s_{LL})$ are obtained from S by glueing the branch of concl sion A to the borders σ and τ, respectively. We show by case analys s that $\mathbf{ribbon}(\Lambda, s)$ is planar, and has a unique external border, which r sits all the conclusions of Λ.

[A] When the two borders σ and τ are different, planarity of both $\mathbf{ribbon}(\Theta, s_{LR})$ and $\mathbf{ribbon}(\Theta, s_{LL})$ implies that the surface S is not connected. More, S has two disconnected components S_1 and S_2, with the branch A in one component, say S_1, and the borders σ and τ in the other component S_2. The surface $\mathbf{ribbon}(\Lambda, s)$ is the result of glueing A in S_1 and $A^\perp$ in S_2. This shows that $\mathbf{ribbon}(\Lambda, s)$ is planar. By our correctness criterion, each border of the surface $\mathbf{ribbon}(\Theta, s_{LR})$ or $\mathbf{ribbon}(\Theta, s_{LL})$ visits either all the conclusions of Θ, or the internal border of exactly one ∇-link. Call ν the border of A in S_1. We claim that ν does not visit any conclusion, nor any internal border of a ∇-link. We proceed by contradiction. Suppose that the border ν visits a conclusion of S_1; then, by the last remark on $\mathbf{ribbon}(\Theta, s_{LR})$ or $\mathbf{ribbon}(\Theta, s_{LL})$, neither σ nor τ visits the internal border of a ∇-link in S_2; thus, two borders of $\mathbf{ribbon}(\Theta, s_{LR})$ are external; this contradicts the hypothesis that Θ is a proof-net. Suppose now that ν visits the internal border of a ∇-link; then, for the same reason as above, neither σ nor τ visits the internal border of a ∇-link in S_2, or a conclusion of S_2; it follows that the external border of $\mathbf{ribbon}(\Theta, s_{LL})$ is the residual of the border σ after glueing τ and ν together; the border visits no conclusion of Θ; according to the correctness criterion, the proof-net Θ does not have any conclusion; this contradicts the fact that Θ translates as an MLL proof-net. This proves our claim that ν visits no internal border of a ∇-link, and no conclusion of S_1. From this, we conclude easily that just like $\mathbf{ribbon}(\Theta, s_{LR})$, the surface $\mathbf{ribbon}(\Lambda, s)$ has a unique external border, visiting all the conclusions of Λ.

[B] When $\sigma = \tau$, and the surface S has two connected components, we call S_1 the component containing the branch A, and S_{23} the component containing the border $\sigma = \tau$. The surface $\mathbf{ribbon}(\Lambda, s)$ is connected because the proof-structure Λ translates as a MLL proof-net. This ensures that the branch $A^\perp$ appears in S_{23}, not in S_1; and implies that the proof-structure $\mathbf{ribbon}(\Lambda, s)$ is planar. There remains to show that $\mathbf{ribbon}(\Lambda, s)$ contains a unique external border, visiting all the conclusions of Λ. We proceed by case analysis. Either σ visits, or does not visit, the branch with conclusion $A^\perp$ in S. When σ visits $A^\perp$, the surface $\mathbf{ribbon}(\Theta, s_{LR})$ may be deformed into $\mathbf{ribbon}(\Lambda, s)$ by letting the component S_1 "slide" along the border σ of S_{23}, until S_1 reaches the branch $A^\perp$. It follows that, like $\mathbf{ribbon}(\Theta, s_{LR})$, the surface $\mathbf{ribbon}(\Lambda, s)$ has a unique external border visiting all conclusions of Λ.

Now, we treat the case when the border σ does not visit the branch with conclusion $A^\perp$ in S. Let S' denote the surface obtained by cutting (with scisors) the branch $B^\perp$ in the surface S_{23}. By planarity of S_{23} and equality of borders $\sigma = \tau$, the surface S' has two connected components: one component, called S_3, contains the branch with conclusion $B^\perp$; the other component, called S_2, contains the cut-elimination pattern $l_{\mathfrak{R}}, l_\otimes$. The three components S_1, S_2 and S_3 are also the result of cutting (with scisors) the branch $B^\perp$ in $\mathbf{ribbon}(\Theta, s_{LR})$, this resulting in two components S_{12} and S_3; then of cutting (with scisors) the branch A in S_{12}, this resulting in the two components S_1 and S_2. Let σ_1 denote the border of A in S_1, and σ_2 and σ_{12} denote the border of the cut-elimination pattern $l_\otimes$, $l_{\mathfrak{R}}$ in S_2 and S_{12} respectively. The surface $\mathbf{ribbon}(\Theta, s_{RR})$ is obtained by glueing the border σ_{12} with the border of $A^\perp$ in $S_{12} + S_3$. Connectedness of $\mathbf{ribbon}(\Theta, s_{RR})$ implies that $A^\perp$ appears in the component S_3, not in the component S_{12}. Now, let τ_A and τ_B denote the borders of $A^\perp$ and $B^\perp$ in the component S_3, respectively. We claim that τ_A and τ_B are different, and prove it as follows: the surface S is the result of glueing σ_2 in S_2 with τ_B in S_3; if τ_A and τ_B were equal in S_3, the border σ would visit $A^\perp$, contradicting our hypothesis. Now, the surfaces $\mathbf{ribbon}(\Theta, s_{LR})$ and $\mathbf{ribbon}(\Theta, s_{RR})$ are obtained by pasting (with glue) the borders σ_{12} in S_{12} with the borders τ_B and τ_A in S_3, respectively. It follows from this and the inequality $\tau_A \neq \tau_B$ and an argument similar to case [A] that the border σ_{12} visits no internal border of a ∇-link, and no conclusion of Θ. A fortiori, the border σ_1 of A in S_1, which is (in a sense) a segment of the border σ_{12}, visits no internal border of a ∇-link, and no conclusion of Θ. We conclude easily that the surface $\mathbf{ribbon}(\Lambda, s)$ which is obtained by glueing the conclusion $A^\perp$ in S_{23} to the border σ_1 in S_1, has a unique external border, visiting all the conclusions of Λ.

[C] When $\sigma = \tau$, and the surface S is connected, we may suppose by symmetry, wlog. that the surface S' obtained by cutting (with scisors) the branch B of the $\otimes$-link $l_\otimes$ in $\mathbf{ribbon}(\Theta, s_{RR})$ is also connected, and that the two borders σ' and τ' of the cut-elimination pattern $l_{\mathfrak{R}}, l_\otimes$ are equal in S'. Removing the cut-link connecting B and $B^\perp$ in S induces the same surface (denoted T) as removing the cut-link connecting A and $A^\perp$ in S', or as removing the cut-elimination pattern $l_{\mathfrak{R}}, l_\otimes$ from the surface $\mathbf{ribbon}(\Theta, s_{XY})$ for any $X, Y \in \{L, R\}$. The equality $\sigma = \tau$, alternatively the equality $\sigma' = \tau'$, implies that the surface T has two connected components. We call T_1 the component of the conclusion B and T_2 the component of the conclusion $B^\perp$, and claim that T_1 is also the component of the conclusion A and T_2 the component of the conclusion $A^\perp$. Indeed, consider the $\{\nabla_3\}$-free MNL switching s' of Θ obtained by replacing in s every switching position ∇_3 by the switching position ∇_L; let T' be the surface obtained from $\mathbf{ribbon}(\Theta, s'_{LR})$ or $\mathbf{ribbon}(\Theta, s'_{RR})$

by cutting (with scisors) the two branches A and B of the $\otimes$-link $l_\otimes$. Because Θ translates as an MLL proof-net Θ^*, the surface T' has three components: one component contains A, the other component contains B, and the last component contains both $A^\perp$ and $B^\perp$. The surface T is obtained by replacing some of the positions ∇_L in s' by the position ∇_3 in s. Consequently, the two formulas $A^\perp$ and $B^\perp$ which appear in the same component of T', appear a fortiori in the same component of T. The component T_2 is this component of $A^\perp$ and $B^\perp$; while T_1 is the component of A and B. Let σ_A and σ_B denote the respective borders of the conclusions $A^\perp$ and $B^\perp$ in T_2.

Now, the surface **ribbon**(Θ, s_{LR}) is obtained from S by gluing the branch of conclusion A to the border σ. The surface S is connected, and the surface **ribbon**(Θ, s_{LR}) is planar. So, the border σ visits the conclusion A in S. On the other hand, after cutting (with scisors) the branch B in S, the border σ becomes the border σ' of B in T_1. From these two facts, it follows that the conclusions A and B lie on the same border σ' of the component T_1. Glue these two conclusions A and B together in T_1, and call T' the resulting surface. The operation divides the border σ' of T_1 into two borders of T'. The borders may be denoted σ'_1 and σ'_2 in such a way that (1) the surface **ribbon**(Θ, s_{LR}) is obtained by glueing $B^\perp$ in T_2 to σ'_1 in T', and (2) the surface **ribbon**(Θ, s_{LL}) is obtained by glueing $B^\perp$ in T_2 to σ'_2 in T'. The correctness criterion, together with an argument similar to case [A] implies that each border σ'_1 and σ'_2 visits exactly one internal border of a ∇-link in T'; and that the border σ_B of $B^\perp$ in T_2 visits no internal border of a ∇-link, and no conclusion of Θ. Now, the surface **ribbon**(Θ, s_{RR}) is obtained by glueing the conclusion $A^\perp$ in T_2 to σ'_1 in T'. It follows from the correctness criterion that ($\bigstar$) the border σ_A of $A^\perp$ in T' visits no internal border of a ∇-link, and no conclusion of Θ.

We claim that the two border σ_A and σ_B coincide in T_2. Suppose not: $\sigma_A \neq \sigma_B$. In that case, the external border of **ribbon**(Θ, s_{LR}) is the residual of σ_A after glueing $B^\perp$ and σ'_2. It follows that the external border of **ribbon**(Θ, s_{LR}) visits no conclusion of Θ by our previous result ($\bigstar$). We conclude from our correctness criterion that Θ has no conclusion, which contradicts the fact that Θ translates as a MLL proof-net Θ^*. This establishes the claim: $\sigma_A = \sigma_B$. We are nearly done. Recall that the border σ_A visits no internal border of a ∇-link, and no conclusion of Θ. It follows that **ribbon**(Θ, s_{LR}) may be deformed into **ribbon**(Λ, s) by "sliding" A along σ_A in T_2, until it reaches $A^\perp$. This

proves that **ribbon**(Λ, s) is planar, has a unique external border, which
visits all its conclusions.   □

Remark. — Try this alternative (but wrong!) definition of MNL proof-
net: relax the condition on internal and external borders, and consider
the class of MNL proof-structures Θ translating as an MLL proof-net Θ^*,
and whose surface **ribbon**(Θ, s) is planar for every topological switching
s. It happens that the class is not closed under cut-elimination, as the
proof-structure below of conclusion $\vdash (B \odot A) \multimap (A \odot B)$ illustrates.

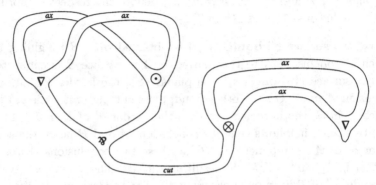

It should be noted that in the figure above, we use a topological notation
for proof-structures, adapted from our notation for switchings. This is
discussed in the appendix, section 8.10.

8.9 The planarity vs. the long trip criterion for MNL

Here, we reformulate our definition of MNL topological proof-nets in
three different ways. The first formulation is topological, but *emancipated* of all reference to MLL, in V. Michele Abrusci's style. We call
the second formulation *intermediate* because it prepares the third formulation, in which any reference to the topology disappears. Planarity
is replaced by a *well-bracketing* condition on the ∇-links of the switched
proof-structures. We benefit from the fact that this third formulation
appears already in [3] and characterizes Abrusci-Ruet proof-nets, to conclude that our planarity criterion coincides with the long trip criterion
for MNL.

Switched surface (2). — Here, we want to extend the definition of
section 8.8, and define a surface **ribbon**(Θ, s) for every MNL proof-
structure Θ and MNL switching s, instead of $\{\nabla_L, \nabla_R\}$-free switchings.
This is easy. The surface **ribbon**(Θ, s) is defined as before, except that

every ∇-link l is replaced by the ribbon diagram of its switching position ∇_L, ∇_R or ∇_3:

The previous definitions of *index*, of *external* and *internal* borders are extended in the obvious way: only ∇-links in position ∇_3 increase the index of a border of **ribbon**(Θ, s).

The emancipated criterion. — An alternative characterization of topological MNL proof-nets follows, which does not mention commutative linear logic.

Lemma 8.16 *An MNL proof-structure Θ is a topological MNL proof-net iff for every MNL switching s:*

(i) *the surface* **ribbon**(Θ, s) *is planar and has a unique external border σ,*

(ii) *σ contains all the conclusions.*

Note that the formulation is very close to Abrusci and Ruet definition 8.12 of an MNL proof-net, except that bilaterality is replaced here by planarity.

The intermediate criterion. — The next criterion makes the first step towards a non topological reformulation of our topological criterion, definition 8.13. Consider an MNL proof-structure Θ whose MLL-translation is an MLL proof-net Θ^*. Obviously, every ∇_3-free switching s of Θ defines a surface **ribbon**(Θ, s) homeomorphic to the disk. The positions of each ∇-link l of Θ may be indicated on the unique border σ of **ribbon**(Θ, s):

- by an opening bracket $(_l$
- by a closing bracket $)_l$.

in such a way that the segment of σ put inside brackets $(_l ...)_l$ coincides with the internal border of the surface **ribbon**$(\Theta, s + (l \mapsto \nabla_3))$. Then, a necessary and sufficient condition for Θ to be a topological proof-net is that, for every ∇_3-free switching s of Θ:

1. the brackets $(_l$ and $)_l$ may be pasted together in **ribbon**(Θ, s) in such a way that the surface remains planar,
2. no conclusion of Θ appears inside brackets.

The well-bracketing criterion. — Now, we make topology disappear entirely from the intermediate criterion, by reformulating the planarity condition of point (i) as a well-bracketing condition on $(_l \ldots)_l$.

Lemma 8.17 *A MNL proof-structure Θ is a topological MNL proof-net iff:*

1. *its MLL translation Θ^* is an MLL proof-net,*

and for every ∇_3-free switching s of Θ:

2. *the brackets $(_l$ and $)_l$ are well-bracketed on the border* **trip**(Θ, s) *of* **ribbon**(Θ, s),
3. *no conclusion of Θ appears inside brackets.*

The planarity and the long trip criteria coincide. — The series of conditions in lemma 8.17 is already mentioned in [3], theorem 2.20, where it characterizes Abrusci-Ruet MNL proof-nets. We conclude that

Theorem 8.18
The topological MNL proof-nets coincide with the Abrusci-Ruet MNL proof-nets.

Remark. — The remark by Mogbil and Puite about bilaterality (see the end of section 8.7) adapted to our topological setting, indicates that the planarity condition of lemma 8.16 may be replaced by the hypothesis that the proof-structure Θ translates as a MLL proof-net Θ^*. Indeed, a topological argument shows that in that case, the surface **ribbon**(Θ, s) is planar for every MNL switching s. Suppose not: there exists a MNL switching s making **ribbon**(Θ, s) non planar. Let s' denote the ∇_3-free switching obtained by switching as ∇_L (or ∇_R) all ∇-links switched ∇_3 in the switching s. The surface **ribbon**(Θ, s') is homeomorphic to the disk because Θ^* is a MLL proof-net, and the MNL switching s' is ∇_3-free. Lemma 8.17 indicates that there exist two ∇-links l_1 and l_2 switched as ∇_3 in the MNL switching s such that the surface **ribbon**(Θ, s'') is already non planar, when one alters s' into $s'' = s' + (l_1, l_2 \mapsto \nabla_3)$. We leave the reader check that the surface **ribbon**(Θ, s'') has a unique border σ, of index 2, which visits all the conclusions of Θ. This contradicts the other hypothesis of lemma 8.16 (there exists a unique external border, which visits all conclusions) and we conclude that **ribbon**(Θ, s) is planar, for every MNL switching s.

We also leave the reader check (hint: by a counter-example in McyLL) that the planarity condition is necessary in our definition 8.13 of topological proof-net, despite the fact that we assume that Θ translates as MLL proof-net Θ^*. The difference with lemma 8.16 is that MNL switchings are restricted here to topological (that is: $\{\nabla_L, \nabla_R\}$-free) switchings.

8.10 Appendix: is MNL an embedded logic?

In this article, we advocate that *switchings* are better expressed as topological objects, than as graphs. One may go further, and declare boldly that proofs *themselves* are topological objects, from which switched surfaces are deduced by topological surgery. From that perspective, the MLL proof π of $\vdash A^\perp \,\mathcal{B}\, A$ defines a surface homeomorphic to the annulus.

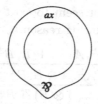

Each of the switching positions $\mathcal{B}_L$ and $\mathcal{B}_R$ of the $\mathcal{B}$-link indicates to cut (with scisors) the annulus π from one border σ_1 to the other border σ_2. In each case, one obtains a surface homeomorphic to the disk. Except for inessential details in the presentation of proofs (ribbon diagrams vs. simplicial complexes) this topological presentation may be found in [14]. It may be worth stressing that the topology of proofs is understood *internally*. In particular, neither the proof theory, nor the topology, reflects the fact that the annulus π may be embedded in several ways in the ambient space, forming all kinds of "twisted knots" like:

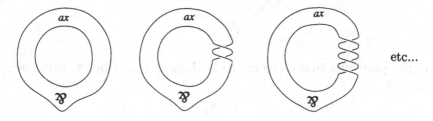

The idea of representing a proof as a surface embedded in an ambient space appears in [6] where Arnaud Fleury interprets the exchange rule as a "braided" permutation, and introduces a "twist" operation on formulas, inspired by similar operations in tortile tensor categories [11].

In the resulting "embedded logic" $\overline{\text{MLL}}$, every embedding of the annulus π in the ambient space happens to be a particular proof of the formula $\vdash A^\perp \,\wp\, A$. More generally, a $\overline{\text{MLL}}$ proof is either constructed sequentially, or characterized geometrically (this is the correctness criterion) as a proof-structure embedded in space, whose switchings are all homeomorphic to the disk. Similarly, one defines an embedded version $\overline{\text{McyLL}}$ of McyLL, whose proofs π are the proofs of $\overline{\text{MLL}}$ verifying the extra condition that π is planar, and has a unique external border visiting all conclusions.

In contrast, there does not seem to exist any satisfactory embedded version of MNL, for the following reason. Consider the MNL proof

$$
\cfrac{\cfrac{\vdash A^\perp, A \qquad \vdash B, B^\perp}{\cfrac{\vdash A^\perp, A \otimes B, B^\perp}{\cfrac{\vdash A^\perp, (A \otimes B) \,\wp\, B^\perp}{\vdash A^\perp \triangledown((A \otimes B) \,\wp\, B^\perp)} \triangledown} \wp} \otimes}{} \tag{8.8}
$$

As in the case of the annulus, there may be several way to embed the proof in ambient space. We choose one of them, which we draw below.

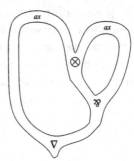

In this particular embedding of the MNL proof (8.8), the switching position s

$$\triangledown \mapsto \triangledown_3 \qquad \wp \mapsto \wp_R \qquad \otimes \mapsto \otimes_L$$

induces a surface admitting a "twist" between the formulas $A^\perp$ and $(A \otimes B) \,\Re\, B^\perp$.

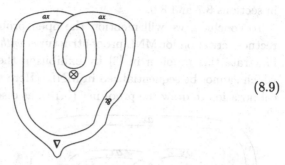

$$(8.9)$$

So, the switched surface, seen as embedded in ambient space, is not planar. More generally, there exists no embedding of (8.8) able to induce only planar MNL switching surfaces. The phenomenon is a consequence of the *see-saw* rule of non-commutative logic, which says that every proof of $\vdash A \,\Re\, B$ is also a proof of $\vdash A \nabla B$. This principle is fine when the topology of proofs is understood internally, but becomes problematic when the topology of proofs is embedded in an ambient space — at least in our ribbon presentation. Typically, the see-saw rule justifies the last ∇-introduction rule of the derivation tree (8.8) which implies in turn that the surface (8.9) is not planar.

8.11 Conclusion

In their correctness criteria [2, 3] Abrusci and Ruet characterize McyLL and MNL proof-nets without mentioning commutative MLL. This conveyed the hope for a theory of McyLL and MNL "emancipated" from any reference to MLL. In this article, we choose to step back, and understand McyLL and MNL as commutative MLL + *a planarity condition:*

- MLL + planarity of proof-nets, for McyLL,
- MLL + planarity of *switched* proof-nets, for MNL.

One reason is that cut-elimination of McyLL and MNL follows essentially from planarity — and its preservation by cut-elimination in MLL. Another reason is that the switching positions $\Re_L$ and $\Re_R$ are internalized in MLL by the "linear" distributivity formulas below, see [5, 1]:

$$A \otimes (B \,\Re\, C) \multimap (A \otimes B) \,\Re\, C, \qquad A \otimes (B \,\Re\, C) \multimap B \,\Re\, (A \otimes C).$$

In contrast, there exists (today) no such internal justification in McyLL or MNL for the "emancipated" criteria formulated in [2, 3] and recalled in sections 8.7 and 8.9.

To conclude, we will mention the open problem of designing a correctness criterion for MNL proof-structures *with cuts*. Abrusci and Ruet illustrate this problem in [3] by exhibiting the MNL proof-net (8.10) which cannot be sequentialized in MNL. (Here again, we use a topological notation to draw the proof-net (8.10), as discussed in the appendix.)

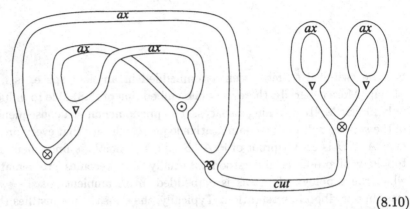

$$(8.10)$$

Finding a satisfactory solution may require to alter MNL — as cyclic linear logic was altered into planar logic in section 8.5. For what matters is not the details of the logic, but its relationship to a geometric (or computational) property of proofs, preserved by cut-elimination.

Acknowledgments

The author would like to thank Jean-Yves Girard for his reaction to a first version of this work, Paul Ruet and Roberto Maieli for suggestions and remarks, and Robin Cockett for mentioning the related work of Robert Schneck.

Bibliography

[1] S. Abramsky and R. Jagadeesan. Games and Full Completeness for Multiplicative Linear Logic. *Journal of Symbolic Logic* (1994), vol. 59, no. 2, 543–574.

[2] V. M. Abrusci, Non-commutative proof-nets. In *Advances in Linear Logic*. Cambridge University Press, 1995.

[3] V. M. Abrusci and P. Ruet. Non-commutative logic I : the multiplicative fragment. *Annals of Pure Appl. Logic*, 101 (1):29-64, 2000.

[4] R. Blute, P-J Scott. The Shuffle Hopf Algebra and Noncommutative Full Completeness, *Journal of Symbolic Logic 63*, pp. 1413-1435, 1998.

[5] V. Danos and L. Regnier, The structure of multiplicatives. *Arch. for Math. Logic*, 28(3):181-203, 1989.

[6] A. Fleury, *La règle d'échange*. Thèse de doctorat, U. Paris VII, Nov. 1996.

[7] J.-Y. Girard. Linear logic. *Theoretical Computer Science*, 50:1-102, 1987.

[8] J.-Y. Girard. Towards a geometry of interaction. *Contemp. Math.*, 92:69-108, 1989.

[9] J.-Y. Girard. Proof nets: a parallel syntax for proofs. *Logic and Algebra*, eds. Ursini and Agliano, Marcel Dekker, New York 1996.

[10] J.-Y. Girard. *On the meaning of logical rules: syntax vs. semantics*. In Berger, U. and Schwichtenbergm H. editors, *Computational logic*, pages 215 - 272. Heidelberg. Springer Verlag, NATO series F 165. 1999.

[11] A. Joyal, R. Street, D. Verity. Traced Monoidal Categories, *Mathematical Proceedings of the Cambridge Philosophical Society*, 119, 1996.

[12] R. Maieli. A new correctness criterion for multiplicative non-commutative proof-nets. Archive for Mathematical Logic, 2002.

[13] R. Maieli and P. Ruet. *Non-commutative logic III : focusing proofs*. Preprint Institut de Mathématique de Luminy, vol. 4, 2000.

[14] F. Métayer. Implicit Exchange in Multiplicative Linear Logic. *Mathematical Structures in Computer Science*, 2000.

[15] V. Mogbil and Q. Puite. A bilateral-free notion of modules for non-commutative logic. *Manuscript*, 2003.

[16] M. Nagayama and M. Okada. A graph-theoretic characterization theorem for multiplicative fragment of Non Commutative Linear Logic. *Electronic Notes in Theoretical Computer Science 3* (1996). Full version in *Journal of Theoretical Computer Science* 294(3):551–573, 2003.

[17] M. Nagayama and M. Okada. A new correctness criterion for the proof-nets of Non Commutative Multiplicative Linear Logics. To appear in *Journal of Symbolic Logic*.

[18] V. Novàk. Cyclically ordered sets. *Czech Math. J.*, 32(107):460-473, 1982.

[19] P. Ruet. *Non commutative logic and concurrent constraint programming*. PhD thesis, Paris 7, 1997.

[20] P. Ruet. Non-commutative logic II : sequent calculus and phase semantics. *Mathematical Structures in Computer Science*, 10(2), 2000.

[21] R. R. Schneck. Natural deduction and coherence for non-symmetric linearly distributive categories. *Theory and Applications of Categories*, Vol. 6, No. 9, 1999.

[22] D. N. Yetter. Quantales and (non-commutative) linear logic. *Journal of Symbolic Logic*, 55(1), 1990.

Part three
Invited Articles

9

Bicategories in Algebra and Linguistics

Jim Lambek

McGill University

To Saunders Mac Lane on his 90th birthday.

Abstract

Reporting on applications of bicategories to algebra and linguistics led me to take a new look at multicategories and polycategories: to replace free monoids by free categories and to introduce a new notation for Gentzen's cuts. This makes it clear that the equations holding in a multi- or polycategory are just those of the 2-category which contains it. Thus, a polycategory is almost the same as a 2-category whose underlying 1-category is freely generated by a graph, except that the class of 2-cells need not be closed under composition, but only under planar cuts.

9.0 Summary of contents

In Section 9.1 we point out that multicategories, slightly generalized, will do for bicategories what they originally did for monoidal categories, i.e. bicategories with one object. At the same time we introduce a new notation for Gentzen's "cut", to present it as a special case of composition in a 2-category.

In Section 9.2 we look at adjunctions in 2-categories and bicategories, with the aim of studying those bicategories in which each 1-cell has both a left and a right adjoint, namely compact noncommutative *-autonomous categories with several 0-cells.

In Section 9.3 we give a short exposition of some applications of bicategories to linguistics that were developed by Claudia Casadio and the present author. These touch on three deductive systems: the syntactic calculus, classical bilinear logic and compact bilinear logic.

325

In Section 9.4 we take a new look at polycategories, which are to classical bilinear logic as multicategories are to the syntactic calculus. Equations in a polycategory are explained by viewing the latter as contained in a 2-category.

In Section 9.5 we show that polycategories in the new sense will do for the linear bicategories of Cockett, Seely and Koslowski what multicategories can do for Bénabou's original bicategories.

In Section 9.6 we study adjoints in polycategories and show that, in a polycategory with residual quotients and "zero" 1-cells, every 1-cell has both a left and a right adjoint.

9.1 Multicategories recalled

A good part of my book "Lectures on rings and modules" [16] was devoted to the residuated bicategory of bimodules, although, at the time, I did not know what a bicategory was. Later I learned from Bénabou [4] that a bicategory resembles a 2- category in having 0-cells (in my case rings $R, S, \cdots$), 1-cells (bimodules $_R A_S : S \longrightarrow R$) and 2-cells (bimodule homomorphisms $f : {_R}A_S \longrightarrow {_R}A'_S$), except that composition of 1-cells (the tensor product $_R A_S \otimes {_S}B_T$) satisfies the usual identity laws ($R \otimes A \cong A \cong A \otimes S$) and associative law only up to coherent isomorphism. All these properties of the tensor product of bimodules may be derived from Bourbaki's [5] universal property, which stipulates a bilinear mapping $m_{AB} : AB \longrightarrow A \otimes B$ such that, for each bilinear mapping $f : AB \longrightarrow C$ into a bimodule $_R C_T$, there is a unique homomorphism $g : A \otimes B \longrightarrow C$ such that $g \circ m_{AB} = f$.

Influenced by an early collaboration with George Findlay, I was particularly interested in the fact that the bicategory of bimodules was *residuated*, there being canonical isomorphisms

$$\text{Hom}\,(A \otimes B, C) \cong \text{Hom}\,(A, C/B) \cong \text{Hom}\,(B, A \backslash C),$$

where

$$_R(C/B)_S = \text{Hom}_S(B, C),$$

$$_S(A \backslash C)_T = \text{Hom}_R(A, C).$$

To explain the universal properties of $\otimes$ (*tensor*), $/$ (*over*) and $\backslash$ (*under*) in general bicategories, I introduced the concept of a multicategory [17][†]

[†] The cited paper contains some mistakes, but these do not affect the concept of a multicategory.

[18]. A multicategory consisted of multilinear maps

$$f : A_1 \cdots A_m \longrightarrow B,$$

which might be viewed as context-free rules in grammar (I have reversed the usual arrow to reflect the hearer's point of view), as deductions in logic (called "sequents" by Gentzen, though here without his structural rules: interchange, contraction and weakening) or as multisorted operations in algebra (where one might write $f a_1 \cdots a_m \in B$, the a_i being variables or indeterminates of type A_i).

Originally, I had assumed that the left side of a multilinear map lives in the free monoid generated by a set, but one may as well let it live in the free category generated by a graph, so that f becomes an arrow from $S \xleftarrow{A_1} \cdots \xleftarrow{A_m} T$ to $S \xleftarrow{B} T$. The step of replacing a monoid by a category, while obvious in the bimodule example, was taken in linguistics by Brame [6] (who may not, however, agree with my [24] interpretation of his ideas). It does not make sense in any logical system which admits the interchange rule.

Among the multilinear maps are the identities $1_A : A \longrightarrow A$, as in a category, but composition is replaced by a more restricted notion, called "cut" by Gentzen:

$$\frac{f : \Lambda \longrightarrow A \quad g : \Gamma A \Delta \longrightarrow C}{g \circ \Gamma f \Delta : \Gamma \Lambda \Delta \longrightarrow C},$$

where capital Greek letters denote *strings* in the free monoid or *chains* in the free category, say

$$\Lambda = R \xleftarrow{L_1} \cdots \xleftarrow{L_k} S.$$

(At one time, I had written $g \,\&\, f$ for this cut, gaining simplicity at the cost of sacrificing information.)

There are the expected identity and associative laws for the cut, but also a kind of commutative law: if $f : \Lambda \longrightarrow A$, $g : \Lambda' \longrightarrow A'$ and $h : \Gamma A \Delta A' \Theta \longrightarrow C$ then

$$(h \circ \Gamma f \Delta A' \Theta) \circ \Gamma \Lambda \Delta g \Theta = (h \circ \Gamma A \Delta g \Theta) \circ \Gamma f \Delta \Lambda' \Theta.$$

This and similar equations will become clear if we think of the multicategory as being contained in a 2-category (see Section 9.4 below). It should be pointed out that Γ and Δ here do not denote terms of the multicategory, but merely serve to remove the ambiguity of the notation $g \& f$ by indicating where f is substituted into g. However, in the 2-category they may be interpreted as horizontal compositions of 1-cells.

In earlier papers [18, 19] I found it useful to pass to an internal language, where the 1-cells are thought of as sorts and variables of each sort are admitted. Thus, to each operation $f : A_1 \cdots A_m \longrightarrow B$ there is associated a term $fa_1 \cdots a_m$ of sort B, where the a_i are terms, e.g. variables, of sort A_i. Then $1_A : A \longrightarrow A$ gives rise to the term $1_A x = x$, where x is a variable of sort A. If $\vec{x}, \vec{c}$ and $\vec{d}$ are appropriate strings of variables,

$$g \circ \Gamma f \Delta : \Gamma \Lambda \Delta \longrightarrow C$$

gives rise to the term

$$(g \circ \Gamma f \Delta)(\vec{c}\vec{x}\vec{d}) = g\vec{c}f\vec{x}\vec{d},$$

which results by substituting $f\vec{x}$ for x in $g\vec{c}x\vec{d}$. The associative, commutative and identity laws can now be proved, provided we identify two operations f and f' whenever $f\vec{x} = f'\vec{x}$ is provable in the language. The variables may be called "indeterminates" in algebra and "assumptions" in logic.

Bourbaki's universal property for the tensor product stipulates a multilinear map $m_{AB} : AB \longrightarrow A \otimes B$ such that, for every $f : \Gamma AB\Delta \longrightarrow C$, there exists a unique $g : \Gamma A \otimes B\Delta \longrightarrow C$ such that $g \circ \Gamma m_{AB}\Delta = f$, equivalently that

$$g\vec{c}m_{AB}ab\vec{d} = f\vec{c}ab\vec{d}.$$

Similar universal properties may be given for the operations "over" and "under". Thus, one stipulates a multilinear map $e_{CB} : (C/B)B \longrightarrow C$ such that, for every $f : \Lambda B \longrightarrow C$ there exists a unique $g : \Lambda \longrightarrow C/B$ such that $e_{CB} \circ gB = f$, equivalently that

$$e_{CB}g\vec{l}b = f\vec{l}b.$$

It is now easy to prove that the operations $\otimes, /$ and $\backslash$ are bifunctors and that $\otimes$ is associative up to coherent isomorphism. In particular, one can derive Mac Lane's famous pentagonal condition [28]. See e.g. [28], where it was unnecessarily assumed that the multicategory has only one 0-cell.

Following Gentzen, one may reformulate the rules for tensor, over and under as introduction rules on the left and on the right:

$$\frac{\Gamma AB\Delta \longrightarrow C}{\Gamma A \otimes B\Delta \longrightarrow C} \; , \qquad \frac{\Gamma \longrightarrow A \quad \Delta \longrightarrow B}{\Gamma\Delta \longrightarrow A \otimes B} \; ,$$

$$\frac{\ulcorner C \Delta \longrightarrow \Delta \quad \Lambda \longrightarrow B}{\ulcorner C/B \Lambda \Delta \longrightarrow D} \quad, \quad \frac{\Lambda B \longrightarrow C}{\Lambda \longrightarrow C/B}.$$

The rules for \ are obtained from those for / by taking the mirror image on each side of the arrow. All these rules are subject to appropriate equations.

While these introduction rules incorporate some cuts, no further cuts are necessary. A categorical version of Gentzen's cut elimination theorem asserts the following:

Proposition 9.1 *Given a multicategory $\mathcal{M}$, one may construct the free residuated tensored multicategory $F(\mathcal{M})$ generated by $\mathcal{M}$ without using any cuts or identities, except those in $\mathcal{M}$.*

For example, $1_{A \otimes B}$ and $1_{C/B}$ may be constructed as follows:

$$\frac{\dfrac{A \longrightarrow A \quad B \longrightarrow B}{AB \longrightarrow A \otimes B}}{A \otimes B \longrightarrow A \otimes B} \qquad \frac{\dfrac{C \longrightarrow C \quad B \longrightarrow B}{(C/B)B \longrightarrow C}}{C/B \longrightarrow C/B}$$

where we introduce $\otimes$ first on the right and then on the left, but we introduce / first on the left and then on the right. Similar rules hold for the cartesian product and coproduct.

9.2 Adjoints in bicategories

I have lately become interested in adjoints in 2-categories and bicategories. For the usual definition of a 2-category see [29]. For the present purpose, a 2-category may be described as having 0-cells (objects), 1-cells (arrows) and 2-cells (transformations). The first two constitute a category and the last act as arrows between 1-cells subject to a *vertical composition*

$$\frac{f : A \to B \quad g : B \to C}{g \circ f : A \to C}$$

and identity arrows $1_A : A \to A$ rendering the 1-cells $S \to R$ objects of a category. The 2-cells can be composed with 1-cells and behave like natural transformations in the familiar 2-category of categories: given

$$T \underset{B}{\overset{A}{\underset{\Longrightarrow}{\rightrightarrows}}} S \underset{G}{\overset{F}{\underset{\Longrightarrow}{\rightrightarrows}}} R$$

one has the commutative diagram

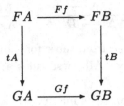

$$
\begin{array}{ccc}
FA & \xrightarrow{\;Ff\;} & FB \\
\downarrow{\scriptstyle tA} & & \downarrow{\scriptstyle tB} \\
GA & \xrightarrow{\;Gf\;} & GB
\end{array}
$$

which may be interpreted mnemonically as describing the *naturality* of t. Its diagonal defines the *horizontal* composition

$$tf = tB \circ Ff = Gf \circ tA.$$

Moreover, 1-cells *distribute* over composition of 2-cells:

$$F(g \circ f) = Fg \circ Ff, \quad (g \circ f)H = gH \circ fH.$$

In the same spirit,

$$F1_A = 1_{FA} = 1_F A.$$

The exchange property [29]

$$ug \circ tf = (u \circ t)(g \circ f)$$

may be deduced. Conversely, one presentation of 2-categories may be deduced from the traditional one by defining

$$tA = t1_A, \quad Ff = 1_F f.$$

An *adjunction* between 1-cells $F : R \longrightarrow S$ and $U : S \longrightarrow R$ in a 2-category is given by transformations $\eta : 1_R \longrightarrow UF$ and $\varepsilon : FU \longrightarrow 1_S$ such that

$$U\varepsilon \circ \eta U = 1_U, \quad \varepsilon F \circ F\eta = 1_F.$$

Proposition 9.2 *Adjoints in a 2-category are unique up to isomorphism.*

While this is well-known, I have never seen a proof and shall produce one here, as it is a little tricky and because the same proof will also serve for the analogous result for polycategories in Section 9.6 below.

Proof : Suppose, for example, that U has another left adjoint F' given by $\eta' : 1_R \longrightarrow UF'$ and $\varepsilon' : F'U \longrightarrow 1_S$. We claim that the composite transformations

$$\varphi = \varepsilon F' \circ F\eta', \quad \psi = \varepsilon' F \circ F'\eta$$

are inverse to one another. Here, for example is a proof that $\psi \circ \varphi = 1_F$:

$$
\begin{aligned}
\psi \circ \varphi &= \psi \circ \varepsilon F' \circ F\eta' \\
&= \varepsilon F \circ FU\psi \circ F\eta' \quad \text{by naturality of } \varepsilon \\
&= \varepsilon F \circ FU\varepsilon' F \circ FUF'\eta \circ F\eta' \quad \text{by distributivity of } FU \\
&= \varepsilon F \circ FU\varepsilon' F \circ F(UF'\eta \circ \eta') \quad \text{by distributivity of } F \\
&= \varepsilon F \circ FU\varepsilon' F \circ F(\eta'UF \circ \eta) \quad \text{by naturality of } \eta' \\
&= \varepsilon F \circ F(U\varepsilon' \circ \eta'U)F \circ F\eta \quad \text{by distributivity of } F \\
&= \varepsilon F \circ F\eta \quad \text{since } U\varepsilon' \circ \eta'U = 1_U \\
&= 1_F.
\end{aligned}
$$

$\square$

The notion of adjunction has been generalized to bicategories, e.g. by Kelly [13] and Street and Walters [31]. When is $F : R \longrightarrow S$ left adjoint to $U : S \longrightarrow R$? We require 2-cells $I_R \longrightarrow U \otimes F$ and $F \otimes U \longrightarrow I_S$ such that the composite 2-cell $U \longrightarrow I_R \otimes U \longrightarrow (U \otimes F) \otimes U \longrightarrow U \otimes (F \otimes U) \longrightarrow U \otimes I_S \longrightarrow U$ is the identity on U and similarly for the analogous 2-cell $F \longrightarrow F$. For example, we may ask: when does a bimodule ${}_R U_S$ have a left adjoint ${}_S F_R$? (Note that I_R is the bimodule ${}_R R_R$.) I was surprised to find the answer in the exercises to Section 4.1 in my 1966 book: ${}_R U_S$ has a left adjoint ${}_S F_R$ if and only if U_S is finitely generated and projective, and then $F = S/U$.

I have become particularly interested in bicategories in which each 1-cell has both a left and a right adjoint. Such bicategories are called *compact*, following Kelly [13][†]. For expository purposes, let me now confine attention to 2-categories with one object in which all 2-cells are just partial orders.

A *pregroup* is a partially ordered monoid with two operations $(-)^\ell$ and $(-)^r$ satisfying

$$
a^\ell a \longrightarrow 1 \longrightarrow aa^\ell, \quad aa^r \longrightarrow 1 \longrightarrow a^r a
$$

for each element a, the arrow denoting the partial order. In the discrete case, when the arrow denotes equality, a pregroup is just a group. More

---

[†] Kelly used the word "compact" for symmetric monoidal categories in which each object has a right adjoint. These are the compact *-autonomous categories of Barr [2], and the concept was generalized to bicategories by Street and Walters [31].

generally, in the cyclic[‡] case, when $a^\ell = a^r$, a pregroup is just a partially ordered group. My favorite example [21, 23] of a non-cyclic pregroup is the monoid of unbounded monotone mappings $f : \mathbb{Z} \longrightarrow \mathbb{Z}$ under composition with elementwise order. Then adjoints may be defined thus:

$$f^r(y) = \max\{x \in \mathbb{Z} | f(x) \le y\}$$

$$f^\ell(y) = \min\{x \in \mathbb{Z} | y \le f(x)\}.$$

For example, let $f(x) = 2x$, then

$$f^r(y) = [y/2], \quad f^\ell(y) = [(y+1)/2].$$

We note that

$$f^{rr}(y) = 2y + 1 \ne f(y);$$

but, of course, $f^{r\ell} = f = f^{\ell r}$. Other examples of pregroups are provided by all natural languages, as we shall illustrate with English in the next section.

9.3 Linguistic applications

We shall look at three possible applications of bicategories to linguistics, see [9]. For expository purposes, we first confine attention to partially ordered monoids:

(1) *Residuated monoids*, namely partially ordered monoids with two operations / and \ such that

$$a \cdot b \longrightarrow c \ \text{ iff } \ a \longrightarrow c/b \ \text{ iff } \ b \longrightarrow a\backslash c.$$

(2) *Grishin algebras*, namely residuated monoids with a dualizing element 0 such that

$$0/(a\backslash 0) = a = (0/a)\backslash 0.$$

It is convenient to write

$$0/a = a^\ell, \quad a\backslash 0 = a^r$$

and one can introduce a second associative operation + by defining

$$a + b = (b^\ell \cdot a^\ell)^r = (b^r \cdot a^r)^\ell$$

[‡] The word "cyclic" in this context is due to Yetter [34].

such that

$$a + 0 = a = 0 + a.$$

One can then prove Grishin's mixed associative laws [12]:

$$c(a + b) \longrightarrow ca + b, \quad (a + b)c \longrightarrow a + bc.$$

(3) *Pregroups* as defined above. These may also be viewed as *compact* Grishin algebras, in which

$$a + b = a \cdot b, \quad 0 = 1.$$

In the corresponding deductive systems, the arrow is not restricted to be a partial order, but equality of arrows is usually ignored. The above partially ordered monoids will then give rise to the following deductive systems:

(1) The *syntactic calculus*, introduced in [15] to study sentence structure.
(2) *Classical bilinear logic* [1, 20] which was pioneered by Claudia Casadio [7] for grammatical investigations.
(3) *Compact bilinear logic*, recently proposed by me [26] for linguistic applications.

Each of these deductive systems becomes a monoidal category (that is, a bicategory with one object), once attention is paid to equality between arrows:

(1) A residuated monoidal category.
(2) A noncommutative *-autonomouns category.
(3) A compact noncommutative *-autonomous category.

As already pointed out, one may remove the restriction to one object. In Linguistics, this step has already been taken by Brame [6].

The idea common to the linguistic applications of these three systems is this: one assigns to each word, say of English, one or more *syntactic types*, namely elements of the free residuated monoid, Grishin algebra or pregroup generated by a partially ordered set of basic types and then calculates the type or types of any string of words.

We shall illustrate this idea by looking at a single English sentence:

whom had she kissed–?

The dash at the end represents Chomsky's trace and is introduced for comparison only.

In (1), the words of this sentence are assigned the types

$$(q'/(q/o))((q_1/p_2)/\pi_3)\pi_3(p_2/o) \longrightarrow q'.$$

The basic types employed here are:

$$q' = \text{question,}$$
$$q = \text{yes-or-no question,}$$
$$q_1 = \text{yes-or-no question in the present tense,}$$
$$o = \text{object,}$$
$$p_2 = \text{past participle,}$$
$$\pi_3 = \text{third person singular pronoun.}$$

The partial order on the set of basic types required for this example stipulates $q_1 \longrightarrow q \longrightarrow q'$.

Although the method advocated here received a belated acceptance by a small group of linguists, I came to reject it myself for various reasons, one being the following. When a person hears the words *whom has*, she may calculate the type of this short string to be $(q'/(p_2/o))/\pi_3$, but the formal proof of this, carried out in the syntactic calculus, is fairly long and, when put to paper, may occupy a quarter page. I had a strong feeling that this kind of calculation could not reflect the pychological reality of how people analyze speech.

In (2), the successive types for the same sentence are

$$(q' + o^{\ell\ell}q^{\ell})(q_1 + p_2^{\ell} + \pi_3^{\ell})\pi_3(p_2 + o^{\ell}) \longrightarrow q'.$$

Here the type of *whom has* is easily calculated algebraically to be $q' + o^{\ell\ell}p_2^{\ell} + \pi_3^{\ell}$; the calculation makes repeated use of the mixed associative laws.

In (3), the distinction between $+$ and $\cdot$ disappears and the successive types of the given sentence are simply

$$(q'o^{\ell\ell}q^{\ell})(q_1\ p_2^{\ell}\pi_3^{\ell})\pi_3(p_2 o^{\ell}) \longrightarrow q',$$

where the underlining indicates the cancellations

$$q^{\ell}q_1 \longrightarrow q^{\ell}q \longrightarrow 1, \quad \pi_3^{\ell}\pi_3 \longrightarrow 1, \quad p_2^{\ell}p_2 \longrightarrow 1, \quad o^{\ell\ell}o^{\ell} \longrightarrow 1.$$

Here the type of the initial segment *whom has* is immediately seen to be $q'o^{\ell\ell}p_2^{\ell}\pi_3^{\ell}$.

The reader should not be misled by this single example that only $/$ and $(-)^\ell$ are useful in grammar and not $\backslash$ and $(-)^r$. For example, in the sentenc

$$she\ kissed\ me$$

the transitive verb has types

$$\pi_3\backslash(s_2/o),\quad \pi_3^r + s_2 + o^\ell,\quad \pi_3^r s_2 o^\ell$$

in the three systems respectively, where

$$s_2 = \text{statement in the past tense.}$$

I believe that a good approximation to English grammar can be obtained by working with the *free* pregroup generated by a partially ordered set of basic syntactic types. For a better approximation, however, freeness must be abandoned. For example, it is difficult to justify the well-formedness of

$$people\ she\ knows\ like\ pizza$$

by the methods outlined above. The problem here is that there is no place for the type of the missing pronoun *whom*. To get around this, one may have to admit grammatical rules not listed in the dictionary, in other words, one may have to work with a pregroup which is not freely generated by its basic types.

9.4 A new look at polycategories

If we adopt the slogan: don't ignore equality between deductions (also known in linguistics as derivations, productions or rewrite rules), a *production grammar* (also known as a semi-Thue system or rewrite system) is just a 2-category whose underlying 1-category is the free monoid generated by a set. Some greater generality is achieved if we allow the free category generated by a graph instead. In fact, for context-free grammars this generality has been advocated by Brame [6].

We recall that a multicategory is essentially a context-free grammar dealing with deductions of the form

$$f : A_1 \cdots A_m \longrightarrow B,$$

where juxtaposition on the left represents the tensor product, attention being paid to equality between deductions. In the presence of Gentzen's structural rules (interchange, contraction and weakening), these deductions are Gentzen's sequents for intuitionistic logic and the tensor product is just conjunction.

Gentzen also devised a deductive system for dealing with classical logic. Its sequents have the form

$$f : A_1 \cdots A_m \longrightarrow B_1 \cdots B_n,$$

where juxtaposition on the left stands for conjunction and juxtaposition on the right stands for disjunction, although, in place of juxtaposition he had used commas on both sides. One may wonder why he did not use a comma on the left and a semicolon, say, on the right, to suggest the two different interpretations? We shall take advantage of his daring notation to embed polycategories, our categorical version of his system (in the absence of his structural rules) into 2-categories.

Polycategories may be regarded as underlying the grammar of Claudia Casadio [7], where juxtaposition on the left and on the right of a deduction represent the tensor and the cotensor, its De Morgan dual, respectively. Compact polycategories, in which the tensor and cotensor are identified, are then essentially production grammars, hence 2-categories whose underlying 1-category is freely generated. Polycategories are like production grammars, except that composition of 1-cells is restricted to cuts.

Cuts in polycategories have the form

$$\frac{f : \Lambda \longrightarrow \Gamma A \Delta \quad g : \Phi A \Psi \longrightarrow \Theta}{\Gamma g \Delta \circ \Phi f \Psi : \Phi \Lambda \Psi \longrightarrow \Gamma \Theta \Delta}$$

subject to the restriction that Γ or Φ is empty and Δ or Ψ is empty. Thus there are four cases:

Case 1. Φ and Ψ are empty and the conclusion is $\Gamma g \Delta \circ f : \Lambda \longrightarrow \Gamma \Theta \Delta$.

Case 2. Φ and Δ are empty and the conclusion is $\Gamma g \circ f \Psi : \Lambda \Psi \longrightarrow \Gamma \Theta$.

Case 3. Γ and Δ are empty and the conclusion is $g \circ \Phi f \Psi : \Phi \Lambda \Psi \longrightarrow \Theta$.

Case 4. Ψ and Γ are empty and the conclusion is $g \Delta \circ \Phi f : \Phi \Lambda \longrightarrow \Theta \Delta$.

The four cases may be illustrated by the following planar diagrams respectively:

$$
\begin{array}{cccc}
\Lambda & \Lambda & \Lambda & \Lambda \\
\overline{\Gamma\, A\, \Delta} & \overline{\Gamma \underline{A} \Psi} & \underline{\Phi\, \overline{A}\, \Psi} & \underline{\Phi\, \underline{A}\, \Delta} \\
\underline{\Theta} & \Theta & \Theta & \Theta
\end{array}
$$

where the 2-cells f and g are represented by horizontal lines.

In case the reader is still skeptical, here is a formal definition to convince her that a polycategory can be embedded into a 2-category with additional 2-cells.

Definition 9.3 A *polycategory over a graph*[†] has 0-cells, 1-cells, 2-cells and equations between 2-cells:

- its 0-cells are the nodes of the graph;
- its 1-cells are the arrows of the free category generated by the graph;
- its 2-cells are certain arrows between 1-cells, Γ and Δ, assuming that Γ and Δ have the same source and target;
- among the 2-cells are all the identity arrows $1_A : A \to A$, where A is an arrow of the graph;
- the set of 2-cells is closed under the four kinds of cuts listed above;
- its equations are precisely those which hold in the 2-category obtained by allowing all identity 2-cells and arbitrary composition of 2-cells, provided we interpret a cut with premises $f : \Lambda \to \Gamma A\Delta$ and $g : \Phi A\Psi \to \Theta$ as the composition of $\Gamma g\Delta$ and $\Phi f\Psi$, as suggested by the above notation.

It was in order to get a grip on the possible equations between deductions that I had suggested the idea of a polycategory in [17]. However, I did not take the trouble to spell out exactly what equations had to hold. This was done by Szabo [32], although he allowed too many cases of the cut for the substructural system of bilinear logic studied here. Polycategories were also investigated by Velinov [33], who considered many variations, even the compact case. A detailed set of equations that meet my approval were presented by Cockett and Seely [1992], who were using polycategories to introduce the tensor and cotensor into what they then called "weakly distributive categories". In fact, they obtained an equivalence between the category of polycategories and the category of weakly distributive categories.

I believe that the present method of inferring all equations between deductions from those valid in 2-categories is new. (We shall ignore here another approach I have been exploring, which replaces the operations that had proved useful in multicategories by binary relations.) We shall look at a few examples of such equations. (A list of five such equations will be found in Cockett and Seely [1992]. I have not checked whether

---

[†] By a *graph* is here understood what graph theorists call an "oriented multigraph".

these five equations imply all the equations that can be inferred from those of 2-categories.)

Example 9.4

$$\frac{\Lambda \xrightarrow{f} \Gamma A \quad A\Delta \xrightarrow{g} \Phi B\Psi \quad B \xrightarrow{h} \Theta}{\Lambda\Delta \longrightarrow \Gamma\Phi\Theta\Psi}$$

There are two ways of deriving the conclusion, depending on whether we first compose g with f or h with g. These are represented by the two sides of the equation

$$\Gamma\Phi h\Psi \circ (\Gamma g \circ f\Delta) = \Gamma(\Phi h\Psi \circ g) \circ f\Delta,$$

which is justified by associativity and distributivity in a 2-category. Note that the intermediate term

$$\Gamma(\Phi h\Psi \circ g) \circ f\Delta$$

does not live in the polycategory, but in the embedding 2-category.

Example 9.5

$$\frac{\Lambda \xrightarrow{f} \Gamma A\Delta \quad A \xrightarrow{g} \Phi B\Psi \quad B \xrightarrow{h} \Theta}{\Lambda \longrightarrow \Gamma\Phi\Theta\Psi\Delta}$$

Here we have the equation

$$\Gamma\Phi h\Psi\Delta \circ (\Gamma g\Delta \circ f) = \Gamma(\Phi h\Psi \circ g)\Delta \circ f,$$

which is also justified by associativity and distributivity.

Example 9.6

$$\frac{\Phi \xrightarrow{f} A \quad \Psi \xrightarrow{g} B \quad \Gamma A\Delta B\Lambda \xrightarrow{h} \Theta}{\Gamma\Phi\Delta\Psi\Lambda \longrightarrow \Theta}$$

Here the equation

$$(h \circ \Gamma A\Delta g\Lambda) \circ \Gamma f\Delta\Psi\Lambda = (h \circ \Gamma f\Delta B\Lambda) \circ \Gamma\Phi\Delta g\Lambda$$

may be reduced by associativity and distributivity to showing that

$$A\Delta g \circ f\Delta\Psi = f\Delta B \circ \Phi\Delta g,$$

which follows from naturality of f.

These examples should support the claim that the equations holding in a polycategory are precisely those which hold in the 2-category which contains it. In retrospect, the same is true for a multicategory. Perhaps *a polycategory should have been called a "sesqui-category"! The algebraic derivations of the equations in the three examples above become redundant if one relies instead on the planar diagrams which illustrate how the conclusion is obtained:*

$$
\frac{\overset{\Lambda}{\overline{\Gamma \underline{A} \;\; \Delta}}}{\underset{\Theta}{\overline{\Phi \underline{B} \Psi}}}
\qquad
\frac{\overset{\Lambda}{\overline{\Gamma \;\; \underline{A} \;\; \Delta}}}{\underset{\Theta}{\overline{\Phi \underline{B} \Psi}}}
\qquad
\frac{\Gamma \overset{\Phi}{\underline{A}} \Delta \overset{\Psi}{\underline{B}} \Lambda}{\Theta}
$$

where the horizontal lines represent the deductions f, g and h: two deductions are identified if they give rise to the same diagram.

A final example will illustrate the behaviour of the identity arrow.

Example 9.7

$$
\frac{\Lambda \xrightarrow{\;f\;} \Gamma A \Delta \quad A \xrightarrow{\;1_A\;} A}{\Lambda \xrightarrow{\;f\;} \Gamma A \Delta}
$$

Here we have

$$
\Gamma 1_A \Delta \circ f = 1_{\Gamma A \Delta} \circ f = f.
$$

9.5 Polycategories and linear bicategories

After composing the first draft of this paper, I was presented with a copy of the article by Cockett, Koslowski and Seely [11], in which they developed the notion of "linear bicategory" and studied "linear adjoints", a generalization of adjoints in the original bicategories of Bénabou.

One purpose of multicategories had been to introduce the tensor product $\otimes$ and the corresponding identity 1-cells I_R into a bicategory so that their properties can be proved instead of having to be postulated. Polycategories will do the same for linear bicategories in helping to introduce also the cotensor $\oplus$ and the corresponding zero 1-cells O_R. This program had in fact been carried out by Cockett and Seely [1992], although they had presented polycategories more directly than here.

We recapitulate the definitions of these operations in a present style polycategory:

- $\otimes$ is given by $m_{AB} : AB \longrightarrow A \otimes B$ such that, for each $f : \Gamma AB\Delta \longrightarrow \Theta$, there exists a unique $g : \Gamma A \otimes B\Delta \longrightarrow \Theta$ such that $g \circ \Gamma m_{AB}\Delta = f$.
- I_R is given by $i_R : 1_R \longrightarrow I_R$ such that, for each $f : \Gamma\Delta \longrightarrow \Theta$, there exists a unique $g : \Gamma I_R\Delta \longrightarrow \Theta$ such that $g \circ \Gamma i_R\Delta = f$. Here 1_R

denotes the identity arrow $R \longrightarrow R$ in a 2-category, that is, the empty chain between Γ and Δ in $\overset{\Gamma}{\longleftarrow} R \overset{\Delta}{\longrightarrow}$.

- $\oplus$ is given by $n_{AB} : A \oplus B \longrightarrow AB$ such that, for each $f : \Theta \longrightarrow \Gamma AB\Delta$, there exists a unique $g : \Theta \longrightarrow \Gamma A \oplus B\Delta$ such that $\Gamma n_{AB}\Delta \circ g = f$.
- O_R is given by $j_R : O_R \longrightarrow 1_R$ such that, for each $f : \Theta \longrightarrow \Gamma\Delta$, there exists a unique $g : \Theta \longrightarrow \Gamma O_R \Delta$ such that $\Gamma j_R \Delta \circ g = f$.

A *residuated polycategory* has residual quotients $/$ and $\backslash$, the first of which is introduced as follows:

- $/$ is given by $e_{AB} : (A/B)B \longrightarrow A$ such that, for each $f : \Gamma B \longrightarrow \Delta A$, there exists a unique $g : \Gamma \longrightarrow \Delta A/B$ such that $\Delta e_{AB} \circ gB = f$.
- For $\backslash$ one takes the mirror image of each side of the arrow.

One may also consider residual differences $\dot{-}$ (*less*) and $\dot{\top}$ (*from*). For a discussion of these see [20].

Gentzen style introduction rules for $\otimes$ and I_R take the following form, while those for $\oplus$ and O_R may be obtained by reversing the arrows:

$$\frac{\Gamma AB\Delta \longrightarrow \Theta}{\Gamma A \otimes B\Delta \longrightarrow \Theta} \ , \quad \frac{\Gamma \longrightarrow \Phi A \quad \Delta \longrightarrow B\Psi}{\Gamma\Delta \longrightarrow \Phi A \otimes B\Psi} \ ,$$

$$\frac{\Gamma\Delta \longrightarrow \Theta}{\Gamma I_R \Delta \longrightarrow \Theta} \ , \quad 1_R \longrightarrow I_R.$$

The introduction rules for $/$ have the form

$$\frac{\Gamma A\Delta \longrightarrow \Theta \quad \Lambda \longrightarrow B}{\Gamma A/B\Lambda\Delta \longrightarrow \Theta} \quad \frac{\Gamma B \longrightarrow \Delta A}{\Gamma \longrightarrow \Delta A/B},$$

while those for $\backslash$ are obtained by taking the mirror image on each side of the arrow.

Here, for example, is how one may construct the arrow

$$(A \oplus B) \otimes C \longrightarrow A \oplus (B \otimes C),$$

representing one of Grishin's mixed associative laws:

$$\frac{\dfrac{\dfrac{A \longrightarrow A \quad B \longrightarrow B}{A \oplus B \longrightarrow AB} \quad C \longrightarrow C}{(A \oplus B)C \longrightarrow A(B \otimes C)}}{\dfrac{A \oplus B)C \longrightarrow A \oplus (B \otimes C)}{A \oplus B) \otimes C \longrightarrow A \oplus (B \otimes C}}$$

central role played by the premises of $\xi \vdash \sigma$ leads to the notion of **partial design** of a behaviour Γ.

A partial design $\mathfrak{D}'$ is a "part of a design" $\mathfrak{D} \in G$: a subtree that has the same base, but where some of the premises may be missing.

A typical example of partial design is a *slice* of a design $\mathfrak{D}$: a subtree of $\mathfrak{D}$ obtained selecting in all negative rules at most one premise. An extreme example of partial designs are the empty ones ($\mathfrak{Fid}$, *Skunk*). G^p denotes the set of all designs (total and partial) of a behaviour G.

Now we can express the fact that all the partial designs

$$\frac{\vdots}{\xi \vdash \sigma} (\xi, \{I\})$$

included in

$$\frac{\vdots \quad \vdots \quad \vdots}{\xi \vdash \sigma} (\xi, \mathcal{P}_f(\mathbf{N}))$$

lead essentially to the same proof.

It becomes natural to introduce on G^p a **partial equivalence relation** (i.e. asymmetric and transitive relation) $\cong$ which separates the *partial* designs with respect to normalization. The key is that the equivalence relation identifies the closed nets normalizing into $\mathfrak{D}ai$ and those normalizing into $\mathfrak{Fid}$.

Since designs and counter-designs (proofs and counter-proofs) have the same status, we need to consider the partial equivalence on $G^{p\perp}$ induced by normalization, and come back by bi-orthogonal... Two partial designs are in the same class, if their reactions against equivalent partial counter-designs are the same.

As an example, the trivial equivalence relation is the one that identifies all *proper*[†] designs of G^p ($\cong^\perp$ distinguishes all proper designs of $G^{p\perp}$).

In a sequent of behaviours $X \vdash P$, we want again that two partial equivalent designs (saying $\mathfrak{D}$ and $\mathfrak{D}'$) react in the same way against two equivalent designs of X (saying $\mathfrak{E}$ and $\mathfrak{E}'$) ; that means that $[\![\mathfrak{D}, \mathfrak{E}]\!]$ and $[\![\mathfrak{D}', \mathfrak{E}']\!]$ produce two equivalent designs in P.

On compound behaviours, the equivalence relations we are interested in must conserve the properties of the connective. For example, the behaviour $P \oplus Q$ is the union of the two distinct behaviours P, Q. The equivalence relation keeps distincts designs coming from distinct be-

---

[†] Any positive design distinct from $\mathfrak{D}ai$ and $\mathfrak{Fid}$.

Proposition 9.8 *In a polycategory any existing adjoints are unique up to isomorphism.*

When can we infer that adjoints exist?

Proposition 9.9 *In a residuated polycategory with zero 1-cells, every 1-cell $A : S \longrightarrow R$ has both a left and a right adjoint:*

$$A^\ell = O_S/A \quad , \quad A^r = A\backslash O_R.$$

Proof : To show the existence of left adjoints, for example, we have to define

$$\varepsilon_A : (O_S/A)A \longrightarrow 1_S \quad , \quad \eta_A : 1_R \longrightarrow A(O_S/A)$$

and verify that

$$A\varepsilon_A \circ \eta_A A = 1_A \quad , \quad \varepsilon_A(O_S/A) \circ (O_S/A)\eta_A = 1_{O/A}.$$

We define $\varepsilon_A = j \circ e_{OA}$ by the cut

$$\frac{e_{OA} : (O/A)A \longrightarrow O \quad j : O \longrightarrow 1}{\varepsilon_A : (O/A)A \longrightarrow 1}$$

and $\eta_A : 1 \longrightarrow A(O/A)$ as the unique $g : 1 \longrightarrow A(O/A)$ such that

$$Ae_{OA} \circ gA = f,$$

where $f : A \longrightarrow AO$ is the unique arrow such that

$$Aj \circ f = 1_A :$$

$$\frac{A \xrightarrow{f} AO \quad O \xrightarrow{j} 1}{A \longrightarrow A} \quad , \quad \frac{1 \xrightarrow{g} A(O/A) \quad (O/A)A \xrightarrow{e_{OA}} O}{A \longrightarrow OA}.$$

Then

$$A\varepsilon_A \circ \eta_A A = Aj \circ Ae_{OA} \circ gA = Aj \circ f = 1_A.$$

To show the other equation to be proved, we recall that 1_{A^ℓ} is the unique $h : A^\ell \longrightarrow A^\ell$ such that

$$e_{OA} \circ hA = e_{OA}.$$

Hence this equation has to be verified when

$$h = \varepsilon_A A^\ell \circ A^\ell \eta_A.$$

Indeed

$$
\begin{aligned}
e_{OA} \circ hA &= e_{OA} \circ \varepsilon_A A^\ell A \circ A^\ell \eta_A A \\
&= \varepsilon_A O \circ A^\ell A e_{OA} \circ A^\ell \eta_A A \quad \text{by naturality of } \varepsilon_A \\
&= jO \circ e_{OA} O \circ A^\ell f \quad \text{by definition of } \varepsilon_a \text{ and } \eta_A
\end{aligned}
$$

Now, by Lemma 9.11 below, we may replace jO by Oj, hence this

$$
\begin{aligned}
&= e_{OA} \circ A^\ell A j \circ A^\ell f \quad \text{by naturality of } e_{OA} \\
&= e_{OA} \circ A^\ell (1_A) \quad \text{by definition of } f \\
&= e_{OA} \circ 1_{A^\ell A} = e_{OA}
\end{aligned}
$$

$\square$

Corollary 9.10 *In a residuated polycategory with zero 1-cells, the zeros are dualizing: for any 1-cell $A : S \longrightarrow R$,*

$$
(O_S/A)\backslash O_R \cong A \cong O_S/(A\backslash O_R).
$$

Proof: If A^ℓ is left adjoint to A, then both A and $A^{\ell r}$ are right adjoints of A^ℓ, hence $A^{\ell r} \cong A$ by Proposition 9.8. Similarly $A^{r\ell} \cong A$. $\square$

Lemma 9.11 *In a polycategory with zero 1-cells, $jO = Oj$.*

Proof By the universal property of $j : O \longrightarrow 1$, any 2-cell $f : OO \longrightarrow 1$ gives rise to a unique $g : OO \longrightarrow O$ such that $j \circ g = f$. Now take $f = j \circ Oj$, then $g = Oj$. But, by naturality of j, $j \circ Oj = j \circ jO$, hence $g = jO$. $\square$

9.7 Postscript

This article is an elaboration of a talk at the 1999 category conference in Coimbra. Its major aim was to explain my idea of what the equations of a polycategory should be. I had introduced this concept in 1969 without spelling out these equations. In the mean time, attempts to produce such equations axiomatically were made by several authors, though not in agreement with one another. While the axioms provided by Cockett and Seely are "sound", "completeness" with respect to the present treatment remains to be shown: the equations of a polycategory should be those that ensure its embedding into a 2-category to be faithful.

Acknowledgements

This research was supported by grants from NSERC and SSHRC.

Bibliography

[1] V.M. Abrusci (1991), *Phase semantics and sequent calculus for pure noncommutative classical linear propositional logic*, J. Symbolic Logic **56**, 1403-1451.

[2] M. Barr (1979), **-Autonomous categories*, Springer LNM **752**.

[3] M. Barr (1995), *Non-symmetric *-autonomous categories*, Theoretical Computer Science **139**, 115-130.

[4] J. Bénabou (1967), *Introduction to bicategories*, Springer LNM **47**, 1-77.

[5] N. Bourbaki (1948), *Algèbre multilinéaire*, Hermann, Paris.

[6] M. Brame (1984, 1985, 1987), *Recursive categorical syntax and morphology* I, II, III, Linguistic Analysis **14**, 265-287; **15**, 137-176; **17**, 147-185.

[7] C. Casadio (1997), *Unbounded dependencies in non-commutative logic*, in: Proc. Conference Formal Grammars, ESSLLI, Aix en Provence.

[8] C. Casadio (2001), *Non-commutative linear logic in linguistics*, Grammars **3/4**, 1-19.

[9] C. Casadio and J. Lambek (2002), *A tale of four grammars*, Studia Logica, **71**, 315-329.

[10] J.R.B. Cockett and R.A.G. Seely (1997), *Weakly distributive categories*, J. Pure & Applied Algebra **114**, 133-173.

[11] J.R.B. Cockett, J. Koslowski and R.A.G. Seely (2000), *Introduction to linear bicategories*, Math. Structures in Computer Science, **10**, 165-203.

[12] V.N. Grishin (1983), *On a generalization of the Ajdukiewicz-Lambek system*, in: *Studies in non-commutative logics and formal systems*, Nauka, Moscow, 315- 343. English translation in: V.M. Abrusci and C. Casadio (eds), *New Perspectives in Logic and Formal Linguistics*, Bulzoni Editore, Roma 2002, 9-27.

[13] G.M. Kelly (1972), *Many variable functorial calculus* I, Springer LNM **281**, 66-105.

[14] S.C. Kleene (1952), *Introduction to metamathematics*, Van Nostrand, New York.

[15] J. Lambek (1958), *The mathematics of sentence structure*, Amer. Math. Monthly **65**, 154-169.

[16] J. Lambek (1968, 1976, 1986), *Lectures on rings and modules*, Blaisdell, Waltham Mass.; Ginn, New York, N.Y.; Chelsea, New York, N.Y.

[17] J. Lambek (1969), *Deductive systems and categories* II, Springer LNM **87**, 76- 122.

[18] J. Lambek (1989), *Multicategories revisited*, Contemporary Math. **92**, 217- 239.

[19] J. Lambek (1993), *Logic without structural rules*, in: K. Došen and P. Schroeder-Heister (eds.), *Substructural Logics*, Studies in Logic and Computation **2**, Oxford Science Publications, 179-206.

[20] J. Lambek (1993), *From categorial grammar to bilinear logic*, ibid., 207-237.

[21] J. Lambek (1994), *Some Galois connections in elementary number theory*, J. Number Theory **47**, 371-377.

[22] J. Lambek (1995), *Bilinear logic in algebra and linguistics*, in: J.-Y.

Girard et al. (eds), *Advances in Linear Logic*, London Math. Soc. Lecture Notes Series **222**, Cambridge University Press.

[23] J. Lambek (1995), *Some lattice models of bilinear logic*, Algebra Universalis **34**, 541-550.

[24] J. Lambek (1999), *Deductive systems and categories in linguistics*, in: H.J. Ohlbach and U. Reyle (eds), *Logic, language and reasoning*, Kluwer Academic Publishers, Dordrecht, 279-294.

[25] J. Lambek (1999), *Bilinear logic and Grishin algebras*, in: E. Orlowska (ed.), *Logic at work*, Essays dedicated to the memory of Helena Rasiowa, Physica-Verlag, Heidelberg, New York, 604-612.

[26] J. Lambek (1999), *Type grammars revisited*, in: A. Lecomte, F. Lamarche and G. Perrier (eds), *Logical Aspects of Computational Linguistics*, Springer LNAI **1582**, 1-27.

[27] F.W. Lawvere (1973), *Metric spaces, generalized logic, and closed categories*, Rend. Sen. Mat. E Fis. Milano **43**, 135-166.

[28] S. Mac Lane (1963), *Natural associativity and commutativity*, Rice University Studies **49**, 28-46.

[29] S. Mac Lane (1971), *Categories for the working mathematician*, Springer-Verlag, New York, N.Y.

[30] K.I. Rosenthal (1994), *∗-autonomous categories of bimodules*, J. Pure & Applied Algebra **97**, 188-201.

[31] R. Street and R.F.C. Walters (1978), *Yoneda structures on 2-categories*, J. Algebra **50**, 350-379.

[32] M.E. Szabo (1975), *Polycategories*, Communications in Algebra **3**, 663-698.

[33] Y. Velinov (1988), *An algebraic structure for derivations in rewriting systems*, Theoretical Computer Science **57**, 205-224.

[34] D.N. Yetter (1990), *Quantales and (non-commutative) linear logic*, J. Symbolic Logic **55**, 41-64.

10

Between Logic and Quantic: a Tract

Jean-Yves Girard

Institut de Mathématiques de Luminy

Abstract

We present a quantum interpretation of the perfect part of linear logic by means of quantum coherent spaces. In particular this yields a novel interpretation of the reduction of the wave packet as the expression of η-conversion, a.k.a, extensionality.

Acknowledgements: this work has been essentially carried out in October 2002, and issued as privately circulated notes in French. The sources were my recent *ludics*, [9], that I was trying to make "quantic" for a couple of years, in relation to my much older "geometry of interaction", [6], an explanation of logic in terms of Hilbert space operators. In Spring 2002, I got a definite jolt from the work of Selinger, [11], in particular his handling of "density matrices". This final version benefited from discussion with colleagues interested in the interface with quantum physics, Ctirad Klimčík, Thierry Paul, and Richard Zekri. It also benefited from the intercession of St Augustine, an output of the discussions led inside the informal group LGC "la Logique comme Géométrie du Cognitif", whose aim is to reconsider various philosophical and methodological issues that were fumbled by the "linguistic turn" of last century, see the page http://www.logique.jussieu/www.joinet.

10.1 Introduction

10.1.1 What is the question?

¿From the beginning, it has been clear that something should be clarified between logic and quantic, that there was a logico-physical puzzle.

346

In such a delicate situation, the main question was to find the *right* question.

10.1.1.1 The punishment of nature

According to Herodotus (VII,35), a tempest destroyed the military bridges built by Xerxes over the Hellespont; he decided to punish nature and to have the sea whipped. To some extent, this is what logicians wanted to do to quantum physics, to punish it for being "against common-sense". Among the untold things was surely the idea of a complete schizophrenia between nature and spirit: our beautiful minds were harboured by the wrong world and this was a mere accident. The logical accounts of quantum phenomena were contrived *on purpose*, as in the notorious *quantum logic*; the subliminal message being: "quantum or not, just a matter of encoding".

10.1.1.2 The failure of quantum logic

We remember what happened to quantum logic —or worse, we no longer remember. Technically speaking, the idea of replacing Boolean algebras with the lattice of closed subspaces of a Hilbert space is obviously wrong: there is a fine negation (the orthogonal complement), but nothing like a decent conjunction, in other words there is no simple account of the intersection of two spaces in terms of operators: $\pi \cap \pi' = \pi \cdot \pi'$ only when π, π' commute. Worse, the expulsion of the Hilbert space in favor of abstract "orthomodular lattices" didn't bring much water in this desert. Viewed from a distance, there was a methodological mistake. Boolean algebras are the truth values of classical logic, they are used as *semantics*, the external world, in opposition with *syntax*, which deals with us, as observers. Quantum logic wanted to keep the opposition semantics/syntax, and, inside the same mould, slightly alter the semantics, from something simple (Boolean algebras) to something artificial (the closed subspaces of a Hilbert). But if there is something that the quantum world refuses, this is this simple minded view of an external reality. The logician Frege thought that any expression was denoting something; but the word "impulsion", denotes nothing in quantum physics, worse, if we want it to denote something, we are performing an irreversible damage. In other words we cannot make a separation between the world and its observation.

This explains the failure of quantum logic. There is little to say about other attempted interpretations, for instance via Kripke models, which are sort of branching parallel universes. These structures are so floppy that they give us back what we want to see in them: they are indeed Loyola models, they obey *perinde ac cadaver*.

10.1.2 Augustine vs. Thomas

We reverse the paradigm. We don't consider quantum as "immoral", we no longer try to "tame" it through some do-it-yourself logic. On the contrary, we consider the quantum world as nice, natural, welcoming. So nice indeed that logic should be interpreted, given a new space of freedom, inside the quantum world. This program forces me to say a few words as to the opposition between *essentialism* and *existentialism*, between Thomas and Augustine, the respective fathers of these two opposite conceptions.

10.1.2.1 Logical essentialism

Logic is surely born essentialist. And the essentialist interpretation is still overwhelmingly dominant. Take for instance Tarski's definition of truth: "$A \wedge B$ is true iff A is true *and* B is true". The essence of conjunction is primitive, all you can do is to express conjunction in terms of a meta-conjunction... We can say the same about a subtler logician, Kreisel, who proposed to reinterpret Brouwer's existentialist paradigms inside a formal system *given in advance*. To sum up: there are preexisting logical *laws*. Logical artifacts, proofs, models,... are constructed accordingly to the law. The reward for obeying the law is that everything goes right.

10.1.2.2 Logical Augustinism

The weak point of essentialism is that, if everything goes right, it means that something could go wrong, but how is it possible when the artifacts always follow the law? The Augustinian[†] approach would be to admit that artifacts like proofs are anterior to logical declarations. Such was the viewpoint of the intuitionistic school (Kolmogoroff, Heyting: proofs as functions), and it seems that Gödel shared this opinion. However, the technical contents remained low.

Quantum is rather on the Augustinian side. An electronic spin is neither up or down w.r.t. a given axis, say $\vec{Z}$. If we only admit spins in these specific states, then we follow the logical laws governing boolean operations. But nature may shuffle the cards, tilt the gyroscopes, so

---

[†] Augustine proposed to define Good end Evil, not as absolute manicheist essences, but through their interaction.

that our would-be boolean has no definite spin on axis $\vec{Z}$. In an essentialist approach, this is illegal, immoral, and the measure of the value on this axis is simply forbidden. But we know from quantum physics that this measurement can take place, and that it involves the process known as the *reduction of the wave packet*, see section 10.5.3 for a logical discussion. Anyway, it is plain that the quantum world follows no rule.

10.1.3 The input of intuitionistic and linear logics

10.1.3.1 Functional interpretations

Around 1930, an alternative explanation of logic was presented by Kolmogoroff and Brouwer's pen-holder, Heyting. Proofs were basically functions, e.g., "a proof of $A \Rightarrow B$ is a function from proofs of A to proofs of B". This (sloppy) definition supposed that, somewhere, lived functions which were anterior to logic. In the late sixties, the Curry-Howard isomorphism expounded the categorical aspects of logic (proofs as morphisms) "a proof of $A \Rightarrow B$ is a morphism from A to B". These interpretations gave more and more importance to the proof, seen as a program, independent of logic. The original essentialist pattern was eventually turned upside down: a proof of A becomes a *program* enjoying *specification A*.

10.1.3.2 Locativity

It is obvious that the same program can enjoy distinct specifications, this is known as *subtyping*. We shall encounter subtyping in this paper, namely the subtyping **Bool** $\subset$ **Spin** (a Boolean, i.e., an electronic spin up or down w.r.t. the axis $\vec{Z}$, is a spin, i.e., a general electronic spin). What is specific about **Spin** is that it contains as many *isomorphic* copies **Bool**$_{\vec{A}}$ of **Bool** as we want (one for each point of the unit sphere S^2); the isomorphism is not difficult to explain as a *spiritual* property; essentialism considers things *as they should be*. However, the fact that **Spin** is the union of all **Bool**$_{\vec{A}}$ cannot be explained in this way. This has to do with things *as they are*, with their precise *location*, with their physical incarnation, so to speak. In Augustinian words, the objects come with a precise location, and isomorphism is the result of an accident —or rather a voluntary *delocation*. Locativity can embody spiritualism, whereas the converse is wrong[†].

[†] Witness the failure of all attempts at axiomatising subtyping. Such a thing shouldn't even be tried, since an axiomatisation keeps a distance between ob-

10.1.3.3 Linear logic and actions

The technical input of linear logic, see, e.g., [7], was to replace proofs as functions with *proofs as actions*. In the linear implication $A \multimap B$, the premise is destroyed. This *perfect* (or perfective) aspect of linear logic is an essential novelty, in harmony with quantum phenomena, typically the fact that a measurement alters the current state. Linear logic contains also *imperfect* connectives, which are more "classical". They are not studied in this paper: they require infinite dimension but, since this work crucially depends on the convergence of the trace, their study has been postponed.

10.1.3.4 Polarity

Why is the implication $\forall \exists \Rightarrow \exists \forall$ wrong? The usual answer is that in $\forall x \exists y$ the y depends on x, whereas in $\exists y \forall x$, y is independent of x... This "answer" is as original as Tarski's definition of truth; it would be more honest to say "it is like this, period". I propose an explanation, based on the concept of *polarity* (positive/negative). This major divide gradually emerged from computer science in the years 1990, especially in the work of Andreoli on proof-search, [1]. This notion roughly separates:

Positive	/	Negative
$\oplus, \otimes$	/	$\&, \bindnasrepma$
active	/	passive
$\underset{\rightarrow}{\lim}$	/	$\underset{\leftarrow}{\lim}$
synchronous	/	invertible
ℓ^1	/	ℓ^∞
explicit	/	implicit
object	/	subject
wave	/	measurement

The basic discovery of Andreoli is that operations of the same polarity commute. When polarities differ, we only have post-commutation of positive: a group $+-$ can be replaced with a group $-+$ (like in usual life, it is easy to *postpone* a decision). This is why $\forall \exists \Rightarrow \exists \forall$ is wrong and $\exists \forall \Rightarrow \forall \exists$ is correct.

ject and subject, hence treats objects up to isomorphism and cannot make sense of an inclusion.

10.1.3.5 *Program of work and achievements*

Intuitionism brought "proofs as functions", linear logic proposed "proofs as actions". We propose to refine this paradigm into "proofs as quantum actions": by this me mean that a proof of an implication $A \multimap B$ is any sort of transformation mapping "waves of type A" into "waves of type B", among which we include pure unitary transformations as well as pure measurements. Following a successful logical pattern, such transformations should also be seen as "waves of type $A \multimap B$", not as sort of "super-operators", like in Selinger's paper [11].

Hence proofs will be interpreted by operators. These operators should contain as particular cases, the usual "density matrices" and also the usual wave transformations and wave reductions, also expressed by hermitian operators. The only essentialist (i.e., "pulled out of a hat") concession is the choice of various finite-dimensional Hilbert spaces, but this is only because our formulas diverge in infinite dimension, otherwise we would once for all fix a separable Hilbert space. The basic duality is expressed by the formula:

$$u \overset{\downarrow}{\sim} v \quad \Leftrightarrow \quad 0 \leq \operatorname{tr}(uv) \leq 1 \tag{10.1}$$

It is to be remarked that logic will define various orderings between hermitians, and that a proper symmetry, such as the flip $\sigma(x \otimes y) = y \otimes x$ might be declared positive. This is because our framework embodies not only waves, but also "negative" (in the sense of polarity) artifacts, i.e., wave transformations $h \rightsquigarrow uhu^*$.

The extension to infinite dimension, in relation to the bosonic or fermionic behaviour of the imperfect (non-linear, traditional) part of logic, is very exciting. But it seems that it deserves another treatment.

10.2 Commutative coherent spaces

Coherent spaces are usually presented in terms of a *web*, i.e., a carrier **X** together with a reflexive and symmetric relation on **X**, its *coherence*. We shall replace this *essentialist* approach, in which the coherence relation is primitive with an alternative existentialist, *Augustinian*, in which coherence is the result of interaction. The starting remark is that we are basically interested in *cliques*, i.e., coherent subsets of the carrier, and that the negation deals with anti-cliques, i.e., incoherent subsets, so that a clique and an anti-clique intersect on at most one point[†].

[†] The question of an Augustinian approach to related notions such as *hypercoherences*, [3], is still open.

10.2.1 Revisiting coherent spaces

Definition 10.1 *Let* X *be a set; two subsets* $a, b \subset$ X *are polar when their intersection is at most a singleton. In notations*

$$a \mathrel{\underset{\sim}{\downarrow}} b \quad \Leftrightarrow \quad \sharp(a \cap b) \leq 1 \tag{10.2}$$

We define the polar $\sim A$ *of a set* $A \subset \wp(\mathrm{X})$ *of subsets of* X *by:*

$$b \in \sim A \quad \Leftrightarrow \quad \forall a \in A \quad a \mathrel{\underset{\sim}{\downarrow}} b \tag{10.3}$$

A coherent space *with* carrier X *is a subset* $X \subset \wp(\mathrm{X})$ *equal to its bipolar. Equivalently, a coherent space is the polar of some subset; moreover the map* $X \rightsquigarrow \sim X$ *is an involution of coherent spaces with carrier* X*, the* (linear) negation.

The fact that we make heavy use of Hilbert spaces prompts us to adapt the terminology and notations of linear logic: orthogonality, $\perp$ and $A^\perp$ are replaced with polarity, $\mathrel{\underset{\sim}{\downarrow}}$ and $\sim A$.

Let X be a coherent space with carrier X:

1. X contains the empty set and all singletons $\{x\}$ ($x \in$ X); in particular, X is not empty.
2. If $a' \subset a \in X$, then $a' \in X$: this is because $\sharp(a' \cap b) \leq \sharp(a \cap b)$.
3. If $a \subset$ X and $a \notin X$, there are $x, y \in X, x \neq y$ such that $\{x, y\} \notin X$: if $b \in \sim X$ is such that $\sharp(a \cap b) \geq 2$, let x, y be two distinct elements of $a \cap b$.
4. If $x, y \in$ X are distinct, then $\{x, y\} \notin X$ iff $\{x, y\} \in \sim X$: obviously $\{x, y\}$ cannot belong to both, moreover, if $\{x, y\} \notin \sim X$, this means that some $a \in X$ contains two distinct points of $\{x, y\}$.

This suggests the following definition:

Definition 10.2 *If* X *is a coherent space with carrier* X*, we define a binary relation on* X*, coherence:*

$$x \mathrel{\bigcirc_X} y \quad \Leftrightarrow \quad \{x, y\} \in X \tag{10.4}$$

By what precedes, coherence w.r.t. $\sim X$, $\bigcirc_{\sim X}$, *is identical to incoherence w.r.t.* X:

$$x \mathrel{\asymp_X} y \quad \Leftrightarrow \quad x = y \vee x \mathrel{\not\bigcirc_X} y \tag{10.5}$$

The following proposition establishes the equivalence between definition 10.1 and definition 10.2, the original definition of coherent spaces.

Proposition 10.1 *Let X be a coherent space with carrier $\mathbb{X}$, and let $a \subset \mathbb{X}$. Then $a \in X$ iff a is a clique w.r.t. the coherence of X, namely, if $\forall x, y \in a \quad x \subset_X y$.*

proofgirard: Immediate. □

10.2.2 Perfection vs. imperfection

Logic can be interpreted in coherent spaces: a formula become a coherent spaces and its proofs become elements (cliques) in it, see, e.g., [7]. Originally, coherent spaces were intended as an explanation of *intuitionistic* logic. The main achievement was to interpret intuitionistic (*imperfect*, see below) implication $X \Rightarrow Y$ in two equivalent ways: either by means of functions from X to Y or by means of a coherent space $X \Rightarrow Y$. $X \Rightarrow Y$ turned out to be a compound operation, made out of two primitives, $\multimap$ and $!$:

$$X \Rightarrow Y = !X \multimap Y \qquad (10.6)$$

The linear implication $\multimap$ is *causal*, in the sense that, in a linear implication, the premise cannot be reused: $X \multimap Y$, enables one, given (a clique in) X, to get (a clique in) Y, but the premise is destroyed. If one wants to reuse the premise, one has to say something like "forever X", which involves the construction of the *exponential* $!X$.

The main achievement of linear logic was not quite to change logical connectives and rules, but to distinguish a primal *linear* layer, in which things are *performed* once for all, that one should therefore called *perfect*, in analogy with linguistics: perfect tenses are used to denote a punctual, well-defined action; in French, English, this is limited to the past, in Russian, this is more systematic. Perfect connectives come as *dual* pairs, $\oplus/\&$, $\otimes/\aleph$; duality means that each pair is swapped by linear (perfect) negation, e.g., $\sim(X \otimes Y) = \sim X \aleph \sim Y$. The most important connective is not part of this official list: *linear* (perfect) implication $X \multimap Y$ is indeed $\sim X \aleph Y$.

Imperfection corresponds to general statements, e.g., mathematical theorems, or to repetitive actions. James Bond movies often have imperfect titles "Diamonds are forever", "You only live twice" (compare to perfect titles like "Gunfight at the OK Corral" !). Imperfect implication $\Rightarrow$ does not correspond to linear maps, but rather to analytical maps, see

[8]†. Mathematically speaking, imperfection deals with infinity, whereas perfection can reasonably live in a small (finite) world. This has a consequence for this paper: *quantum coherent spaces* make a heavy use of the *trace* which (basically) lives in finite-dimensional spaces. This means that we shall forget the imperfect connectives !/? which would involve infinite dimension and concentrate on the perfect $\oplus/\&$, $\otimes/\mathfrak{P}$ /—∘. Since this paper is basically concerned with the relation logic/quantum, this is not a major restriction: perfection is rather "quantum" whereas imperfection is more "classical".

10.2.3 Perfect connectives

The basic perfect connectives are divided into *additives* and *multiplicatives*; additives make use of disjoint *unions* (later: direct sums), multiplicatives make use of cartesian *products* (later: tensor products).

10.2.3.1 Additives

Assume that the respective carriers $\mathbb{X}, \mathbb{Y}$ of X, Y are disjoint (if not, do the obvious thing!). Then we define $X \oplus Y$, "Plus", and $X \& Y$, "With", both with carrier $\mathbb{X} \cup \mathbb{Y}$:

Definition 10.3 $\oplus$ *and* $\&$ *are defined by the dual definitions:*

$$X \oplus Y = X \cup Y \tag{10.7}$$

$$X \& Y = \{a \cup b; a \in X, b \in Y\} \tag{10.8}$$

Proposition 10.2 $X \oplus Y$ *and* $X \& Y$ *are coherent spaces; their respective negations are* $\sim X \& \sim Y$ *and* $\sim X \oplus \sim Y$.

proofgirard: Everything eventually amounts at showing that the spaces $X \oplus Y$ and $\sim X \& \sim Y$ are swapped by negation. Any $c \subset \mathbb{X} \cup \mathbb{Y}$ can uniquely be written $c = a \cup b$, with $a \subset \mathbb{X}$, $b \subset \mathbb{Y}$. $c = a \cup b \in \sim(X \oplus Y)$ iff $c \stackrel{|}{\smile} a'$ and $c \stackrel{|}{\smile} b'$ for all $a' \in X$, all $b' \in Y$, i.e., iff $c \in \sim X \& \sim Y$, which shows that $\sim(X \oplus Y) = \sim X \& \sim Y$.

From this we deduce that $X \oplus Y \subset \sim\sim(X \oplus Y) = \sim(\sim X \& \sim Y)$. But if $c = a \cup b \in \sim(\sim X \& \sim Y)$, one of a, b must be empty: if $x \in a \subset \mathbb{X}, y \in b \subset \mathbb{Y}$, then $\{x, y\} \in \sim(\sim X \& \sim Y)$, and $\neg(c \stackrel{|}{\smile} \{x, y\})$.

---

† And $!X$ is a sort of symmetric (co)-algebra, much *bosonic* in spirit... but this is beyond the scope of this paper.

Let us say that $c = b$; then c meets any $a' \cup b'$ ($a' \in \sim X$, $b' \in \sim Y$) on at most one point, which means that $b \downharpoonleft b'$, and that $c \in Y$. From this, $X \oplus Y = \sim(\sim X \& \sim Y)$. $\qquad \square$

The coherence relations related to additives work as follows: if $x, x' \in X$, then $x \mathbin{\supset\!\!\!\subset} x'$ w.r.t. $X \oplus Y$ or $X \& Y$ iff they were coherent w.r.t. X, similarly for $y, y' \in Y$. The connectives differ as to the coherence between $x \in X$ and $y \in Y$:

$X \oplus Y$: incoherent, $x \asymp y$.
$X \& Y$: coherent, $x \mathbin{\supset\!\!\!\subset} y$.

10.2.3.2 Multiplicatives

Assume that the respective carriers of X and Y are X and Y. Then we define $X \otimes Y$, "Times", and $X \bindnasrepma Y$, "Par", both with carrier $X \times Y$; we start with the essentialist version (via coherence), and later discuss the possibility of an existentialist version. The following abbreviations are useful: $x \frown y$ for $x \mathbin{\supset\!\!\!\subset} y \wedge x \neq y$, $x \smile y$ for $x \asymp y \wedge x \neq y$ (equivalently, $x \frown y \Leftrightarrow x \not\asymp y$, $x \smile y \Leftrightarrow x \not\mathbin{\supset\!\!\!\subset} y$).

Definition 10.4 *The respective coherences of "Times" and "Par" are as follows:*
in $X \otimes Y$, $(x, y) \mathbin{\supset\!\!\!\subset} (x', y')$ iff $x \mathbin{\supset\!\!\!\subset} x'$ and $y \mathbin{\supset\!\!\!\subset} y'$.
in $X \bindnasrepma Y$, $(x, y) \frown (x', y')$ iff $x \frown x'$ or $y \frown y'$.

The two definitions are clearly dual; "Par" is an artificial creation, the dual of "Times". Indeed "Par" is a contrived way to speak of *linear implication*, $X \multimap Y = \sim X \bindnasrepma Y$, and $X \bindnasrepma Y$ is better understood as $\sim X \multimap Y$ or $\sim Y \multimap X$. The coherence on $X \multimap Y$ is obviously given by: $(x, y) \frown (x', y')$ iff $x \mathbin{\supset\!\!\!\subset} x' \Rightarrow y \frown y'$.

Definition 10.5 *A function φ from (cliques of) X to (cliques of) Y is linear when it preserves all disjoint unions: if a_i are disjoint cliques in X whose union is still a clique, then*

$$\varphi\Big(\bigcup_i a_i\Big) = \bigcup_i \varphi(a_i)$$

The following result is elementary, but essential:

Theorem 10.3 *If A is a clique in $X \multimap Y$ and a is a clique in X, define*

$$[A]a := \{y \in Y; \exists x \in a \quad (x, y) \in A\} \tag{10.9}$$

Then $[A]a$ is a clique in Y and the map $a \rightsquigarrow [A]a$ is linear.
Moreover, any linear function φ from X to Y is of the form $[A]\cdot$, with
a unique A given by:

$$A = \{(x,y); y \in \varphi(\{x\})\} \qquad\qquad (10.10)$$

proofgirard: See the literature, e.g., [7]. The crucial point in the proof is
the fact that in (10.9), the x such that $(x,y) \in A$ is indeed unique. $\quad\square$

Example 10.6 *Since linearity is a preservation property, the identity*
map is surely linear. The clique in $X \multimap X$ associated to it is the set
$\Delta_X = \{(x,x); x \in X\}$. This set is not the graph $\{(a,a); a \in X\}$ of the
function, it is much smaller, and depends only on the carrier X.

The theorem establishes a link between the cliques of the coherent space
$X \multimap Y$ and the linear functions from X to Y; we could as well take
linear functions ψ from $\sim Y$ to $\sim X$, using $b \rightsquigarrow b[A]$:

$$b[A] := \{x \in X ; \exists y \in b \ (x,y) \in A\} \qquad\qquad (10.11)$$

and

$$A = \{(x,y) ; x \in \psi(\{y\})\} \qquad\qquad (10.12)$$

Let us now try an Augustinian definition of multiplicatives. There is no
problem as long as $\otimes$ is concerned:

Proposition 10.4 $X \otimes Y = \{c ; \exists a \in X \ \exists b \in y \ c \subset a \times b\}.$

proofgirard: If $a \in X$, $b \in Y$, then $a \times b \in X \otimes Y$, and if $c \subset a \times b$, we
still have $c \in X \otimes Y$. Conversely, if $c \in X \otimes Y$, let a, b be the respective
projections of c on X and Y; then $c \subset a \times b$ and $a \in X$, $b \in Y$. $\quad\square$

But there is nothing of the like for the connective "Par", or equivalently,
linear implication. However the following is true:

Proposition 10.5 $X \multimap Y = \{A ; \forall a \in X \ [A]a \in Y\}.$

proofgirard: Trivial reformulation of theorem 10.3. $\quad\square$

Moreover, $[A]a$ is characterised as the unique subset of Y such that:

$$\sharp([A]a \ \cap \ b) = \sharp(A \ \cap \ a \times b) \qquad\qquad (10.13)$$

for any $b \in \sim Y$, so what is the problem? Following the existentialist
pattern, existence (here: objects, functions) must be anterior to essence

(here: logical declarations). This means that we should be able to define $[A]a$ for any subsets $A \subset$-$\mathbf{X} \times \mathbf{Y}$, $a \subset \mathbf{X}$, in such a way that (10.13) holds for all $b \subset \mathbf{Y}$. But this is clearly impossible: we have an explicit definition of $[A]a$ in (10.9), and it is plain that (10.13) is satisfied iff the x such that $(x, y) \in A$ is unique. Our construction is essentialist in the sense that $[A]a$ is defined only when A, a "obey the law".

You may think that I am gilding the lily, asking for some fancy purity criteria... And this is correct as long as we stay with usual (commutative) spaces: everything can be handled in terms of a well-defined set of atoms (the singletons $\{x\}, x \in \mathbf{X}$). But imagine that the atoms are no longer well-defined (no canonical base in a vector space), or, worse, that there are no atoms at all (e.g. in a von Neumann algebra of type distinct from I). By the way, in what follows (PCS, QCS), there will be no direct, manageable, account of the tensor product in the style of proposition 10.4, and our only hope will be the linear implication.

10.2.4 Probabilistic coherent spaces

In proceeding towards quantum, we must replace qualitative features with quantitative ones. Here it is the place to remark that my first glimpse of linear logic came from *quantitative* domains, [5], see also [2], soon replaced with qualitative domains and coherent spaces. Indeed there is something quantitative in coherent spaces, namely the unicity of the x in (10.9), which is behind (10.13).

The idea will therefore to replace $\wp(\mathbf{X})$ —the subsets of the carrier $\mathbf{X}$— with the space $\Re(\mathbf{X})$ of all functions $\mathbf{X} \xrightarrow{f} \mathbb{R}^+$. We have in mind that, instead of saying whether or not $x \in \mathbf{X}$ belongs to a set, we rather give a probability, which would mean $0 \leq f \leq 1$; incoherence between two atoms x, y now means that their mutual weights $f(x), f(y)$ are such that $f(x) + f(y) \leq 1$, which amounts to a mutual exclusion, in case f is a characteristic function. But this is only a basic intuition: once for all, forget about coherence, or any limitation of the values to the interval $[0, 1]$.

10.2.4.1 The bipolar theorem

Definition 10.7 *Let $\mathbf{X}$ be a finite set; two functions $f, g : \mathbf{X} \to \mathbb{R}^+$ are polar when:*

$$\sum_{x \in \mathbf{X}} f(x) \cdot g(x) \leq 1 \qquad (10.14)$$

We define the polar of a set of positive functions as in definition 10.1, and a probabilistic coherent space (PCS) as a set of positive functions equal to its bipolar.

(10.14) is obviously inspired from (10.2), since, when f, g are the characteristic functions of the subsets $a, b \subset \mathbb{X}$, then $\sum_{x \in \mathbb{X}} f(x) \cdot g(x) = \#(a \cap b)$.

Theorem 10.6 (Bipolar theorem) *Let X be a PCS with carrier $\mathbb{X}$; then*

(i) *X is non-empty (in fact, $0_{\mathbb{X}}$ belongs to X).*

(ii) *X is closed and convex.*

(iii) *X is downward closed.*

Conversely every subset of $\Re(\mathbb{X})$ enjoying (i)-(iii) is a PCS.

proofgirard: That every PCS enjoys (i)-(iii) is a trifle. Conversely, assume that X enjoys (i)-(iii) and that $f \notin X$. $\Re(\mathbb{X})$ is a closed convex subset of the real Banach space $\mathbb{R}^{\mathbb{X}}$. By Hahn-Banach, there is a linear form φ such that $\varphi(X) \leq 1$, $\varphi(f) > 1$. This linear form can be identified with an element of $\mathbb{R}^{\mathbb{X}}$, i.e., a real-valued function h: $\varphi(g) = \sum_{x \in \mathbb{X}} g(x) \cdot h(x)$. Define the positive h' by $h'(x) = \sup(h(x), 0)$. Obviously $\sum_{x \in \mathbb{X}} h'(x) \cdot f(x) \geq \sum_{x \in \mathbb{X}} h(x) \cdot f(x) > 1$. If $g \in X$, then $\sum_{x \in \mathbb{X}} h'(x) \cdot g(x) = \sum_{x \in \mathbb{X}} h(x) \cdot g'(x)$ with $g'(x) = 0$ if $h'(x) = 0$, $g'(x) = g(x)$ otherwise; $g' \leq g \in X$, hence $g' \in X$ by (iii), and $\sum_{x \in \mathbb{X}} h'(x) \cdot g(x) \leq 1$. This shows that $h' \in \sim X$, but $\neg(h' \mathrel{\underset{\sim}{\llcorner}} f)$, hence $f \notin X$. $\square$

10.2.4.2 Additives

As before, additives are defined in case the carriers $\mathbb{X}$ and $\mathbb{Y}$ of X, Y are disjoint, as a PCS with carrier $\mathbb{X} \cup \mathbb{Y}$. If $f \in \Re(\mathbb{X})$, $g \in \Re(\mathbb{Y})$, I can define $f \cup g \in \Re(\mathbb{X} \cup \mathbb{Y})$ in the obvious way. The set

$$X \mathbin{\&} Y := \{f \cup g; f \in X, g \in Y\} \tag{10.15}$$

is the polar of $\sim X \cup \sim Y$ (modulo the obvious abuse which identifies $f \in \Re(\mathbb{X})$ with $f \cup 0_{\mathbb{Y}} \in \Re(\mathbb{X} \cup \mathbb{Y})$, so that $\sim X \cup \sim Y$ is indeed short for $\{f \cup g; f \in \sim X, g \in \sim Y, f = 0 \vee g = 0\}$). On the other hand $X \cup Y$ is not a PCS; $X \oplus Y$ must be defined as $\sim \sim (X \cup Y)$, with no hope of removing the double negation. The bipolar theorem 10.6 yields:

Proposition 10.7

$$X \oplus Y = \{\lambda f \cup (1 - \lambda)g; f \in X, g \in Y, 0 \leq \lambda \leq 1\} \tag{10.16}$$

proofgirard: The right-hand side is the convex hull of $X \cup Y$. It obviously enjoys conditions (i)-(iii). □

10.2.4.3 Multiplicatives

As before, multiplicatives are defined as PCS with carrier $\mathbb{X} \times \mathbb{Y}$, where $\mathbb{X}$ and $\mathbb{Y}$ are the respective carriers of X, Y. But, in contrast with section 10.2.3.2, the definition is really Augustinian.

Definition 10.8 *If* $\Phi \in \Re(\mathbb{X} \times \mathbb{Y})$, *if* $f \in \Re(\mathbb{X})$, *then one defines* $[\Phi]f \in \Re(\mathbb{Y})$ *by:*

$$([\Phi]f)(y) = \sum_{x \in \mathbb{X}} \Phi(x,y) \cdot f(x) \qquad (10.17)$$

This makes sense because $\mathbb{X}$ is finite.

Theorem 10.8 *The map* $\Phi \rightsquigarrow [\Phi] \cdot$ *is a bijection from* $\Re(\mathbb{X} \times \mathbb{Y})$ *onto the set of linear maps from* $\Re(\mathbb{X})$ *to* $\Re(\mathbb{Y})$. Φ *can be retrieved from its associated linear map* $\varphi = [\Phi] \cdot$ *by means of:*

$$\Phi(x,y) = \varphi(\delta_x)(y) \qquad (10.18)$$

proofgirard: A linear map satisfies $\varphi(\lambda f + \mu g) = \lambda \varphi(f) + \mu \varphi(g)$, hence it is determined by its values on the δ_x, this is the explanation of equation (10.18). Everything is straightforward. □

In the basic case (subsets) this didn't work: if Φ and f are characteristic functions, $[\Phi]f$ need not be one (again the unicity of the x in (10.9)).

Definition 10.9 *If* X, Y *are PCS with respective carriers* $\mathbb{X}, \mathbb{Y}$, *one defines the PCS* $X \multimap Y$, *with carrier* $\mathbb{X} \times \mathbb{Y}$, *as the set of all* Φ *such that* $[\Phi] \cdot$ *maps* X *to* Y.

Example 10.10 *The characteristic function* $\Delta_\mathbb{X}$ *of the diagonal belongs to* $X \multimap X$; *in fact* $[\Delta_\mathbb{X}]f = f$.

$X \multimap Y$ is obviously the polar of $\{f \times g; f \in X, g \in \sim Y\}$, this why it is a PCS. It could as well be defined as the set of all Φ such that $\cdot[\Phi]$ (whose definition is easy to figure out) sends $\sim Y$ to $\sim X$.
¿From $\multimap$, "Par" and "Times" follow, e.g., $X \otimes Y := \sim(X \multimap \sim Y)$, equivalently, $X \otimes Y = \sim\sim\{f \times g; f \in X, g \in Y\}$. The bipolar theorem characterises this set as a closure under certain operations, but this is

not very manageable. Should we try to prove associativity of "Times", it is much simpler to first establish it for the dual connective γ.

Proposition 10.9 *"Par" is commutative, associative, and distributes over "With".*

proofgirard: Let us prove, for instance, that "Par" is associative. For this, we pretend that cartesian product is really[†] associative, so that we can write $\mathbb{X} \times \mathbb{Y} \times \mathbb{Z}$ as the common carrier of $X \,\gamma\, (Y \,\gamma\, Z)$ and $(X \,\gamma\, Y) \,\gamma\, Z$. We use the possibility of expressing "Par" in two ways, by means of $[\cdot]\cdot$ or $\cdot[\cdot]$. We get $X \,\gamma\, (Y \,\gamma\, Z) = \{A; \; \forall f \in \sim X \; \forall h \in \sim Z \;\; h[[A]f] \in Y\}$, whereas $(X \,\gamma\, Y) \,\gamma\, Z = \{A; \; \forall h \in \sim Z \; \forall f \in \sim X \;\; [h[A]]f \in Y\}$. Everything amounts to checking that $h[[A]f] = [h[A]]f$, which is obvious.

Similarly, if we want to prove that "Par" distributes over "With", by establishing an isomorphism between $X \,\gamma\, (Y \,\&\, Z)$ and $(X \,\gamma\, Y) \,\&\, (X \,\gamma\, Z)$, we express "Par" in terms of $[\cdot]\cdot$ (and not in terms of $\cdot[\cdot]$, which is suitable for distribution on the left). □

10.3 Generalisations

10.3.1 Köthe spaces

The restriction to finite carriers ensures the convergence of (10.18). In the case of infinite carriers, one can liberalise the definition so as to accept the value $+\infty$. One can also use *Köthe spaces*: the objects are functions from a carrier $\mathbb{I}$ to $\mathbb{R}$, and polarity is defined by:[‡]

$$f \underset{\sim}{\perp} g \quad \Leftrightarrow \quad \sum_{i \in \mathbb{I}} |f(i) \cdot g(i)| \leq +\infty \qquad (10.19)$$

This is what Ehrhard did in [4]; in that case, (10.18) does not always make sense. However, everything works fine, as long as one does not try to "change the basis", i.e., as long as one "stays commutative".

10.3.2 Continuous carriers: an interesting failure

There seems to be an alternative way to accommodate infinite carriers, namely, to consider $\mathbb{X}$ as a *measure space*, typically the segment $[0, 1]$ with Lebesgue measure. It will turn out that this attempt fails, but

---

[†] Cartesian product, like "Par", is associative only up to isomorphism.
[‡] The formula defines in fact what I call Fin_C, see definition 10.11.

sometimes a wrong idea is far more interesting than a "correct" one. We only sketch the definitions:

Carriers: measure spaces $(\mathbb{X}, \mu)$, $\mathbb{X}$ for short.

Objects: functions $\mathbb{X} \xrightarrow{f} \mathbb{R}^+$ which are essentially bounded, i.e.,
$f \in \mathcal{L}^\infty(\mathbb{X}, \mathbb{R}^+)$.

Polarity: $f \overset{\downarrow}{\sim} g \quad \Leftrightarrow \quad \int_{\mathbb{X}} f \cdot g \, d\mu \leq 1$.

Application: given Φ, f with respective carriers $\mathbb{X} \times \mathbb{Y}$ and $\mathbb{X}$, define $[\Phi]f$, with carrier $\mathbb{Y}$, by $([\Phi]f)(y) = \int_{\mathbb{X}} \Phi(x, y) \cdot f(x) \, d\mu(x)$.

The map $\Phi \rightsquigarrow [\Phi] \cdot$ associates to each Φ with carrier $\mathbb{X} \times \mathbb{Y}$ a linear map sending objects with carrier $\mathbb{X}$ to objects with carrier $\mathbb{Y}$. Unfortunately, this map is far from being surjective, the typical example being given by the identity map (in case $\mathbb{X} = \mathbb{Y}$). The obvious candidate for this is still Δ, the characteristic function of the diagonal, see example 10.10. But this function is likely to be almost everywhere null. This is where we fail, and we shall meet the same obstacle when dealing with QCS. This failed attempt introduced an important novelty, namely that the basic duality should be seen as an *integral* (remember that we started with an intersection). Since, following Connes, the non-commutative integral is a trace, this explains the role played by the trace in a QCS.

10.3.3 Banach spaces

In [8], I introduced *coherent Banach spaces* as an explanation for logic. These spaces were complex because of the use of analytic functions in the imperfect case; they were also infinite dimensional, which forced one to be careful with problems of *reflexivity*. Here we restrict our discussion to real, finite dimensional, Banach spaces.

Norms Banach spaces are normed: X is a finite dimensional real Banach space, and $\sim X$ is its dual, with dual norm, so that the identification $X = \sim \sim X$ makes sense. But what is this norm for? The answer is that the norm measures *incoherence*, what corresponds to cliques of a coherent space, to objects of a PCS, is now translated as a vector of norm ≤ 1.

Additives The underlying space is a direct sum $X \oplus Y$, only the norms differ:

$$\|f \oplus g\|_{X \oplus Y} \;=\; \|f\|_X + \|g\|_Y \tag{10.20}$$

$$\|f \oplus g\|_{X \& Y} \;=\; \sup(\|f\|_X, \|g\|_Y) \tag{10.21}$$

The two choices are dual.

Multiplicatives $X \multimap Y$ is the space of linear maps from X to Y, endowed with the usual supremum norm. $X \otimes Y$ is the tensor product, endowed with the usual tensor norm, defined as $\|a\|_{X \otimes Y} = \inf\{\sum_i \|x_i\|_X \cdot \|y_i\|_Y\}$, the infimum being taken over all decompositions $a = \sum_i x_i \cdot y_i$. Again the two choices are dual.

Polarity Certain norms are defined via supremum, this is the case for $\&$ and $\multimap$ (i.e., $\invamp$), others in terms of sums ($\oplus, \otimes$). The choice of supremum corresponds to coherence, the choice of sum to incoherence. This distinction is a major divide of logic, known as *polarity*, see the introduction: supremum is negative (observation-like), sum is positive (object-like).

Semi-norms There is *a priori* no room for semi-norms in this picture. In usual mathematics, a semi-norm behaves like a norm on a quotient space. However this is wrong in the case of logic, especially if we want to accommodate Augustinian features such as subtyping. The subtyping $X \subset Y$ means that, on the same underlying vector space $\mathbb{E}$, we can have more "coherent" objects, i.e., that the unit ball increases. In other words, $\| \cdot \|_Y \leq \| \cdot \|_X$: the norm decreases. It can decrease up to 0 on certain vectors, and this explains why semi-norms naturally occur.

Positivity PCS were made of positive functions, hence they were an ordered structure. The same is true of real Köthe spaces, which are spaces of sequences. With Banach spaces, things are different, since there is *a priori* no distinguished basis. However, remark the following property:

Proposition 10.10 $f \in \mathbb{R}^{\mathbb{X}}$ *belongs to* $\Re(\mathbb{X})$ *iff for all* $g \in \Re(\mathbb{X})$ *the "scalar product"*
$\sum_{x \in \mathbb{X}} f(x) \cdot g(x)$ *is positive.*

proofgirard: Immediate.   □

This means that positivity itself can be defined in Augustinian style. We shall make a heavy use of this when dealing with QCS... although QCS are spaces of hermitian operators, which come with a natural ordering (positive hermitians), we shall not content ourselves with the "standard" notion of positivity. This can be very easily understood: if $\|a\| = 0$ and $a \neq 0$, then it is reasonable to assume that a can be identified with 0, which means that $0 \leq a$ and $0 \leq -a$. The a and $-a$ cannot both be positive hermitians.

10.3.4 The bipolar theorem, revisited

We shall complete our preliminary works with an alternative version of the bipolar theorem 10.6 which requires some care. The setting is as follows: $\mathbb{E}$ is a finite-dimensional Euclidian space, equipped with the bilinear form $\langle \cdot \mid \cdot \rangle$. Polarity is defined by means of:

$$x \underset{\sim}{\downarrow} y \quad \Leftrightarrow \quad 0 \leq \langle x \mid y \rangle \leq 1 \tag{10.22}$$

The question is to determine bipolars.

Theorem 10.11 (Bipolar theorem) *A subset $C \subset \mathbb{E}$ is its own bipolar iff the following hold:*

(i) $0 \in C$.

(ii) C *is closed and convex.*

(iii) *If $nx \in C$ for all $n \in \mathbb{N}$, then $-x \in C$.*

(iv) *If $x, y \in C$ if $\lambda, \mu \geq 0$ and $\lambda x + \mu y \in C$, then $\lambda x \in C$.*

proofgirard: (i) and (ii) are immediate. (iii): if $nx \in C$ for $n \in \mathbb{N}$, and $z \in {\sim}C$, then $\langle x \mid z \rangle \in [0, 1/n]$ for $n \in \mathbb{N}$, hence $\langle -x \mid z \rangle = \langle x \mid z \rangle = 0 \in [0, 1]$.

(iv): if $z \in {\sim}C$, then $0 \leq \langle \lambda x + \mu y \mid z \rangle \leq 1$, $0 \leq \langle \lambda x \mid z \rangle$, $0 \leq \langle \mu y \mid z \rangle$, hence $0 \leq \langle \lambda x \mid z \rangle \leq 1$. By the way remark that (iv) yields a sort of converse to (iii): if $x, -x \in C$, then $nx + n(-x) = 0 \in C$, hence $nx \in C$. We now prove the converse, and assume that C enjoys (i)-(iv); let C^+ be the cone $\bigcup_{n \in \mathbb{N}} n \cdot C \; (= \bigcup_{\lambda \in \mathbb{R}^+} \lambda \cdot C)$. Then we can reformulate (iv) as:

$$C = C^+ \cap (C - C^+) \tag{10.23}$$

Assume that $b \notin C$, then, by (10.23), we have to consider two cases:

$b \notin C^+$: using Hahn-Banach, one can find a vector $d \in \mathbb{E}$ such that
$\langle b \mid d \rangle < 0 \leq \langle c \mid d \rangle$ for all $c \in C$. By condition (ii) the subset

$I = \{c; \forall n \in \mathbb{N} \ nc \in C\}$ is a vector space, moreover, $\langle \cdot \mid d \rangle$ vanishes on I, so that we can write $C = I \oplus C'$, with $C' = I^{\perp} \cap C$. C' is compact: if we embed $\mathbb{E}$ in the projective space, C' has a compact closure, and its boundary corresponds to the lines $\mathbb{R} \cdot a$ which are included in C'. But there is no such line (all of them have been gathered in I): the boundary is empty, and C' is compact. From this, $\langle \cdot \mid d \rangle$ is bounded on C', hence on C, so $\langle b \mid d \rangle < 0 \leq \langle c \mid d \rangle \leq \lambda$. By rescaling d we can assume that $\lambda = 1$, in which case $d \in \ {\sim} C$, and $b \notin {\sim}{\sim} C$.

$b \notin C - C^{+}$: the same Hahn-Banach yields a vector $d \in \mathbb{E}$ such that $\langle p \mid d \rangle \leq 1 < \langle b \mid d \rangle$, for all $p \in C - C^{+}$. Assume that $\langle c \mid d \rangle < 0$ for some $c \in C$; then $-nc \in C - C^{+}$ for $n \in \mathbb{N}$ and the values $\langle -nc \mid d \rangle$ cannot be bounded by 1. From this we deduce that $0 \leq \langle c \mid d \rangle \leq 1 < \langle b \mid d \rangle$ for all $c \in C$. As above, $d \in \ {\sim} C$, and $b \notin {\sim}{\sim} C$.

$\square$

10.3.5 Norm and order

With the notations of theorem 10.11, in particular, $D = \ {\sim} C, C^{+} = \bigcup_{n \in \mathbb{N}} n \cdot C$:

Definition 10.11 *The domain* Fin_C *of* C *is the vector space* $C^{+} - C^{+}$ *generated by* C.

Proposition 10.12 $\mathrm{Fin}_C = (D \cap (-D))^{\perp}$.

proofgirard: If $c \in C, d \in D \cap (-D)$, then $\langle c \mid d \rangle = 0$, and the same remains true for $c \in \mathrm{Fin}_C$, the linear span of C, so that $\mathrm{Fin}_C \subset (D \cap (-D))^{\perp}$. Conversely, if $c \notin \mathrm{Fin}_C$ there is a vector $d \in (\mathrm{Fin}_C)^{\perp}$ such that $\langle c \mid d \rangle \neq 0$. But $(\mathrm{Fin}_C)^{\perp} = C^{\perp} \subset D \cap (-D)$, hence $c \notin (D \cap (-D))^{\perp}$. $\square$

In other words, the domain of C is the orthogonal of the null space of ${\sim} C$.

Definition 10.12 *The domain* Fin_C *is naturally equipped with a semi-norm* $\| \cdot \|_C$ *and a preorder* $\preccurlyeq_C$:

$$\|x\|_C = \sup\{|\langle x \mid d \rangle| \ ; \ d \in D\}$$
$$x \preccurlyeq_C y \iff \forall d \in D \ \langle x \mid d \rangle \leq \langle y \mid d \rangle$$

Let $\cong_C$ be the equivalence associated with $\preccurlyeq_C$.

Proposition 10.13 *The zero space 0_C of $\|\cdot\|_C$ is identical to the zero class modulo $\cong_C$.*

proofgirard: Obvious. □

In particular, $\mathrm{Fin}_C/0_C$ is a partially ordered Banach space.

Proposition 10.14

1. C^+ *is the set of positive elements modulo $\preccurlyeq_C$.*
2. $0_C = C^+ \cap (-C^+) = C \cap (-C)$.
3. *The unit ball w.r.t. $\|\cdot\|_C$ is $(C - C^+) \cap (C^+ - C)$.*

proofgirard: (i) and (iii) come respectively from the cases "$b \notin C^+$" and "$b \notin C - C^+$" in the proof of theorem 10.11. (ii) is immediate. □

The next properties are more or less reformulations of what we already established.

(i) The partial order $\preccurlyeq_C$ is continuous w.r.t. $\|.\|_C$: if $x_n \preccurlyeq_C y_n$ and $(x_n), (y_n)$ are Cauchy sequences w.r.t. $\|\cdot\|_C$ with limits x, y, then $x \preccurlyeq_C y$.

(ii) If $0 \preccurlyeq_C x \preccurlyeq_C y$, then $\|x\|_C \leq \|y\|_C$.

(iii) If $x \in \mathrm{Fin}_C$, then there exists $y, z \succcurlyeq_C 0$ such that $x = y - z$ and $\|y\| \leq \|x\|$.

Now what is the relation between norm and order w.r.t. C and norm and order w.r.t. $\sim C$? The question is not to establish any new result, everything has been said, but to look for symmetries $C/ \sim C$. We consider successively: equivalence, positivity, semi-norm.

Equivalence

$$x \cong_C y \quad \Leftrightarrow \quad \forall x', y'(x' \cong_{\sim C} y' \Rightarrow \langle x \mid y \rangle = \langle x' \mid y' \rangle) \qquad (10.24)$$

The introduction of the domain Fin_C, i.e., the fact of considering a *partial* (non-reflexive) equivalence relation (PER) is responsible for this symmetrical formulation.

Positivity

$$x \in C^+ \quad \Leftrightarrow \quad \forall y(y \in (\sim C)^+ \Rightarrow \langle x \mid y \rangle \geq 0) \qquad (10.25)$$

The relation $\preccurlyeq_C$ is a preorder on the domain Fin_C. I don't know how to call a transitive relation enjoying *weak reflexivity*:

$$x \preccurlyeq y \Rightarrow x \preccurlyeq x \wedge y \preccurlyeq y \qquad (10.26)$$

"partial preorder" conflicts with the use of "partial" in "partial order". I therefore propose to call it a "POR" (like we say "a PER").
The next result generalises the familiar decomposition of a hermitian as a difference $u = u^+ - u^-$ of two positive hermitians, see the default choices in section 10.3.6.

Theorem 10.15 *Given $x \in \mathbb{E}$ there are unique $x^+ \in C^+$ and $x^- \in (\sim C)^+$ such that $x = x^+ - x^-$ and $\langle x^+ \mid x^- \rangle = 0$.*

proofgirard: Let x^+ be the projection of x on the convex set C, and let $x^- := x - x^+$. It is well-known that x^- is the unique y such that $\langle y \mid x - y \rangle \geq \langle y \mid z \rangle$ for all $z \in C$. This last condition is easily shown to be equivalent to $y \in (\sim C)^+$ and $\langle y \mid x - y \rangle = 0$. $\qquad \square$

Semi-norm

$$nfx_C = \inf\{\lambda \; ; \; \forall y \in C^+ \quad |\langle x \mid y \rangle| \leq \lambda nfy_{\sim C}\} \qquad (10.27)$$

It is not the case that $|\langle x \mid y \rangle| \leq \|x\|_C \cdot \|y\|_{\sim C}$ for all $x \in \text{Fin}_C$, $y \in \text{Fin}_{\sim C}$. I am not sure that one should spend too much time on this, since the choice of the norm makes sense for us only for *positive* elements, as a way of defining *coherence*.

Proposition 10.16 *If $C \subset D$, then:*
$$\begin{array}{ccc} \text{Fin}_C & \subset & \text{Fin}_D \\ \preccurlyeq_C & \subset & \preccurlyeq_D \\ \cong_C & \subset & \cong_D \\ \| \cdot \|_C & \geq & \| \cdot \|_D \end{array}$$

The last inequality can be understood by extending $\| \cdot \|_C$ into a *total* function with values in $[0, +\infty]$.

10.3.6 Quantum coherent spaces

Let $\mathbb{X}$ be a finite-dimensional (complex) Hilbert space; let $\mathbb{E} = \mathcal{H}(\mathbb{X})$ be the set of *hermitian* (self-adjoint) operators on $\mathbb{X}$. $\mathbb{E}$ is a real vector space (whose dimension is the square of the dimension of $\mathbb{X}$) naturally endowed with the scalar product

$$\langle u \mid v \rangle := \operatorname{tr}(uv) \tag{10.28}$$

which makes it an Euclidian space: $\operatorname{tr}(uv) = \operatorname{tr}(vu) = \overline{\operatorname{tr}(uv)}$, $\operatorname{tr}(u^2) > 0$ for $u \neq 0$. Two hermitians are said to be *polar* when $0 \le \langle u \mid v \rangle \le 1$.

Definition 10.13 *A* quantum coherent space *(QCS) with carrier $\mathbb{X}$ is a subset of $X \subset \mathcal{H}(\mathbb{X})$ equal to its bipolar.*

Theorem 10.11 yields a characterisation of QCS. Some default choices are given by:

Example 10.14

Negative default: *N consists of all positive hermitians of norm ≤ 1. N^+ therefore consists in all positive hermitians; on N^+, $\| \cdot \|_N$ coincides with the usual (supremum) norm $\| \cdot \|_\infty$.*

Positive default: *P consists of all positive hermitians of trace ≤ 1. P^+ therefore consists in all positive hermitians; on P^+, $\| \cdot \|_P$ coincides with the usual trace norm $\|u\|_1 = \operatorname{tr}(\sqrt{uu^*})$.*

Hilbert-Schmidt default: *H consists of all positive hermitians of Hilbert-Schmidt norm less than 1. H^+ therefore consists in all positive hermitians; on H^+, $\| \cdot \|_H$ coincides with the usual Hilbert-Schmidt norm $\sqrt{\operatorname{tr}(uu^*)}$. This choice is self-dual: $\sim H = H$.*

In fact, $P = \sim N$; one basically uses $|\operatorname{tr}(uv)| \le \|u\|_\infty \cdot \|v\|_1$, and, for $u, v \ge 0$, $\operatorname{tr}(uv) = \operatorname{tr}(\sqrt{u}\,v\sqrt{u}) \ge 0$ and $\operatorname{tr}(uxx^*) = \langle u(x) \mid x \rangle$.

10.4 Additives

10.4.1 Basics of quantum physics

Let us recall a few basics of quantum mechanics; we stay in finite dimension to avoid technical problems.

(i) The state of a system is represented by a *wave function*, i.e., a vector x of norm 1 in some Hilbert space $\mathbb{X}$.

(ii) A *measurement* is a hermitian operator Φ on $\mathbb{X}$. To say that the *value* of x w.r.t. Φ is λ is the same as saying that $\Phi(x) = \lambda x$. This means that, under normal conditions, there is no value at all. Moreover, if Φ, Ψ do not commute, they are likely to have no common eigenvector, so x cannot have a value w.r.t. both of them, as in the famous *uncertainty principle*. For instance the Pauli matrices (see *infra*) which measure the spin along the axes $\vec{X}, \vec{Y}, \vec{Z}$, do not commute: if the spin is $+1/2$ along the axis $\vec{Z}$, then it is completely undetermined along $\vec{X}$.

(iii) The process of measurement is a Procustus's bed, it forces the system to "have a value". This means, that, after a measurement, the wave function x is replaced with an eigenvector x' of Φ. This process is non-deterministic: in fact, if $\mathbb{X}$ is split as the direct sum of the eigenspaces of Φ: $\mathbb{X} = \Sigma_\lambda \mathbb{X}_\lambda$, so that $x = \Sigma_\lambda x_\lambda$, then x' is one of the components x_λ, up to renormalisation (multiplication by $1/\|x_\lambda\|$), and the probability of the transition $x \rightsquigarrow x_\lambda/\|x_\lambda\|$ is $\|x_\lambda\|^2$. This process is known as the *reduction of the wave packet*, *reduction* for short.

(iv) In this pattern, wave functions make sense up to multiplication by any element of the unit circle. Typically, when we deal with the *spin* of an electron, which is nothing but the quantum analogue of a boolean, a rotation of 2π will replace x with $-x$, without any significant consequence.

(v) *Density matrices* have been introduced by von Neumann; they take care of the scalar indetermination of wave functions, they also take care of the probabilistic aspect of measurement. A *density operator* is a positive hermitian of trace 1. Density matrices form a compact convex set, whose extremal points are operators of the form xx^*, where x is a vector of norm 1, i.e., a wave function, uniquely determined up to multiplication by a scalar of modulus 1. When one performs a measurement, xx^* is replaced with $\Sigma_\lambda x_\lambda x_\lambda^*$: this density operator is a "mixture", a convex combination of extremal points $x_\lambda x_\lambda^*/\|x_\lambda\|^2$, with coefficients $\|x_\lambda\|^2$ which correspond to the respective probabilities of each transition.

(vi) One can iterate measurements, this means, apply this process to an arbitrary density operator, not necessarily extremal. Concretely, this means that, if we write our density matrix u as a "matrix" $(u_{\lambda\mu})$ w.r.t. the decomposition $\mathbb{X} = \Sigma_\lambda \mathbb{X}_\lambda$ ($u_{\lambda\mu} \in \mathcal{L}(\mathbb{X}_\mu, \mathbb{X}_\lambda)$), then the reduction of the wave packet consists in annihilating the non-diagonal "coefficients" $u_{\lambda\mu}$: after the measurement, the density matrix becomes $v = (u_{\lambda\mu})$, with $v_{\lambda\lambda} = u_{\lambda\lambda}$, $v_{\lambda\mu} = 0$ for $\lambda \neq \mu$.

(vii) The measurement process is irreversible: if $u \rightsquigarrow v$ through measurement, then $\operatorname{tr}(v^2) \leq \operatorname{tr}(u^2)$, i.e., the Hilbert-Schmidt norm decreases[†]. If $\mathbf{X}$ is of dimension n, then the HS norm can vary between 1 (extremal point xx^*) and $1/\sqrt{n}$, which corresponds to $1/n \cdot I$, the "tepid mixture", which conveys no information at all.

10.4.2 Quantum booleans

10.4.2.1 Commutative booleans

With start with 2×2 *matrices*. As long as traditional logic is concerned, there is little to say:

1. The booleans `true,false` are naturally represented by

$$\begin{bmatrix} 1 & 0 \\ 0 & 0 \end{bmatrix}, \begin{bmatrix} 0 & 0 \\ 0 & 1 \end{bmatrix}.$$

2. It is natural to think that a diagonal matrix $\begin{bmatrix} \lambda & 0 \\ 0 & \mu \end{bmatrix}$, with $\lambda + \mu = 1$, $\lambda, \mu \geq 0$ represents a probabilistic boolean.

But, as soon as one "forgets the diagonal", i.e., when one considers "booleans of arbitrary basis", then the three —nay the four— dimensions of space come into the picture.

10.4.2.2 Space-time

Any hermitian can be written $h = 1/2 \begin{bmatrix} t+z & x-iy \\ x+iy & t-z \end{bmatrix}$, i.e., $t.s_0 + x.s_1 + y.s_2 + z.s_3$, where t, x, y, z are real and the s_i are the *Pauli matrices* $1/2 \begin{bmatrix} 1 & 0 \\ 0 & 1 \end{bmatrix}$ $1/2 \begin{bmatrix} 0 & 1 \\ 1 & 0 \end{bmatrix}$ $1/2 \begin{bmatrix} 0 & -i \\ i & 0 \end{bmatrix}$ $1/2 \begin{bmatrix} 1 & 0 \\ 0 & -1 \end{bmatrix}$. Remark that *time t* is nothing but the *trace*, $t = \operatorname{tr}(h)$. As to the determinant, we get $4\det(h) = (t^2 - (x^2 + y^2 + z^2))$, the square of the pseudo-metrics. Remark that $\operatorname{tr}((t.s_0 + x.s_1 + y.s_2 + z.s_3)(t'.s_0 + x'.s_1 + y'.s_2 + z'.s_3)) = tt' + xx' + yy' + zz'$.

For $1 \leq i \neq j \leq 3$, we have the anti-commutations $s_i.s_j + s_j.s_i = 0$.

In order to characterise *positive hermitians*, remember that, modulo a

---

[†] The reduced hermitian is not smaller: the difference has null trace, and can hardly be positive.

unitary transformation, $uhu^* = \begin{bmatrix} \lambda & 0 \\ 0 & \mu \end{bmatrix}$, with $\lambda, \mu \in \mathbb{R}$, so that h is positive iff $\lambda, \mu \geq 0$. In other words, the condition $\det(h) \geq 0$ (vectors in position "time") characterises hermitiens which are either positive or negative. Positive hermitians correspond to the further requirement $\operatorname{tr}(h) \geq 0$, i.e., to the "future cone".

The most general transformation preserving positive hermitians is of the form $h \rightsquigarrow uhu^*$, with $\det(u) = 1$, i.e., $u \in \mathrm{SL}(2)$: such transformations correspond to the familiar positive Lorenz group, which is the group of linear transformations preserving the pseudo-metrics and the future. By the way, remark that the inverse of $u \in \mathrm{SL}(2)$ is given by:

$$\begin{pmatrix} a & b \\ c & d \end{pmatrix}^{-1} = \begin{pmatrix} d & -b \\ -c & a \end{pmatrix} \tag{10.29}$$

Therefore, inversion can be extended into an involutive anti-automorphism of the C^*-algebra $\mathcal{M}_2(\mathbb{C})$ of 2×2 matrices. This anti-automorphism acts on space-time by negating the spacial coordinates.

The positive Lorenz group admits as a subgroup the group $\mathrm{SO}(3)$ of *rotations*, which modify only space: they correspond to trace-preserving transformations, those who are induced by *unitaries*. In other words, $\mathrm{SO}(3)$ admits a double covering by $\mathrm{SU}(2)$, the group of unitary transformations of determinant 1, whose general form is $\begin{pmatrix} a & b \\ -\bar{b} & \bar{a} \end{pmatrix}$, with $a\bar{a} + b\bar{b} = 1$. The rotations of axes $\vec{X}, \vec{Y}, \vec{Z}$ and angle θ are induced by the unitaries $e^{i\theta s_k}$, i.e.,

$$\begin{bmatrix} \cos\theta/2 & i\sin\theta/2 \\ i\sin\theta/2 & \cos\theta/2 \end{bmatrix} \quad \begin{bmatrix} \cos\theta/2 & \sin\theta/2 \\ -\sin\theta/2 & \cos\theta/2 \end{bmatrix} \quad \begin{bmatrix} e^{i\theta/2} & 0 \\ 0 & e^{-i\theta/2} \end{bmatrix}$$

respectively. Remark the "heresy" consisting in dividing an angle by 2, an operation with two solutions... This is why one speaks of a double covering; this is also why a rotation of angle 2π acts on a spin (seen as a wave function) by multiplying by -1.

10.4.2.3 Quantum booleans

"Classical" booleans correspond to projections on two 1-dimensional subspaces which are distinguished by the matricial representation. A *quantum* boolean will therefore be a subspace of dimension 1. By the way, remark that this definition refuses any differentiation between true and false: if the space E is a quantum boolean, its negation is $E^\perp$, period. By the way, remark that, due to problems of commutation, it will be impossible to construct convincing binary connectives. It remains to determine the subspaces of dimension 1, i.e., the matrices of orthogonal projections of rank 1. Those are the hermitian matrices of trace 1 and determinant 0, i.e., the points of space-time $t.s_0 + x.s_1 + y.s_2 + z.s_3$, with $t = 1, x^2 + y^2 + z^2 = 1$, which are therefore in $1 - 1$ correspondence with the sphere S^2. What we have just explained is the natural way to speak of a quantum boolean, which also known to physicists as the *spin* of an electron.

10.4.2.4 Probabilistic quantum booleans

Probabilistic quantum booleans (PQB) are just convex combinations of quantum booleans, i.e., "density matrices", positive hermitians of trace 1. Any PQB can be diagonalised in an orthonormal basis. In which respect is this unique?

(i) The PQB $\begin{bmatrix} 1/2 & 0 \\ 0 & 1/2 \end{bmatrix}$ is diagonal in all bases. This is the extreme form of non-unicity.

(ii) Apart from this case, our boolean can be written $\lambda b + (1 - \lambda)c$, where b, c are quantum booleans and $0 \leq \lambda \leq 1$. λ, b, c are uniquely determined if we require $0 \leq \lambda < 1/2$.

The reduction of the wave packet occurs when we want to measure a boolean, this corresponds to the measurement of a spin in physics. First we must specify an orthonormal basis, and write operators as matrices w.r.t. this base. Say that our PQB corresponds to the matrix $\begin{pmatrix} a & b \\ \bar{b} & c \end{pmatrix}$, then, after measurement, it becomes $\begin{pmatrix} a & 0 \\ 0 & c \end{pmatrix}$, i.e., true with probability a, false with probability $c = 1 - a$.

10.4.2.5 Negation

Specifying an orthonormal basis consists in chosing two orthogonal subspaces of dimension 1, i.e., two quantum booleans π and $1 - \pi$, whose four-dimensional coordinates are therefore $(1, x, y, z)$ and $(1, -x, -y, -z)$. The two vectors $\vec{A} = (x, y, z)$ and $-\vec{A}$ correspond to two opposite directions on the same three-dimensional axis (spin up, spin down). The symmetry w.r.t. the origin comes from the anti-automorphism $\begin{pmatrix} a & b \\ c & d \end{pmatrix} \rightsquigarrow \begin{pmatrix} d & -b \\ -c & a \end{pmatrix}$ of the C^*-algebra $\mathcal{M}_2(\mathbb{C})$ of 2×2

matrices. This transformation corresponds to *negation*. It must be remarked that, since symmetry w.r.t. the origin is of determinant -1, it is not in $SO(3)$, and therefore it is not induced by an element of $SU(2)$.

10.4.2.6 Binary boolean connectives

Whereas negation does not need reduction, binary boolean connectives will badly need it; there are two reasons for that.

(i) We cannot combine non-commuting 1-dimensional projections in a way that will produce another projection.

(ii) Common sense tells us that, if we cannot distinguish between true and false, then we cannot distinguish between conjunction and disjunction.

Hence binary connectives will be probabilistic: they yield a PQB even when the inputs are "pure" quantum booleans. Moreover, they depend on the choice of a basis, and an order of evaluation; I give an example:
$\begin{bmatrix} a & b \\ \bar{b} & c \end{bmatrix} \vee \begin{bmatrix} a' & b' \\ \bar{b}' & c \end{bmatrix} := \begin{bmatrix} a + ca' & cb' \\ c\bar{b} & cc' \end{bmatrix}$. The first argument is "reduced" in the canonical base: true with probility a, in which case the answer is $\begin{bmatrix} 1 & 0 \\ 0 & 0 \end{bmatrix}$, false with probility c, in which case the answer is $\begin{bmatrix} a' & b' \\ \bar{b}' & c \end{bmatrix}$. There is a symmetrical choice which reduces the second argument. But only a real Jivaro will choose the third possibility, which reduces *both* arguments, yielding $\begin{bmatrix} a + ca' & 0 \\ 0 & cc' \end{bmatrix}$, which is in fact symmetrical, since $a + ca' = a' + ca = a + a' - aa'$.

10.4.3 Quantum and additives

10.4.3.1 Basics

Definition 10.15 *If X, Y are QCS with respective carriers $\mathbb{X}, \mathbb{Y}$, one defines the additive combinations $X \oplus Y$ and $X \,\&\, Y$, as QCS of carrier $\mathbb{X} \oplus \mathbb{Y}$.*

$$X \oplus Y = \{\lambda u \oplus (1 - \lambda)v;\ u \in X, v \in Y, 0 \leq \lambda \leq 1\}$$
$$X \,\&\, Y = \{w;\ \mathbb{X}w\mathbb{X} \in X, \mathbb{Y}w\mathbb{Y} \in Y\}$$

As usual, we have identified the subspaces $\mathbb{X}$ and $\mathbb{Y}$ with the associated orthogonal projections.

Proposition 10.17 $\oplus$ *and* & *are swapped by negation.*

proofgirard: Essentially because $\langle u \oplus v \mid u' \oplus v' \rangle = \langle u \mid u' \rangle + \langle v \mid v' \rangle$. $\square$

Remark that $\| \cdot \|_{X \oplus Y}$ and $\| \cdot \|_{X \& Y}$ are not norms. This is because this definition mistreats all hermitians which are not of the form $u \oplus v$. W.r.t. an obvious matricial notation, every hermitian on $X \oplus Y$ can be written $\begin{pmatrix} u & w \\ w^* & v \end{pmatrix}$, with, u, v hermitian. If $w \neq 0$, then this operator has infinite norm in $X \oplus Y$. *A contrario*, its norm w.r.t. $X \& Y$ does not depend on w: the null space $0_{X \& Y}$ contains all $\begin{pmatrix} 0 & w \\ w^* & 0 \end{pmatrix}$.

10.4.3.2 Dimension 2

If X is of dimension 1, then $\mathcal{H}(X)$ is of dimension 1 (isomorphic to $\mathbb{R}$) and the three defaults of example 10.14 coincide, and yield the same QCS, noted 1, which corresponds to the segment $[0, 1]$ of $\mathbb{R}$. The ordering is the usual ordering, and the norm the usual absolute value.

In dimension 2, $\mathcal{H}(X)$ has dimension 4, and there are many choices.

Spin: the positive default. The elements of **Spin** are positive hermitians of trace at most 1. They are not quite PCB, since a PCB is of trace 1, they are sort of "partial PCB". Concretely, if we measure an element $\begin{pmatrix} a & b \\ \bar{b} & c \end{pmatrix}$, it will yield "true" with probability a, "false" with probability c, and nothing with probability $1 - a - c$. This "nothing" is natural from the computational viewpoint: if we assume that the measurement is done through a computing device, then we are likely to wait before getting our probabilistic answer "true" or "false". "Nothing" corresponds to the case of a computing loop, i.e., when we wait too long.

$\sim$Spin: the negative default. The elements of $\sim$**Spin** are positive hermitians of (usual) norm at most 1. They should be understood as "anti"-booleans.

Bool: the "Plus" of two copies of 1. The space $1 \oplus 1$ consists of all diagonal matrices $\begin{pmatrix} a & 0 \\ 0 & c \end{pmatrix}$ such that $0 \leq a, c \leq a + c \leq 1$. This QCS is a subset, a "subtype" of **Spin**. It has a well-defined notion of truth and falsity.

$\sim$Bool: the negation of the former, i.e., $1 \& 1$. It consists in all matrices $\begin{pmatrix} a & b \\ \bar{b} & c \end{pmatrix}$ such that $0 \leq a, c \leq 1$.

Now remark that our construction of **Bool** depends on the choice of a 1-dimensional subspace (corresponding to "true"). This means that, given any vector $\vec{A} \in S^2$, there is a QCS made of "booleans of axis $\vec{A}$", noted **Bool**$_{\vec{A}}$.

Proposition 10.18 Spin $= \bigcup_{\vec{A} \in S^2}$ **Bool**$_{\vec{A}}$.

proofgirard: Obviously **Bool**$_{\vec{A}} \subset$ **Spin**. Conversely, if $h \in$ **Spin**, it can be put in diagonal form $\begin{pmatrix} a & 0 \\ 0 & c \end{pmatrix}$, with $0 \leq a, c \leq a + c \leq 1$, w.r.t. a certain basis $\mathbf{e}, \mathbf{f}$. If $\vec{A}$ is the point of S^2 corresponding to $\mathbf{e}$, then $h \in$ **Bool**$_{\vec{A}}$. $\square$

Corollary 10.16 $\sim$**Spin** $= \bigcap_{\vec{A} \in S^2} \sim$**Bool**$_{\vec{A}}$.

10.4.3.3 Reduction: a discussion

In the next section, we shall deal with multiplicatives and linear implication. In particular, we shall be able to transform a boolean $h \in$ **Spin** into something else by using an element of the QCS **Spin** $\multimap \ldots$, then transform the result by means of another implication... Some of these transformations will behave like negation (wave-like) others will use reduction. We try now to understand to which extent reduction is subjective. For this, we make an impossible thing, we assume that the process of transformation is over, i.e., that in this sequence of successive implications, we have succeeded in "closing the system". This means that there is an ultimate implication with values in **1**. If I compose all my implications, I eventually discover that a sequence of transformation, eventually "closing the system" is exactly an anti-boolean $k \in \sim$**Spin**. The resulting output is objective: $\langle h \mid k \rangle = \text{tr}(hk)$. But the choice of k (the transformations, observations made on h) is highly subjective, we are biased, we are "on the side of k). If we are on the side of k, then put k in diagonal form w.r.t. a basis $\mathbf{e}, \mathbf{f}$. Then $h = \begin{pmatrix} a & b \\ \bar{b} & c \end{pmatrix}, k = \begin{pmatrix} \alpha & 0 \\ 0 & \gamma \end{pmatrix}$, so that $\langle h \mid k \rangle = a\alpha + c\gamma$. If $h' = \begin{pmatrix} a & 0 \\ 0 & c \end{pmatrix}$, then $\langle h \mid k \rangle = \langle h' \mid k \rangle$, i.e., it is as if h had been reduced.

It may be the case that we know that f is a boolean in a certain base (e.g., if f is the result of a measurement). Then we can select this base, in which case $h = \begin{pmatrix} a' & 0 \\ 0 & c' \end{pmatrix}, k = \begin{pmatrix} \alpha' & \beta' \\ \bar{\beta}' & \gamma' \end{pmatrix}$, and we can write

$\langle h \mid k \rangle = a'\alpha' + c'\gamma'$. In that case, we can "reduce" the observer k into $k' = \begin{pmatrix} \alpha' & 0 \\ 0 & \gamma' \end{pmatrix}$ so that $\langle h \mid k \rangle = \langle h \mid k' \rangle$. This shows the extreme subjectivity of reduction.

10.5 Multiplicatives

10.5.1 Linear functionals

Theorem 10.19 *Let* $\mathbf{X}, \mathbf{Y}$ *be finite dimensional Hilbert space. Then* $\mathcal{L}(\mathcal{L}(\mathbf{X}), \mathcal{L}(\mathbf{Y})) \simeq \mathcal{L}(\mathbf{X} \otimes \mathbf{Y})$.

proofgirard: The complex vector space $\mathcal{L}(\mathbf{X})$ is generated by rank 1 endomorphisms xw^*: $xw^*(y) = \langle y \mid w \rangle x$. If $\varphi \in \mathcal{L}(\mathcal{L}(\mathbf{X}), \mathcal{L}(\mathbf{Y}))$, define $\Phi \in \mathcal{L}(\mathbf{X} \otimes Y)$ by

$$\langle \Phi(x \otimes y) \mid w \otimes z \rangle = \langle \varphi(xw^*)(y) \mid z \rangle \tag{10.30}$$

Conversely, given $\Phi \in \mathcal{L}(\mathbf{X} \otimes Y)$, if $f \in \mathcal{L}(\mathbf{X})$, then one defines $[\Phi]f \in \mathcal{L}(\mathbf{Y})$ by:

$$\langle ([\Phi]f)(y) \mid z \rangle = \text{tr}(\Phi \cdot (f \otimes yz^*)) \tag{10.31}$$

so that $[\Phi] \cdot \in \mathcal{L}(\mathcal{L}(\mathbf{X}), \mathcal{L}(\mathbf{Y}))$. □

Corollary 10.17 *If* $\Phi \in \mathcal{H}(\mathbf{X} \otimes Y)$, *if* $f \in \mathcal{H}(\mathbf{X})$, *then* $[\Phi]f \in \mathcal{H}(\mathbf{Y})$. *The map* $\Phi \rightsquigarrow [\Phi] \cdot$ *is a bijection from* $\mathcal{H}(\mathbf{X} \times \mathbf{Y})$ *onto the set of linear maps from* $\mathcal{H}(\mathbf{X})$ *to* $\mathcal{H}(\mathbf{Y})$.

proofgirard: An easy computation shows that $[\Phi^*]f^* = ([\Phi]f)^*$, hence a hermitian Φ sends hermitians to hermitians. Conversely, if φ is a linear map from $\mathcal{H}(\mathbf{X})$ to $\mathcal{H}(\mathbf{Y})$, then φ can be uniquely extended into a $\mathbb{C}$-linear map from $\mathcal{L}(\mathbf{X})$ to $\mathcal{L}(Y)$: $\varphi(u) = 1/2(\varphi(u+u^*)+i\varphi(iu^* - iu))$. Now the $\mathbb{C}$-linear maps obtained in this way are *hermitian*, i.e., $\varphi(f^*) = \varphi(f)^*$, and they are in 1-1 correspondence with hermitians of $\mathcal{H}(\mathbf{X} \otimes Y)$. □

The essential property of $[\varphi] \cdot$ is summarised by the equation

$$\text{tr}(([\Phi]f) \cdot g) = \text{tr}(\Phi \cdot (f \otimes g)) \tag{10.32}$$

Example 10.18 *If* $\sigma_\mathbf{X} \in \mathcal{H}(\mathbf{X} \otimes \mathbf{X})$ *is such that* $\sigma(x \otimes y) = y \otimes x$ *(the "flip"), then* $\langle [\sigma](xw^*)(y) \mid z \rangle = \langle \sigma x \otimes y \mid w \otimes z \rangle = \langle y \otimes x \mid w \otimes z \rangle = \langle y \mid w \rangle \langle x \mid z \rangle = \langle (xw^*)(y) \mid z \rangle$. *Hence* $[\sigma]xw^* = xw^*$ *and by linearity* $[\sigma]f = f$.

Example 10.19 *More generally, let u be any map from* X *to* Y. *Then* $u \otimes u^*$ *maps* $X \otimes Y$ *into* $Y \otimes X$, *and if* σ_{XY} *is the "flip" from* $Y \otimes X$ *to* $X \otimes Y$, *then* $U = \sigma \cdot u \otimes u^* \in \mathcal{H}(X \otimes Y)$. *It is immediate that* $[U]f = ufu^*$.

Example 10.20 *Let* $1_X = E + F$ *be a decomposition of the identity as a sum of orthogonal projections (subspaces). Then* $R = \sigma(E \otimes E + F \otimes F)$ *acts as follows:* $[R]f = EfE + FfF$. *R is a typical reduction operation, it chops off the "non-diagonal" portions* EfF *and* FfE *of* f.

One can wonder what is the status of the identity map of $X \otimes Y$. An easy computation shows that $[1_{X \otimes Y}](u) = \mathrm{tr}(u) \cdot 1_Y$. Not very exciting... But this will help us with our last example:

Example 10.21 *If* X *is of dimension 2, then* $[1_{X \otimes X} - \sigma_X] \begin{pmatrix} a & b \\ \bar{b} & c \end{pmatrix} = \begin{pmatrix} c & -b \\ -\bar{b} & a \end{pmatrix}$, *i.e., acts like* negation. *Remark that* $1_{X \otimes X} - \sigma_X = 2\pi$, *where* π *is the orthogonal projection corresponding to the antisymmetric (one-dimensional) subspace of* $X \otimes X$, *i.e., the space of vectors* $x \otimes y - y \otimes x$.

10.5.2 Connectives

Definition 10.22 *Let* X, Y *be QCS with respective carriers* X, Y. *We define the QCS* $X \multimap Y$, *with carrier* $X \otimes Y$, *as the set of all* Φ *sending* X *to* Y:

$$X \multimap Y \quad = \quad \{\Phi; \, \forall f \in X \, [\Phi]f \in Y\} \tag{10.33}$$

$X \multimap Y$ could as well be defined by

$$X \multimap Y \quad = \quad \{\Phi; \, \forall g \in \sim Y \, g[\Phi] \in \sim X\} \tag{10.34}$$

and also as $\sim \{f \otimes g; f \in X, g \in \sim Y\}$. This last expression shows that $X \multimap Y$ is a QCM. From this we can define $X \,\invamp\, Y = \sim X \multimap Y$ and $X \otimes Y = \sim\sim\{f \otimes g; f \in X, g \in Y\}$. As usual, $\invamp$ is commutative, associative, and distributive over & (all this up to isomorphism).

As usual, "Times" is more difficult to access than "Par". By equation (10.25) (and Hahn-Banach) one can characterise the "positive" cone of a "Times", as the closure of the set of finite sums $\sum_i f_i \otimes g_i$, $f_i, g_i \geq 0$[†].

---

† This is obviously related to *separable mixed states*, see, e.g., [10].

In the same way, (10.27) can be used to determine the semi-norm associated with a "Times".

Remark 10.23 *It is important to remark that multiplicatives force a departure from the standard ordering of hermitians. For instance, assume that X, Y have been equipped with the positive defaults, e.g., $X = Y = $ **Spin**. Then $X \multimap Y$ will declare as positive any hermitian sending positive hermitians to positive hermitians. The most typical example is the flip σ which behaves like the identity map. But σ is a proper symmetry, not a positive hermitian. So $X \multimap Y$ is more liberal as to positivity than expected. This means that, dually, $X \otimes Y$ is more restrictive. In fact, the positive cone of $X \otimes Y$ is the closure of the set of finite sums $\sum_i f_i \otimes g_i$, $f_i, g_i \geq 0$. Most positive hermitians on $\mathbf{X} \otimes \mathbf{Y}$ cannot be obtained in this way: take any orthogonal projection zz^*, where z is not a pure tensor!*

10.5.3 η-expansion and reduction

The question "is a function a graph?" is traditional in logic, and quite scholastic. It is such a long time that people exchange the same arguments; do they actually believe in what they say? There is peculiar form of this question, known as "η-conversion, and limited to the sole identity function. Given a logically compound formula F, then the identity function admits two alternative descriptions, as a proof of $C \multimap C$:

Generic: since C is identical to C, the *identity axiom* maps C into C.

η-expanded: decompose C into components, $A, B, \ldots$, and recompose the identity functions of $A, B, \ldots$, in order to produce an identity function of C.

The two processes are identified by all *honest* interpretations, i.e., interpretations which are not contrived to make a difference between them. This is why, in my own *ludics*, [9], everything was "η-expanded", i.e., the identity was not primitive.

We shall show that η-expansion is wrong, by differentiating the identity from its η-expansion in the case $C = A \oplus B^\dagger$. For simplicity, let us assume that A, B have both carriers of dimension 1. Our two identities respectively correspond to:

The flip: the generic identity map of a space $\mathbf{X}$ of dimension 2. This

---

† But η stays correct in the case of a "Times".

map writes as $\sigma = \begin{pmatrix} 1 & 0 & 0 & 0 \\ 0 & 0 & 1 & 0 \\ 0 & 1 & 0 & 0 \\ 0 & 0 & 0 & 1 \end{pmatrix}$ in any base $\mathbf{e} \otimes \mathbf{e}$, $\mathbf{e} \otimes \mathbf{f}$,

$\mathbf{f} \otimes \mathbf{e}$, $\mathbf{f} \otimes \mathbf{f}$ of $X \otimes X$.

The η-expanded flip: it corresponds to putting together two identities. W.r.t. a specific base (corresponding to the decomposition of C as a direct sum), it writes $\iota = \begin{pmatrix} 1 & 0 & 0 & 0 \\ 0 & 0 & 0 & 0 \\ 0 & 0 & 0 & 0 \\ 0 & 0 & 0 & 1 \end{pmatrix}$.

These two maps are clearly distinct: $[\sigma] \begin{pmatrix} a & b \\ b & c \end{pmatrix} = \begin{pmatrix} a & b \\ b & c \end{pmatrix}$, it is the real identity. On the other hand, $[\iota] \begin{pmatrix} a & b \\ \bar{b} & c \end{pmatrix} = \begin{pmatrix} a & 0 \\ 0 & c \end{pmatrix}$ is a Procustus's identity. It behaves as the identity w.r.t. matrices which already have the right logical form $\begin{pmatrix} a & 0 \\ 0 & c \end{pmatrix}$, and those who don't follow the logical rule, it chops off their anti-diagonal coefficients. Of course, if we remember our basics, ι is quite the reduction of the wave packet, corresponding to the measurement of *spin* along the vertical axis $\vec{Z}$.

In logic, only the identity can be η-expanded, but this is an accident. For instance the negation $\nu = \begin{pmatrix} 0 & 0 & 0 & 0 \\ 0 & 1 & -1 & 0 \\ 0 & -1 & 1 & 0 \\ 0 & 0 & 0 & 0 \end{pmatrix}$ which is such that

$[\nu] \begin{pmatrix} a & b \\ \bar{b} & c \end{pmatrix} = \begin{pmatrix} c & -b \\ -\bar{b} & a \end{pmatrix}$ w.r.t. a given base can be η-expanded into $\nu' = \begin{pmatrix} 0 & 0 & 0 & 0 \\ 0 & 1 & 0 & 0 \\ 0 & 0 & 1 & 0 \\ 0 & 0 & 0 & 0 \end{pmatrix}$; obviously $[\nu'] \begin{pmatrix} a & b \\ \bar{b} & c \end{pmatrix} = \begin{pmatrix} c & 0 \\ 0 & a \end{pmatrix}$: ν' corresponds to a

measurement of the spin along the axis $\vec{Z}$ and a subsequent inversion. To come back to the original question about functions and graphs. In the "commutative" world, every function is bound to be a graph. This is because everything is diagonal in a fixed basis. When the distinguished bases disappear, the "atoms" disappear as well. η-expansion corresponds to the choice of a set of atoms (a basis), the decomposition of a function along this basis, and its recomposition. This process is violently incorrect in a non-commutative setting.

10.5.4 Still to be done

The main challenge is the extension to infinite dimension:

(i) First, the approach is not fully Augustinian, since the carriers $\mathbb{X}, \mathbb{Y}$ are "pulled out of a hat". It would be nicer to fix once for all a separable Hilbert space.

(ii) Second, the imperfect (infinite) part of logic needs to be studied too. It is to be remarked that the exponential $!A$ "forever A" is much bosonic in spirit. In general the question of a possible logical status for the two types of quantum symmetry (fermionic, bosonic) is much exciting.

However, this stumbles on serious problems.

(i) Köthe spaces, as used by Ehrhard, see section 10.3.1, are perfect as an infinite *commutative* Augustinian explanation of logic. One can fix once for all a denumerable index set $\mathbb{I}$ and define polarity by:

$$f \overset{\perp}{\sim} g \quad \Leftrightarrow \quad |\sum_{i \in \mathbb{I}} f(i) \cdot g(i)| \leq 1 \qquad (10.35)$$

But this approach does not allow significant changes of basis, and is inappropriate for quantum.

(ii) Finite-dimensional Hilbert spaces give rise to type I_n *factors*, i.e., "connected" von Neumann algebras. The most trivial generalisation is a type I_∞ factor, i.e., the space $\mathcal{B}(\mathbb{H})$ of *bounded operators* on an infinite-dimensional Hilbert space. The main problem is that such an algebra is *semi-finite*, i.e., trace makes sense, as an element of $[0, +\infty]$, only for *positive* operators. But we badly need equations like $[\sigma_{\mathbb{X} \otimes \mathbb{X}}]\sigma_{\mathbb{X}} = \sigma_{\mathbb{X}}$, which has strictly no meaning from this viewpoint.

(iii) Another direction would be type II_1 factors, typically the famous matricial factor, which harbours a (unique) finite trace. But $\mathrm{tr}(\sigma \cdot 1 \otimes 1) = \mathrm{tr}(\sigma) = 0 \neq \mathrm{tr}(1 \cdot 1) = 1$. The reason for this vanishing of σ is the same as the reason of the vanishing of Δ in section 10.3.2.

What is most likely to happen is the use of a matricial factor of type II_1 together with the replacement of trace with determinant, $\det(1 - uv)$, instead of $\text{tr}(uv)$. But this involves *geometry of interaction*, see [6], and this is quite another story.

10.5.5 *Relation to quantum computing*

Although it is not my primary interest, the relation to quantum computing should be considered. It would be interesting to revisit Selinger's language for quantum computation [11] in the spirit of QCS. However, the use of loops in the style of geometry of interaction may suggest that determinant might be more appropriate. Perhaps more appropriate (because explicitely based on linear logic) is the "quantum lambda-calculus" recently proposed by van Tonder [12].

NON SI NON LA

Bibliography

[1] J.-M. Andreoli and R. Pareschi. Linear objects: logical processes with built-in inheritance. *New Generation Computing*, 9(3 – 4):445 – 473, 1991.

[2] Nuno Barreiro and Thomas Ehrhard. Quantitative semantics revisited (extended abstract). In *Proceedings of the fourth Typed Lambda-Calculi and Applications conference*, volume 1581 of *Lecture Notes in Computer Science*, pages 40–53. Springer-Verlag, 1999.

[3] T. Ehrhard. Hypercoherences: a strongly stable model of linear logic. In Girard, Lafont, and Regnier, editors, *Advances in Linear Logic*, pages 83 – 108, Cambridge, 1995. Cambridge University Press.

[4] T. Ehrhard. On Köthe sequence spaces and linear logic. *Mathematical Structures in Computer Science*, 12:579–623, 2002.

[5] J.-Y. Girard. Normal functors, power series and λ-calculus. *Annals of Pure and Applied Logic*, 37:129 – 177, 1988.

[6] J.-Y. Girard. Geometry of interaction I: interpretation of system F. In Ferro, Bonotto, Valentini, and Zanardo, editors, *Logic Colloquium '88*, pages 221 – 260, Amsterdam, 1989. North-Holland.

[7] J.-Y. Girard. Linear logic, its syntax and semantics. In Girard, Lafont, and Regnier, editors, *Advances in Linear Logic*, pages 1 – 42, Cambridge, 1995. Cambridge University Press.

[8] J.-Y. Girard. Coherent Banach Spaces : a continuous denotational semantics. *Theoretical Computer Science*, 227:275 – 297, 1999.

[9] J.-Y. Girard. Locus Solum. *Mathematical Structures in Computer Science*, 11:301 – 506, 2001.

[10] M. Horodecki, P. Horodecki, and R. Horodecki. Mixed-state entanglement and quantum communication. In *Quantum Information:*

An Introduction to Basic Theoretical Concepts and Experiments, volume 173 of *Springer tracts in modern physics*. Springer Verlag, 2001.

[11] P. Selinger. Towards a quantum computing language. *Mathematical Structures in Computer Science*, 2003.

[12] A. van Tonder. A lambda-calculus for quantum computing. Technical report, Dept of Physics, Brown university, Providence, RI, July 2003.

Printed in the United States
By Bookmasters